NATURE'S LAST STRONGHOLDS

NATURE'S LAST STRONGHOLDS

GENERAL EDITOR

Robert Burton

IN ASSOCIATION WITH THE INTERNATIONAL UNION FOR CONSERVATION OF NATURE AND NATURAL RESOURCES

New York
OXFORD UNIVERSITY PRESS
1991

CONSULTANT EDITOR
Professor Peter Haggett, University of Bristol

Dr John Akeroyd
Italy and Greece

Dr Richard Barnes, Large Animal Research Group, Cambridge
Central Africa

Dr Hans Bibelriether, Federation of Nature and National Parks,
West Germany
Central Europe

Robert Burton
Conserving the World's Wild Places

Sally Collins, The National Trust of South Australia
Australasia, Oceania and Antarctica

Teresa Farino
The British Isles, Spain and Portugal

Graham Drucker, World Conservation Monitoring Center, Cambridge
Northern Africa

Paul Goriup, The Nature Conservation Bureau, Berkshire
The Middle East

Dr Gina C. Green, Oxford Forestry Institute, Oxford
Central America

Dr Michael J.B. Green, International Union for Conservation of Nature
and Natural Resources
The Indian Subcontinent

Dr Claes Grundsten
The Nordic Countries

Brian J. Huntley
Southern Africa

Dr Z.J. Karpowicz, World Conservation Monitoring Center,
Cambridge
Eastern Europe, The Soviet Union

Dr Haruko Laurie
Japan and Korea

Dr John Mackinnon
China and its neighbors

Dr Kathy MacKinnon
Southeast Asia

Dr Alan Moore
South America

Dr Peter Lewis Nowicki, Aménagement-Environment, Lille
France and its neighbors

Dr D. Scott Slocombe, University of Waterloo, Ontario
Canada and the Arctic

Dr Wim J.M. Verheugt
The Low Countries

Dr Peggy Wayburn
The United States

AN EQUINOX BOOK
Copyright © Equinox (Oxford) Limited 1991

Planned and produced by
Equinox (Oxford) Limited
Musterlin House, Jordan Hill Road
Oxford, England OX2 8DP

Published in the United States of America by
Oxford University Press, Inc.,
200 Madison Avenue,
New York, N.Y. 10016

Oxford is a registered trademark of
Oxford University Press

Library of Congress
Cataloging-in-Publication Data

Volume editors	Jill Bailey, Susan Kennedy
Designers	Jerry Goldie, Rebecca Herringshaw
Cartographic manager	Olive Pearson
Picture research manager	Alison Renney
Picture researcher	Linda Proud
Project editor	Candida Hunt
Art editor	Steve McCurdy

ISBN 0-19-520862-5

Printing (last digit):9 8 7 6 5 4 3 2 1

Printed in Spain by Heraclio Fournier SA, Vitoria

INTRODUCTORY PHOTOGRAPHS
Half title: *Rain forest stream, New Zealand (NHPA/John Shaw)*
Half title verso: *Ayers Rock, Australia (Ardea/J.P. Ferrero)*
Title page: *Yellowstone Park, northwestern USA (William Ervin)*
This page: *Kakadu, Northern Territory, Australia (A.N.T. Photo Library/R.W.G. Jenkins)*

Contents

PREFACE
7

PREFACE

TEN THOUSAND YEARS AGO THE WORLD WAS INHABITED BY SMALL scattered bands of hunters and gatherers. Their impact on the environment must have been negligible compared with the effects of natural phenomena – storms, lightning, volcanic eruptions, floods and droughts. As plants and animals were domesticated by humans, the naturally evolved landscapes that covered the Earth began to be replaced by agricultural landscapes; and with more reliable and controllable sources of food, the human population began to expand.

As the pace of human development accelerated, centers of civilization developed in the most fertile and easily cultivable areas of the world. Wilderness was regarded as unproductive and wasteful: forests had to be cleared, grasslands plowed and wetlands drained to accommodate the world's growing populations. Even after the cataclysmic changes of the Industrial Revolution there still appeared to be plenty of wilderness left to be plundered at will for its wealth of natural resources.

In the last three decades the pace of human growth and technological innovation has speeded up so much that human activity now presses on all the world's surviving natural landscapes. Even areas that a few years ago seemed inviolate because they appeared empty and worthless, such as the polar and desert regions, are now exploited for their minerals and their tourist value.

As the scale of global environmental destruction has magnified, however, the seeds of a realization have taken root and blossomed; namely that wilderness areas must be preserved, not merely so that endangered wild animals should survive, or that ecologists and explorers should have their outdoor laboratories and their wild terrain, or even that the remaining small numbers of hunters and gatherers should be allowed their forest homes. It has become increasingly clear that the world's wildernesses are needed in order to sustain the proper working of the physical and chemical processes that keep the world a fit and viable environment for all living organisms.

It is with this in mind that this book sets out to show the importance the world's surviving wildernesses have for the health and continuation of our planet and looks at the steps that have been, and are being, taken to conserve and protect the world's fragile ecosystems from the destruction that threatens to overwhelm them all.

Tracts of boreal forest in northern Canada

A salt-encrusted lake high in the Andes of Bolivia *(overleaf)*

CONSERVING THE WORLD'S WILD PLACES

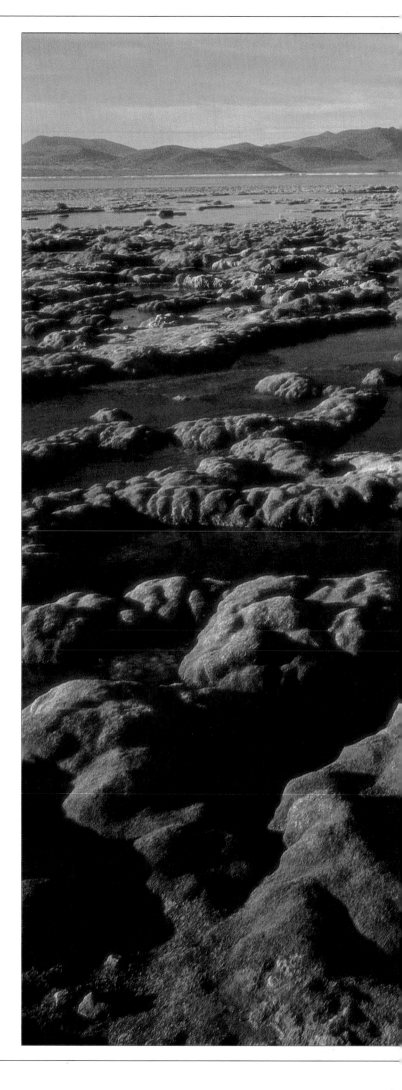

What is Wilderness?

THE WORD "WILDERNESS" HAS A NUMBER OF different meanings. To someone in the city, the tangle of undergrowth that envelops a stretch of wasteland may seem to be a wilderness; for the backpacker, wilderness is an area of open country devoid of human habitation, where access is confined to small trails. The naturalist regards wilderness as an area untouched by human hand, still covered by its original natural vegetation. It is in this last sense that the term wilderness is used in this book.

The world's many varied wildernesses range from dense tropical rainforests to barren treeless deserts, from frozen wastes of ice and snow to sparkling oceans and lakes. The character of a wilderness is determined by the landscape and the vegetation (or lack of it) that covers it. The nature of the vegetation is itself determined by the climate, soils and topography of the area, and to some extent by its past history.

In its turn, the vegetation influences its environment. The vegetation provides shade, cooling the air below and screening out wind; it reduces soil evaporation and slows the runoff of rainwater, allowing it time to soak into the soil. Roots help to anchor the soil against both water and wind, and fallen leaves and dead plants add to its organic and mineral content. Mature wilderness is thus a stable, self-sustaining system.

Wilderness around the world

The major wilderness types broadly follow the changes in latitude, and hence in climate. At the poles there is little precipitation (rain, snow, sleet, hail) and what does fall becomes locked up in ice and snow; this creates polar desert conditions. At slightly lower latitudes, where the ground thaws in summer, lies the tundra, an area of lowgrowing plants that in winter are protected under a blanket of snow. Beyond the tundra is a band of coniferous or boreal forest, which gives way to deciduous forests of species such

A flock of flamingoes feeding on one of the many saline lakes in the Rift Valley of eastern Africa. Large animal populations require extensive wilderness areas to support them.

Adelie penguins in Antarctica The icesheets of the polar oceans are among the world's areas of pristine wilderness. The nutrient-rich waters support many seabirds and marine mammals.

Forests constitute many of the world's wildernesses. They are generally rich in animal species, but they are being felled for timber the world over. These temperate rainforests are in New Zealand.

as oak and beech as winters become warmer. Closer to the Equator, where rainfall is markedly seasonal, there are hard-leaved (sclerophyll) forests, adapted to conserving water. The moist, humid tropics and subtropics support lush evergreen tropical forests.

Where high mountains or distance from the sea reduce the amount of moisture reaching the land, vast expanses of grassland have developed, and where rain is still scarcer there are deserts. By contrast, some coasts experience very heavy rainfall, which can promote the growth of temperate rainforests as lux-

uriant as any tropical forest. The largest wilderness of all is the ocean, which covers some 70 percent of the globe to an average depth of 4 km (2.5 mi).

These great wilderness areas are by no means uniform. There are thousands of lakes and bogs among the dark boreal forests, islands of pines and birches invade the temperate grasslands, and green oases enliven even the largest deserts. Wilderness areas contain a bewildering variety of plants and animals, an intricate web of life whose components are continually changing as they interact with each other and respond to global changes.

Classifying the Wilderness

Although wilderness areas across the planet have enormous diversity, certain general types, such as grassland or tundra, can easily be distinguished. These wilderness types are characterized by their dominant vegetation and are known as biomes.

Classifying the biomes

As part of UNESCO's Man and Biosphere Program, a system for classifying the major biomes of the world according to their dominant vegetation was proposed by Professor Miklos Udvardy. He recognized 14 basic types of biome, ranging from forests and grasslands to deserts, tundra, islands and lakes. By studying the biomes of the world's last surviving wilderness areas and their environments, it is possible to work out where the different biomes would occur if the natural vegetation were allowed to flourish undisturbed. This can help in planning the location of protected areas.

When similar biomes in different parts of the world are compared, it is found that the species differ from one region to another. Grasslands in Africa and North America, for example, are grazed by antelopes, while in South America the main grazers are guanacos and vicunas, relatives of the camel; in Australia kangaroos fulfil this role. The biomes are therefore subdivided into provinces, which are related to geographical locations, but whose boundaries are determined according to the similarities in and differences between their component species. The classification thus takes into account the diversity of species contained in the world's wildernesses.

Diversity in the wilderness

The wilderness is not static but is continuously changing. At a local level, the marsh vegetation fringing lakes and seas traps silt, creating new lands. As mountains gradually rise or are eroded, the landscape and climate change, and the nature of the wilderness alters with it. Plants and animals need to evolve in response to these changes. Some adapt to the new environment, while others migrate to surrounding areas and adopt new ways of life.

Over many millions of years the continents have drifted over the face of the Earth, moving from one latitude to another and one climate to another, coalescing and fragmenting, allowing species to mingle and separate again. In the process of adapting to these changes, a huge diversity of living species has evolved.

Species are the basic unit of diversity. A species is a group of similar living organisms that can interbreed. Individual populations gradually evolve adaptations that make them particularly successful in their local habitat. This can lead to the evolution of local races, or subspecies, and eventually of new species when the isolated population differs so much from other populations that it can no longer interbreed with them. One of the principal aims of conservation is to maintain this wealth of diversity. To achieve this, examples of every biome in every province need to be protected. By protecting a particular biome, the many species of plants, animals and microorganisms within it are also preserved.

Coniferous forest, also known as boreal forest or taiga, is one of the world's major biomes. It extends in a broad belt across the high latitudes of the northern hemisphere, where the growing season is too short for deciduous trees.

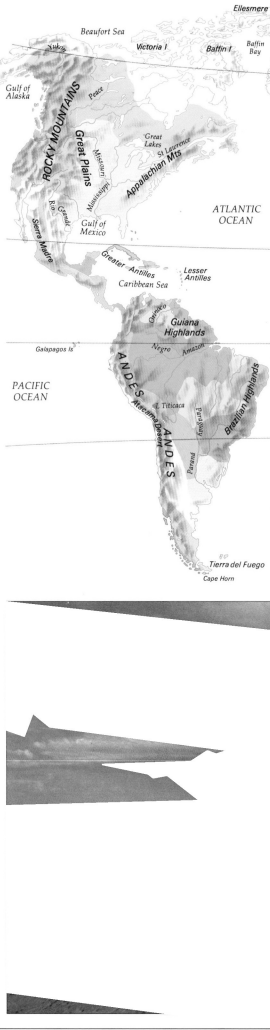

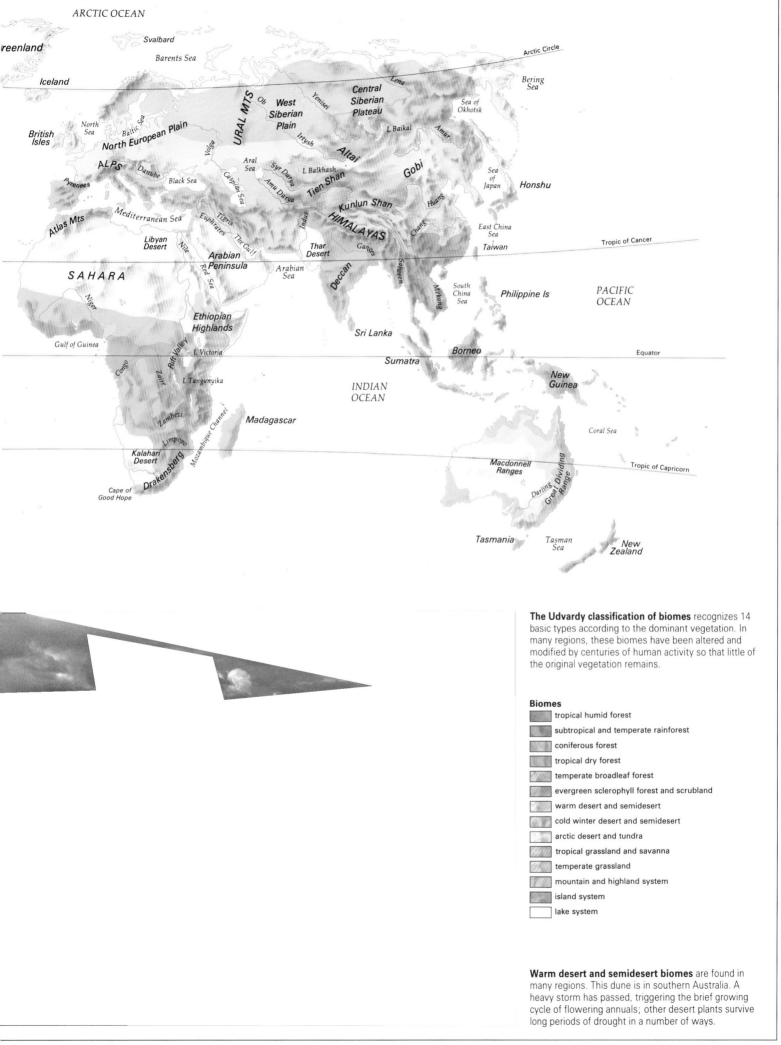

ARCTIC OCEAN

Svalbard

Barents Sea

Greenland

Iceland

Arctic Circle

Bering Sea

British Isles

North Sea

Baltic Sea

North European Plain

URAL MTS

Ob

West Siberian Plain

Yenisei

Central Siberian Plateau

Lena

Sea of Okhotsk

ALPS

Danube

Black Sea

Volga

Irtysh

Aral Sea

Syr Darya

Caspian Sea

L Balkhash

Altai

Gobi

Amur

L Baikal

Sea of Japan

Honshu

Pyrenees

Amu Darya

Tien Shan

Kunlun Shan

Atlas Mts

Mediterranean Sea

Euphrates

Tigris

The Gulf

HIMALAYAS

Chang

Huang

East China Sea

Taiwan

Tropic of Cancer

Libyan Desert

Nile

Arabian Peninsula

Red Sea

Arabian Sea

Thar Desert

Ganges

Indus

Salween

SAHARA

Deccan

Mekong

South China Sea

Philippine Is

PACIFIC OCEAN

Niger

Ethiopian Highlands

Sri Lanka

Gulf of Guinea

Rift Valley

L Victoria

Borneo

Equator

Congo

Zaire

L Tanganyika

Sumatra

INDIAN OCEAN

New Guinea

Zambezi

Mozambique Channel

Madagascar

Limpopo

Coral Sea

Kalahari Desert

Drakensberg

Macdonnell Ranges

Great Dividing Range

Tropic of Capricorn

Cape of Good Hope

Darling

Tasmania

Tasman Sea

New Zealand

The Udvardy classification of biomes recognizes 14 basic types according to the dominant vegetation. In many regions, these biomes have been altered and modified by centuries of human activity so that little of the original vegetation remains.

Biomes

tropical humid forest

subtropical and temperate rainforest

coniferous forest

tropical dry forest

temperate broadleaf forest

evergreen sclerophyll forest and scrubland

warm desert and semidesert

cold winter desert and semidesert

arctic desert and tundra

tropical grassland and savanna

temperate grassland

mountain and highland system

island system

lake system

Warm desert and semidesert biomes are found in many regions. This dune is in southern Australia. A heavy storm has passed, triggering the brief growing cycle of flowering annuals; other desert plants survive long periods of drought in a number of ways.

The Ecology of the Wild

EACH WILDERNESS IS UNIQUE IN ITS LOCATION and in the history of its colonization by plants and animals. The presentday physical and chemical features and the events of the past combine to determine which species are present. Intimate interactions exist between plants and animals and between these living organisms and their environment. Even slight changes in the acidity of the soil or the degree of protection from the wind may significantly affect the distribution of smaller forms of life, and hence of the predators that depend upon them, while an accident of past history may make a species rare in one location but common in another. The study of plant and animal communities and their interactions with each other and with their environment is called ecology.

The place where a particular species or group of species lives is known as its habitat. This term is often used rather loosely, depending on the context. For example, a red squirrel's habitat might be described as deciduous woodland, while that of the caterpillar of a woodland moth as deciduous woodland or as an oak tree. Habitats may be very large or very small: the habitat of the gray whale is the ocean, while the bread mould's habitat is no larger than a loaf of bread.

Ecosystems

Factors such as temperature, availability of light, water and minerals, soil type and drainage all affect the growth of plants, and hence of animals. In turn, the living organisms affect their environment – trees provide shade and trap moisture, the droppings of animals add to the organic matter in the soil, and so on. They also affect each other. For example, many plants use animals to pollinate their flowers and disperse their fruits. The environment, the living communities and the interactions between them are often referred to as an ecosystem.

The basis of every ecosystem is the population of primary producers. These organisms trap the energy that powers the rest of the living community. In most ecosystems the primary producers are the green plants. They make their own organic material from water and minerals taken up from the soil and carbon dioxide taken in from the atmosphere during the process known as photosynthesis. During this process light energy is trapped and converted into chemical energy.

The plants provide herbivores (primary consumers) with a plentiful supply of food. Carnivores (secondary consumers) feed on the herbivores, and other carnivores (tertiary consumers) feed on them. These different levels of feeding are called trophic levels. Each trophic level contains approximately 10 percent less energy than the preceding level, which explains why there are fewer carnivores than herbivores. Some of this energy is lost as heat and carbon dioxide during the metabolic processes of plants and animals. Not all food taken in by animals is digested: some passes out as feces. Energy is also used in searching for food, in courting, mating and other activities. All the energy used by animals comes ultimately from the photosynthesis of plants.

Nutrients follow a similar path. They are returned to the ecosystem in the form of feces or when an animal dies. The organic material is acted on by decomposers such as worms, fungi and bacteria, releasing the nutrients and returning them to the soil. Here they are taken up by plants, and the cycle starts again.

A spread of habitats High mountains, a freshwater stream with marshy banks, and flanking tropical forest in Nepal all provide specific conditions in which different species of animals and plants are able to live.

The community of plants and animals within this stretch of heathland in France depend upon each other, and any disturbance of the soil or drainage by human interference will affect them all. Remove one species and the whole balance of the ecosystem will alter.

After the volcano As the cooled volcanic lava on this crater floor on the island of Réunion wears down to soil, it is colonized by wind-blown seeds of shrubs and trees. A new habitat is created, and the complex combination of temperature and degree of rainfall, soil type and drainage will determine which plants and animals can survive in that particular environment.

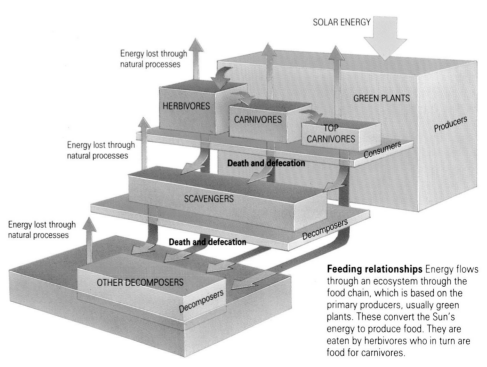

SOLAR ENERGY

Energy lost through natural processes

HERBIVORES

CARNIVORES

TOP CARNIVORES

GREEN PLANTS

Producers

Consumers

Energy lost through natural processes

Death and defecation

SCAVENGERS

Decomposers

Energy lost through natural processes

Death and defecation

OTHER DECOMPOSERS

Decomposers

Feeding relationships Energy flows through an ecosystem through the food chain, which is based on the primary producers, usually green plants. These convert the Sun's energy to produce food. They are eaten by herbivores who in turn are food for carnivores.

15

Tropical Forest

TROPICAL RAINFOREST FORMS THE MOST complex and varied biome in the world. These are the "jungles" of popular imagination: vast areas of giant trees, evergreen and moist. The temperature is constantly warm and the average annual rainfall of more than 2,000 mm (78 in) is spread more or less evenly throughout the year. These conditions mean that the growing season is never interrupted.

Tropical rainforests cover only 6 percent of the Earth's surface, but they contain more than half the world's species of animals and plants. A report by the United States National Academy of Sciences suggests that in 250 ha (618 acres) of rainforest in South America there are some 1,500 species of plants, 125 mammal species, 400 birds, 100 reptiles, 600 amphibians and 150 butterflies.

One of the reasons for this great diversity is the number of different habitats contained within the tropical rainforest. There are several distinct layers of foliage, each with its own community of animals. The tallest trees have large spreading crowns that form the canopy layer, an almost continuous roof of branches extending across the forest and shading the plants below.

The layers beneath the canopy consist of some naturally lowgrowing species, and of the saplings of canopy-forming species. In some forests there is also a dense layer of shrubs, but the lack of light restricts plant growth on the ground. Twining between the various layers are vines and lianas, climbing figs and other trailing plants.

The constant warmth encourages both rapid growth and rapid decomposition. Pockets of debris accumulate in crevices in the bark of the trees, providing footholds for epiphytic plants – mosses, ferns, orchids and bromeliads. Fallen trees and branches rot quickly, their damp wood covered in colorful fungi. There are numerous insects – ants, beetles, cockroaches, and termites feed among the rotting leaves. There are also many mammals, including elephants, deer, gorillas, tapirs and big cats.

The fragile forests
Tropical forests are surprisingly fragile. If a patch of tropical forest is cut down, within a few years the land will be exhausted, because the products of decay are immediately taken up by new growth rather than being incorporated into humus in the soil, which remains thin. When trees are felled and removed the nutrients trapped in the vegetation are lost. Nutrients locked in the underlying rocks are released too slowly to support longterm crop cultivation or ranching, and removal of the foliage cover and the binding mesh of roots leaves the soil vulnerable to erosion. Without the vegetation, which holds water like a vast sponge and acts as a natural reservoir, rivers start to dry up, and at times of heavy rain are subject to flash floods that increase erosion.

Rainforests have been exploited by humans for thousands of years, but they are under increasing threat both from expanding indigenous populations and from ranching and commercial farming. Almost any disturbance of the forest will result in the loss of species. Many animals can survive in the altered forest, but specialized species, such as bearded saki monkeys of the Amazon and lemurs in Madagascar, may disappear. Many small lifeforms cannot tolerate the change in temperature and humidity when the canopy is opened. In Sarawak, Malaysia, for example, the number of species of termites, which are essential for the recycling of dead trees, was halved in selectively logged forests.

It is becoming increasingly clear that tropical forests have a significant effect on local and even global climates, and that they are important for the health of the planet as well as being a largely unexplored natural resource. Yet losses of rainforest are enormous. Brazil and Indonesia will still have large blocks of forest in 50 years' time, but at the present rate some countries such as Costa Rica, Nigeria and Thailand will have lost their forest cover almost completely.

Dense tropical forests cover the highlands of western Malaysia. The heavy year-round rainfall supports a lush growth of tall trees, tangled undergrowth and ferns. These forests are being logged at a devastating rate to satisfy the worldwide demand for hardwood timber.

Mist and cloud swathe the upper slopes of eastern Africa's taller mountains much of the time. Despite the cold temperature at this altitude, these forests are quite luxuriant. Lichens trail from the branches, and epiphytic plants grow in the dense moss cushions on the stunted trees.

Deep in the Brazilian rainforest, little light reaches the forest floor, but where large trees fall the sunlight brings a host of new seedlings to life. In this warm humid climate decomposition is rapid. Fungi abound – leaves and logs are soon rotted down, returning minerals to the soil. Although the vegetation here is dense and luxuriant, the soils are thin. If the forest is felled, the few soil nutrients are soon washed out by the rain and they become barren. The virgin forests contain more than half the world's known species of plants and animals, but this abundance will never be recaptured; secondary growth will be poor in species.

TROPICAL DRY FORESTS

In areas where the annual rainfall drops below 2,000 mm (78 in) and part of the year is dry, the forest becomes dry or seasonal forest. Many of the trees drop their leaves each year to reduce water loss. When less than 500 mm (19.5 in) of rain falls, and the dry season lengthens to more than six months, growth thins to savanna forests with open grassy patches. Local variations in rainfall and the water-holding capacity of the soil create a patchwork of seasonal forest – along rivers, for example, dense, evergreen gallery forest stands out among

drier types of forest nearby. Fire is also important in seasonal forest ecology and contributes to its characteristically this patchy appearance.

Rainforests have received most attention from ecologists and conservationists, but tropical dry forests are also important. They include monsoon forest, which is the dominant biome of large parts of tropical Asia and contains valuable teak and sal forests. Although there are fewer tree species than are found in the rainforests, seasonal forests also have a very rich animal life.

Temperate and Coniferous Forest

Twisted moss-covered trees are characteristic of ancient oak woodlands such as these in Killarney, Ireland. Such woodlands once covered much of western Europe's upland areas, but they were cleared by humans centuries ago, and today only moorlands and heaths survive on many hills.

THICK FORESTS ONCE COVERED LARGE AREAS of the northern hemisphere's temperate zone. Today many of the densely settled parts of North America, Europe and Asia fall within the belt of broadleaf forest. Most of these woodlands are deciduous, with seasonal leaf-fall as an adaptation to a cold winter climate that limits the availability of water. The species composition of the forest varies according to local conditions of soil, attitude and rainfall, so the temperate forest is really a mosaic of communities dominated by different tree species such as beech, lime, maple and oak.

The relatively open structure of the forest encourages plant growth on the forest floor, which is carpeted in spring with the flowers of plants that survive the winter as underground bulbs, corms or rhizomes. In the fall the forest turns yellow, red and gold as the leaf pigments break down before the leaves drop.

Many animals live in these forests. Before human activity drove them into decline bears, lynx and wolves were common, and deer, squirrels, woodpeckers and other animals still thrive. In summer the abundance of caterpillars and other insects attracts temporary visitors such as warblers and flycatchers; fruits and berries provide food for birds migrating south in the fall.

Decomposition of the leaves and rapid weathering of the underlying rocks leads to a buildup of rich soil, which is able to support intensive cultivation when the trees have been cleared. As a result, the natural forest cover of temperate Europe and Asia was reduced to scattered remnants hundreds of years ago; this took place in North America more recently. Much of the temperate forest growing today is secondary or seminatural forest, and only pockets of virgin broadleaf forest have survived.

Coniferous forest

The broad band of forest lying to the north of the temperate broadleaf zone is dominated by coniferous trees, which grade into birch, willow, alder and aspen at the northern extremity. This is known as the boreal forest or taiga. At these latitudes the growing season is short, and is followed by a long winter of heavy snowfall and drying winds. A similar climate is found on the upper slopes of mountains at lower latitudes; here, too, coniferous forest grows.

This boreal forest is not a uniform expanse of trees. Where soils are poor and waterlogged, bogs develop, and near its northern edge the forest begins to merge with lowgrowing tundra vegetation.

Like the broadleaf forest, the coniferous forest is home to many animals, and is an important winter refuge for mammals such as moose and caribou. The range of plants is limited partly because of the

Mature beech trees In contrast to tropical forests, the canopy of deciduous forest harbors a very limited number of mammals; the seasonal loss of leaf cover means that the complex layers of animal communitites are lacking. Consequently most mammals live on the forest floor. The beeches' dense shade discourages the growth of an understorey.

Coniferous forest covers large areas of North America and northern Europe. Boreal forest is by no means a uniform expanse of densely growing conifers. Where the deep shade is interrupted by lakes, birches thrive in the sunlight, and a succession of plant communities forms concentric rings around the water. This rich mixture of habitats supports a varied wildlife, including both residents and seasonal visitors.

dense shade and thick leaf litter of the forest floor; lichens, ferns and shrubby plants predominate.

Conifers are felled in large quantities for timber, pulp and wood chip, but the boreal forests of Canada, Scandinavia and Siberia, which survive in an environment that has been only sparsely settled by humans, have been considerably less affected and modified by human activities than have many other biomes.

Sclerophyll forest and scrubland
In areas that have a Mediterranean-type climate of hot, dry summers and cool, wet winters, trees and shrubs have become adapted to drought. The dry summers often result in fires, and some species are fire resistant – seeds may even need the heat of the flames to stimulate germination. In California this vegetation is known as chaparral, in southern Europe it is called maquis and garigue, in Chile matorral, in South Africa fynbos and in Australia mallee scrub.

The trees may be deciduous, shedding their leaves in the dry season to reduce the loss of water through transpiration. Others such as the cypress, cedar and the evergreen oaks of the Mediterranean have tough evergreen leaves. Their leathery, waxy surface also helps to reduce water loss. Human activities, especially grazing by flocks of sheep and goats, have been destroying the tree cover for centuries. This has resulted in the spread of flowering herbs and aromatic shrubs such as lavender and thyme.

Savanna and Grassland

AROUND THE EDGES OF THE GREAT TROPICAL forests, where the climate becomes too dry to support a closed canopy of trees, there is a biome of parklike savanna. Depending on the amount of rain that falls, tree cover ranges from open woodland with a sward of grass underneath to grassland with scattered trees. This kind of open forest community lies on each side of the Equator, receiving 250–1,300 mm (10–50 in) of rain and having one or two dry seasons each year.

The main savannas are in Africa, where acacias and other tree species form a thorny scrub; in Australia, where the tree cover is eucalyptus; and in South America, where they are known as the llanos in Venezuela and Colombia, and the campos and cerrados in Brazil. Savanna vegetation is maintained by frequent fires, and the trees tend to be resistant to both burning and drought.

The African savanna, in particular, supports a wide range and abundance of large animals. It is the home of herds of browsing and grazing animals such as elephants, giraffes, buffaloes, zebras and many kinds of antelope, as well as flocks of birds and swarms of locusts and termites. Savannas are also visited in winter by many migrant birds. A number of nomadic species migrate long distances, following the seasonal pattern of rains in search of food and water.

Savannas originally covered 20 percent of the Earth's land surface, but they are increasingly being encroached on for crop cultivation and stock-keeping. This leads to destruction of the tree cover and overgrazing of the grass, and eventually the disappearance of some plant species. The trees are felled for firewood and for forage. This destruction has caused a slump in breeding populations of migrant birds in North America and Europe.

Temperate grasslands

Often called steppe, these grasslands cover areas where the annual rainfall is only 250–750 mm (10–30 in) a year, and the land is too dry for trees. They support a wealth of animal life. When a grass shoot is nibbled or burned, several new shoots sprout from buds at the base of the leaves, so the turf actually gets thicker.

Perennial species of grass, which persist throughout the year, may be outnumbered by a variety of colorful "nongrasses" or forbs. After a flush of growth following the melting of the winter snow or after irrigation by the spring rains, the grasses and forbs wither to hay under the summer sun.

The world's great temperate grasslands include the North American prairies, the South American pampas, the steppe of central Europe and Asia, the South African veldt and the Australian downland. They are inhabited by large grazing animals such as the horse, bison, pronghorn, guanaco, kangaroo, saiga and many other antelopes. There are also hares, rabbits, prairie dogs and other ground squirrels, tuco-tucos and viscachas, gerbils, lemmings, mice and voles, as well as many seed-eating birds, grasshoppers, ants and termites.

Scattered acacia trees form oases of shade in the African savanna grassland. The tall dry grasses are readily ignited by lightning during the drought season, and fire spreads rapidly. Nutrients from the ashes are later washed into the soil to feed a new burst of growth, enriching the pasture for large herds of grazing mammals.

Wetland plants of prairie sloughs, or bogs, provide a splash of spring color. Grassland ecosystems are by no means uniform. The miniature water worlds of sloughs and potholes are important breeding grounds for waterfowl, and provide welcome stopover points for birds migrating to and from the tundra in spring and fall.

Virgin grassland is now rare. In North America the bison have been replaced by herds of cattle and native grasses by imported aliens such as meadow grass, a plant of the Eurasian steppes renamed Kentucky blue grass. In parts of New Zealand more than half the tussock grassland species are aliens that have been deliberately introduced.

The principal change to the grassland biome has come from the plow. The grasslands are now the great "bread baskets" of the world, in North America, Argentina and central Asia. In Illinois in the United States only 10 sq km (4 sq mi) remain of the prairies that once covered most of its 145,000 sq km (56,000 sq mi).

EQUILIBRIUM ON THE GRASSLANDS

A natural or seminatural grassland is one that is maintained by the animals that feed on it. Their continual nibbling down of the vegetation keeps out invading woody plants, and preserves the mixture of grasses and forbs that they need for sustenance.

For nearly 12,000 years the North American prairies were maintained by herds of bison that supported the Plains Indians. Some 40 to 60 million bison roamed the plains. The herds broke up and enriched the soil, thus allowing nutritious buffalo grass and blue grama to thrive while the taller, ranker bunch grasses were kept at bay. Grass seeds were distributed by clinging to the thick fur on the heads and forequarters of the bison. In wetter areas, where trees had the chance to invade, they were held in check by bison fraying them with their horns during the mating season. In less than a century the enormous herds of bison were reduced to little over one thousand animals by the European settlers who colonized North America and regarded the bison as a nuisance.

The prairie dog is another animal that is being eradicated because it competes with cattle for grazing. Prairie dogs live in "towns" with populations of millions of individuals. Their grazing on annuals and shrubs clears the ground for perennial grasses and forbs, and their excavations benefit species that colonize open ground. They too maintain the prairie ecosystem.

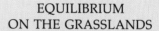

On cold, dry plateaus between the high mountain ranges of central and eastern Asia the steppe becomes so sparse it approaches cold desert. Yet these areas support large herds of grazers – wild asses and antelopes, gazelles, yaks, sheep and even camels.

Deserts and Semideserts

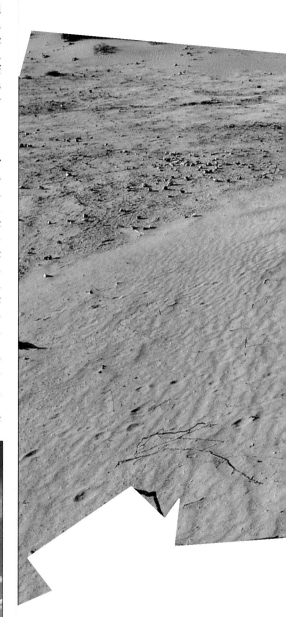

DESERTS FORM WHERE LESS THAN 250 MM (10 in) of rain falls each year. The scant rain is often sporadic, and while some desert areas receive a regular wetting others may receive no rain for years. Without rain, the only source of moisture is dew. As well as being dry, many deserts are very hot, but there are also cool deserts, such as the coastal desert of Baja California, the Gobi in central Asia and the Atacama of Peru and Chile.

With such meager water supplies there is only sparse vegetation. The plants are well adapted to resist drought, with extensive root systems, small leaves and waterproof coatings. Many desert plants can also store water. There is not enough food and water to support a some large population of animals, but there are large animals such as camels, asses and oryx, which travel extensively to find food. Small animals – gerbils, jerboas, lizards and tortoises – live in cool, moist burrows and come out at night to feed.

Unlike other biomes, deserts are spreading, encroaching on the savannas and steppe that border them. Although desert plants and animals are well adapted for survival in the harshest of environments, they are very vulnerable to human interference. Desert nomads survived with their flocks of camels and goats because their numbers were small and they kept on the move, but with increasing populations and periodic failures of the rains, there has been massive overgrazing and destruction of woody vegetation for fuel. Real deserts have their own forms of vitality; the new deserts are barren and lifeless.

Arctic desert and tundra

Polar regions are also in effect deserts. For much of the year what precipitation does fall remains frozen as ice or snow and cannot be used by plants and animals. The Arctic and Antarctic are geographic opposites: the former is an ocean surrounded by land, the latter a continent surrounded by ocean. Both are very cold, but whereas the Arctic lands warm up in summer the huge icecap of the Antarctic prevents warming, and life there is limited to the coastal fringe. The isolation of Antarctica from neighboring continents has also restricted colonization by plants and animals, whereas many hardy species have spread into the Arctic from the northern continents.

The Arctic tundra is a vast expanse of

Light green scrubland, known as mallee, fringes the sandy red deserts of the Australian outback. It is characteristically dominated by dwarf eucalypts and tufts of spinifex grass. Where the land becomes more arid the darker green of drought-tolerant acacias (mulga) are apparent.

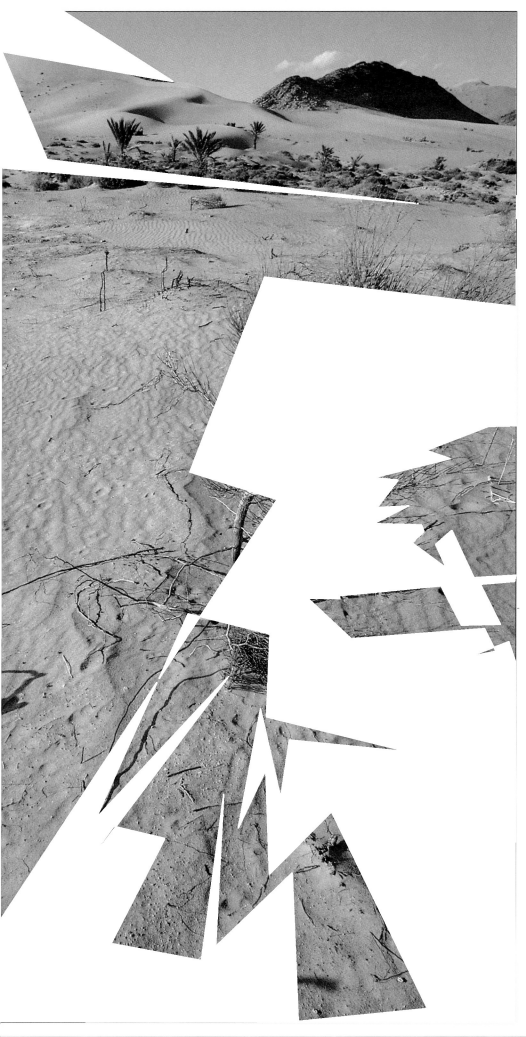

In this polar desert at Polar Bear Pass, Canada, lichens and mosses cover less than 20 percent of the ground, yet they are sufficient to support large herds of migrating caribou.

A drought-resistant acacia tree flowers after rare rains in the Algerian desert. The distribution of desert shrubs and trees reflects the location of scarce underground water supplies.

lowgrowing shrubby vegetation covering a soil that, except for a thin upper layer, is frozen all the year round (permafrost). The plants are typically small, longlived cushion plants; they survive the severe Arctic winters beneath an insulating covering of snow. Lichens and mosses often grow extensively. Small rodents such as lemmings live in tunnels in the ground or under the snow, where temperature fluctuations are not so severe, feeding on underground roots and stems and on seeds. They are food for predators such as Arctic foxes, owls and gyrfalcons. In summer insects breed in the numerous pools of meltwater, providing food for millions of breeding birds, temporary summer migrants to these feeding grounds.

The Antarctic has never been permanently settled, but the Arctic has supported nomadic people for millennia. They were either hunters like the Inuit or herders of reindeer like the Lapps (Saami). The indigenous peoples have traditionally lived in harmony with the environment, but commercial whalers, sealers, trappers and fishermen have destroyed animal populations on both continents. With the growing demand for minerals, oil and gas, these areas of pristine wilderness may also be at risk.

Oceans and Seas

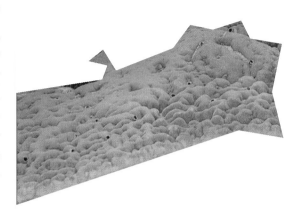

SALT WATER COVERS TWO-THIRDS OF THE world's surface. With an average depth of 4,000 m (13,000 ft), the oceans are the largest biome on Earth. Although they are all interconnected, forming one continuous body of water, the seas and oceans are not uniform. Barriers are created by changes in depth, light, temperature and salinity, creating distinct habitats. Even powerful animals such as whales and dolphins are constrained in their use of the ocean. As on land, some animals migrate from one habitat to another to breed or to feed: the migration of whales, fur seals, sea turtles and eels may cover thousands of kilometers.

The basis of life in the oceans is a vast floating community of microscopic plants and animals, the plankton. They thrive in the warmth and light of the surface waters in such numbers that they provide the sole food supply of many species of filter-feeding fish and also large animals such as the great blue whale and the basking shark.

In addition to the plankton a community of larger animals, the ocean surface drifters, also lives at the surface. These animals, mostly predators, rely on the wind to blow them across the oceans; other predators, such as sea slugs and bubble raft snails, feed on them. Most of these animals are deep blue on top, to camouflage them from the eyes of the seabirds that prey on them.

Below the surface waters there are other predators – tuna, barracuda and carnivorous sharks feed on the filter feeders, and on each other. Squid and octopus hunt at all depths, and sperm whales have been found hundreds of meters down. Near the coasts, marine mammals such as seals and sea otters dive to considerable depths in search of fish or mollusks, and seagulls and diving birds plunder the surface waters. A few birds, such as the albatross, spend their whole life at sea.

In the deep sea, where the temperature is low and light does not penetrate, there are many strange luminous fish and shrimps, generating their own light to attract prey or deceive predators. Many have distensible stomachs in order to make the most of rare opportunities for a meal. Hagfish, crabs and other scavengers help to decompose corpses on the bed of the deep sea, living at pressures that would crush the bodies of humans.

Until recently the great variety of ocean life was threatened only in limited ways, as when particular species such as the Peruvian anchovy, the Atlantic cod and the great whales were overfished. Pollution by chemicals and oil threatens small areas from time to time. There is a new and more serious threat in the form of huge nets extending for tens of kilometers across the Pacific, indiscriminately trapping fish, dolphins, turtles and other animals. These walls of death are putting whole food chains at risk, and could have a devastating impact on ocean life.

Plundering the oceans Chinstrap penguins breed on Antarctica and on subantarctic islands, but may range 1,000 km (620 mi) across the southern ocean in search of food in summer. Many seabirds spend much of their lives at sea; some albatrosses do not touch land for years at a time.

CORAL REEFS

Coral reefs are the rainforests of the sea. They form a highly complex environment, with steep cliffs, a flat top exposed to waves and high light intensities, and numerous nooks and crannies. They support a rich variety of colorful, bizarre and often specialized animal life. They also form a buffer for the shore against storms, and provide a nursery for many economically important fish, mollusks and other forms of life. Coral reefs require clear shallow water and a temperature between 17°C and 35°C (63–95°F).

Reefs are built up from the limestone skeletons of millions of small coral animals known as polyps. They resemble sea anemones, with cup-shaped bodies and a ring of stinging tentacles with which to catch their prey. Other animals contribute to the reef: parrotfish scrape at the corals, producing a coral sand that helps to bind the coral fragments together.

There are many ecological niches in a coral reef: moray eels, nurse sharks and octopus inhabit coral caves; lobsters hide in crevices; and a variety of invertebrates graze the plankton.

Coral reefs are fragile, slow-growing and susceptible to change. Reefs that are devastated by shortterm events such as hurricanes or sudden changes in the water temperature will eventually recover, but many are threatened in the long term by human activities.

Adapting to circumstance Corals have evolved special characteristics to fill the different niches of the coral reef. Near the surface the corals are robust enough to survive pounding by the waves: tiers of coral follow the direction of the waves to give minimum resistance, while widely spaced fingers of *Acropora* coral allow the energy of the waves to dissipate between them.

On the fringes of the sea Guillemots bob on shallow, nutrient-rich waters off a rocky coastline. Cliffs are important nesting sites for seabirds, as they are often free from predators.

Bare beaches of sand seem devoid of life, but the intertidal zone between land and sea, periodically exposed to air by the tidal movements of the water, is a rich habitat for plant and animal life. Many species of mollusks, worms, sea urchins and other invertebrates burrow below the sand when it is uncovered at low tide.

Islands

ISLANDS ARE CURIOSITIES. THEY TYPICALLY have a limited variety of plants and animals, but these may be extremely interesting in two ways, both of which have a bearing on the study of evolution. Relicts of ancient and once widespread forms – so-called "living fossil" species – are often found on islands. The tuatara, for example, the only survivor of a group of lizard-like reptiles that flourished alongside the dinosaurs, survived until a century ago in New Zealand and still lives on some offshore islands. By contrast, islands are also of interest because of their new species that have evolved in isolation quite recently. The diversity of newly evolved species on the Galapagos Islands off the Pacific coast of South America helped the naturalist Charles Darwin (1809–82) to formulate his theory of evolution by natural selection.

There are two kinds of islands: offshore or landbridge islands and oceanic islands. Offshore islands were once connected to the adjacent landmass, but they became separated by rising sea levels at the end of the last ice age about 10,000 years ago. Japan and the British Isles are examples. Their variety of animal and plant life is limited by separation, by their relatively small size, and by the fact that evolutionary processes are slow and new species have not yet had time to develop.

Oceanic islands are formed as a result of volcanic activity and are not directly linked to continental landmasses. Their inhabitants are descended from chance immigrants. Species that make the landfall may find little or no competition, and evolve into new species to fit the novel environment. The most famous are the many kinds of finches (Darwin's finches) that have evolved from the single species that reached the Galapagos Islands from the South American mainland. The number of these endemic species, as they are known, tends to increase the farther an island is from the nearest mainland: in the remote islands of Hawaii 97 percent of the species are found nowhere else.

Island wildlife is especially vulnerable because of the small population of each species and the specialized ways of life that many species adopt. Islands seldom have large native mammals, and in their absence land birds often become flightless. They have no defense when predatory mammals are introduced. Other islands have suffered from the introduction of adaptable species that destroy specialized habitats. An endemic gecko living on Round Island, off Mauritius, is endangered because the destruction of palm trees by introduced goats and rabbits has robbed it of its daytime roosts.

The invasion of settlers and visits by the trading and whaling ships, especially during the 18th and 19th centuries, has left very few islands in a pristine state. Where the vegetation has not been unwittingly destroyed by fire and introduced animals, it has been superseded by cultivation. Conservation of species has been successfully accomplished by taking specimens for captive breeding and propagation, but protection of often pitifully small remnants of the natural landscape may only be achieved if the human and other settlers leave the island altogether.

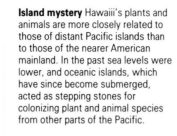

Island mystery Hawaiii's plants and animals are more closely related to those of distant Pacific islands than to those of the nearer American mainland. In the past sea levels were lower, and oceanic islands, which have since become submerged, acted as stepping stones for colonizing plant and animal species from other parts of the Pacific.

Unique species This iguana in the Galapagos Islands is descended from ancestors who reached the islands, 965 km (600 mi) from the South American mainland, by chance.

Islands at risk Coral atolls form as a result of island subsidence. As the island sinks coral reefs that have built up a barrier around it are left encircling a sheltered lagoon – a rich habitat for fish and many other marine animals.

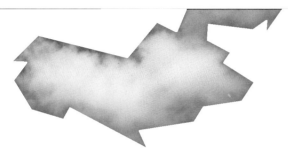

ST HELENA – SAVING SPECIES

St Helena is a small island in the Atlantic Ocean, lying south of the Equator and nearly 2,000 km (1,250 mi) off the west coast of Africa. It is the home of many relict species of plants and animals that survive nowhere else. The island was discovered by the Portuguese in 1502, when it was densely covered in dense vegetation; its subsequent history was a classic case of island degradation.

Goats were introduced as early as 1513; in increasing numbers they ate herbs and seedlings, preventing natural regeneration. Mature trees were cut for firewood, timber and bark for tanning. Crops and weeds were also introduced; European, American and Australian weed species all flourished. Many of its endemic plants were lost before they could be collected by botanists, and skeletons have provided evidence that at least eight endemic species of birds have also become extinct.

Conservation of the remaining endemics – some 24 species of flowering plants and ferns – has come just in time. In 1980 the ebony *Trochoetiopsis melanoxylon* was rediscovered, having been presumed extinct for over a hundred years. It has been successfully brought into cultivation. Propagation has also saved the St Helenian olive and the bastard gumwood, both of which had been reduced to single specimens. The government is now working with conservation organizations to reintroduce other endangered endemic species, which are now being planted as part of St Helena's official forestry policy to ensure their future survival.

Mountains

MOUNTAINS CAN BE LIKENED TO ISLANDS, AS they too form distinct, separate units of wildlife. Their plant and animal species may have evolved in isolation, or may be related to others in distant habitats. The latter are relicts that survived in the harsh mountain environment when conditions changed. Plants that were widespread in Europe when the climate was cooler immediately after the last ice age, for example, are now found only in the Arctic tundra and high up in mountains. Purple saxifrage and moss campion are two such plant species; among birds, the snow bunting and ptarmigan have a similar distribution.

The plant and animal life that can develop on mountains is determined largely by altitude. The air temperature drops by 1°C (1.8°F) for every 180 m (590 ft) rise in altitude; even in the tropics will have a snowcapped peak. As the climate becomes more severe, the vegetation on the mountain slopes changes. This vertical zonation, as it is known, produces quite distinct vegetation zones that often resemble the sequence of biomes from the Equator to the poles.

The vegetation zones follow a similar pattern, though the plant species they contain vary from one region to another. Between the lowlands and the snowcapped mountain peak the vegetation changes from lowland savanna or tropical forest, through temperate to coniferous forest and, above the treeline, through scrub to alpine meadows and then to

The Donkey's Ears peaks at the summit of Mount Kinabalu, the highest mountain in Southeast Asia. Even in the tropics, windswept mountain summits may be too hostile an environment to support more than a few lowgrowing shrubs.

tundra-like vegetation such as lichens and mosses, and eventually to bare rock and snow.

The pattern of these zones is affected by aspect as well as altitude. One side of a mountain may receive more of the Sun's warmth and more rain than the other: in India the forests extend up the slopes of the Himalayas to 3,000 m (10,000 ft); on the Tibetan side the treeline is 1,000 m (3,300 ft) higher.

Survival in the mountains

On the lower slopes of mountains the animal life is plentiful and rich, and similar to that of corresponding biomes elsewhere in the world. Higher up, the inhospitable mountain environment – sparse vegetation, steep, rocky terrain, penetrating cold and chilling winds – makes life difficult for all but the hardiest of animals. Mammals can survive only if they are protected by thick woolly fur, such as that of the alpaca and chincilla of the Andes in South America and angora goats in the mountains of southern central Asia. On steep slopes the sure-footedness of the chamois and ibex is essential; the clumsier Tibetan yak is restricted to less precipitous areas. Snow leopards and wolves are among the few large predators. In summer the alpine

pastures are rich with butterflies; even at high altitudes insects can be surprisingly abundant, blown up from the lowlands and deposited, frozen, to be preyed on by resident insects and spiders. Some species, such as moose, deer and mountain quail, avoid the harshest times of year by retreating to lower altitudes for the winter. Mountain people also follow this way of life, taking their herds up the mountains for summer grazing.

Tussucky native grasses cover the mountain slopes of New Zealand's South Island. Separated from the other continents before mammals had evolved, New Zealand has no native grazing mammals and this ecological niche was filled by birds, many of them flightless.

As the winter snows melt mountain slopes above the treeline are covered in lush meadows and wildflowers. Plants survive the winter covered by a protective layer of snow that insulates them from the cold. During the summer many animals migrate up the mountain to graze the meadows.

Mountain zonation Traveling up a mountain, plant and animal communities change with increasing altitude and falls in temperature. The different zones correspond to the succession of biomes between the Equator and the higher latitudes. Even mountains on the Equator have tundra near the summit.

The remoteness and harsh environment of mountains has ensured their survival as wilderness, refuges for animals such as bears, wolves and eagles after they have been eradicated from lowland areas. Few people live permanently in the mountains, though in some areas there is now a threat from growing numbers of visitors. Trekking and, in particular, skiing activities can inflict major damage on local mountain ecosystems.

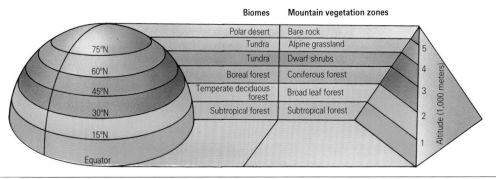

Biomes	Mountain vegetation zones
Polar desert	Bare rock
Tundra	Alpine grassland
Tundra	Dwarf shrubs
Boreal forest	Coniferous forest
Temperate deciduous forest	Broad leaf forest
Subtropical forest	Subtropical forest

Lakes and Wetlands

Large lakes have many habitats for wildlife. Conditions change with depth, as light, temperature and oxygen availability decrease. Over a long period of time, sediments carried into lakes by rivers are trapped in the vegetation at the water's edge, building up the banks and creating new wetlands. Around the edge of the lake the spread of aquatic vegetation creates new habitats such as reed-beds and marshes. Where a lake is large enough to be affected by the prevailing wind, the exposed shore may provide a very different habitat from that of the more sheltered shore.

A large floating community of algae forms the base of the aquatic food chain. The most numerous of these are the diatoms, with their delicate silica shells. The zooplankton – microscopic crustaceans and the larvae of fish, mollusks and other aquatic animals – feed on the algae. These in turn are food for fish, water beetles and other small predators.

Filter feeders are common. They include larvae of mosquitoes and midges, freshwater shrimps and mussels. Worms, snails and other scavengers feed on organic debris. Beetles and snails graze the algae that coat the water weeds.

The water's edge is home to frogs and salamanders and the insect adults of many aquatic larvae. Mammals such as deer and antelope visit the lake to drink. In the spring and fall lakes are important stopover sites for migrating waterfowl: stocked with food and easily recognizable from a distance, the open water offers a safe roosting site.

Apart from sediments and organic material that wash in and out, lakes form virtually closed environments. Like other isolated habitats, over long periods of time they gradually evolve their own unique species. Lake Victoria in eastern Africa contains hundreds of species of small fish called cichlids, all of which have evolved from a single species in the million or so years since the lake was formed. In the Soviet Union Lake Baikal, the world's deepest lake, contains 960 species of animals and 400 species of plants that live nowhere else in the world.

Wetlands

Wetland ecosystems are dominated by water and inhabited by animals and plants with a more or less aquatic way of life. This includes a wide range of habitat types: river, lake, marsh, fen, swamp, bog or mire, saltmarsh, mangrove swamp and estuary. Wetlands are a stage in the natural succession from an aquatic to a terrestrial habitat. They have abundant areas of open water, breeding grounds for insects that in turn attract birds, amphibians and other insect-eaters. The wet mud contains worms, crustaceans and mollusks, and the damp soil is easily probed by beaks and paws.

Wetland ecosystems are dispersed throughout the major biomes, and are some of the most productive habitats. Coastal wetlands – estuaries, saltmarshes and mangroves – are used by marine animals for breeding, and are the nurseries for many commercial species of fish. In many parts of the world there is increasing pressure to convert wetlands into agricultural land to support growing human populations.

Reed beds are important habitats for wildlife. Their shelter conceals breeding birds, small mammals and amphibians from predators.

A glacial lake adds to the beauty of upland scenery. The margins of lakes are never static. Sediment accumulates around the roots of plants growing at the water's edge, creating a swampy habitat that is then colonized by other plants.

A focus for wildlife Dense vegetation surrounds a waterhole in Yala National Park, Sri Lanka. The presence of water allows a wide variety of animals to live in the surrounding area.

MANGROVES

A mangrove swamp is like a forest on stilts. Mangroves grow in tidal waters in the tropics; they thrive in waters where there is too much silt for coral reefs to form. Rotting leaves falling into the water and nutrients from the silt trapped by the mangrove roots feed the microscopic plankton that support a nursery for crustaceans, mollusks and fish. This abundance of food attracts numerous birds such as ibises, pelicans and frigatebirds. They nest in the mangroves, their droppings adding to its fertility.

Tropical coasts support millions of hectares of mangroves, but they are disappearing rapidly in many places. Large areas have been destroyed for building, rice-growing, charcoal burning and aquaculture, even though the fish farms themselves get their stock from mangroves. There is, however, a growing realization that mangroves provide a self-repairing coastal flood barrier, and they are increasingly being replanted and protected.

Storm barrier Mud trapped between mangrove roots gradually builds up the coast.

Changing Biomes

COMPLETELY NATURAL LANDSCAPES ARE able to develop only where the land and its vegetation are left undisturbed for hundreds or even thousands of years. These communities of plants and animals eventually reach a stable equilibrium, and are often called climax communities. If the vegetation and soil are disturbed they pass through a succession of transitional stages before the original landscape is restored. These stages are much more common in our overpopulated world than the pristine natural biomes favored by wilderness enthusiasts.

Humans are not, of course, the only agents of change. There are also natural disturbances such as floods and landslides, earthquakes and volcanic eruptions, fire and hurricanes that destroy the climax vegetation, and sometimes fundamentally alter soil and topography. Volcanic outpourings and retreating ice sheets and glaciers expose new land surfaces, on which soil formation and plant colonization has to start afresh.

Even where there are no obvious signs of disturbance a slow progression may be taking place, such as that around the shores of lakes. Here mud trapped among the roots of plants growing in the shallow water builds up the land, and the shore encroaches on the lake. As the land builds up it drains better: in temperate regions the reed beds and marshes give way to trees such as alder and willow, which can tolerate waterlogged soil. As the leaf litter and decaying tree trunks build up the soil, less tolerant species such as oak and beech can invade. Gradually a woodland community is established, and in time this eventually evolves into the climax vegetation.

Where forests are logged or cleared for cultivation it takes a long time for the original forest to reestablish itself. The species of the deep forest are not adapted to the exposed conditions of cleared land, with its high winds, bright light and wide fluctuations in temperature and humidity. Nutrients may have been leached out of the exposed soil, making it difficult for forest seedlings to become established. Only when dense forest cover returns can the original forest species compete, and replace those of the secondary forest.

Many familiar habitats are the result of human activities arresting the natural process of succession, holding the vegetation in a transitional state. Grasslands are maintained by the grazing of domesticated animals such as cattle and sheep.

Many natural heath and grassland habitats are maintained by periodic fires caused by lightning. Manmade heaths and grasslands have been deliberately fired since prehistoric times, usually to provide fresh new growth for grazing. The fires accelerate the rate at which plant material decomposes for recycling, and can also prevent tree species from becoming established. Misguided attempts to prevent fire in nature reserves have led to the invasion of scrub and woodland.

Succession can be a major problem for the managers of national parks, who may have to intervene to preserve a particular transitional habitat that is rich in wildlife. In order to preserve as many species as possible, the successional stages of the major biomes need to be considered. Succession is as much a part of the natural order as the equilibrium that some climax communities eventually achieve.

Once woodlands have been destroyed they can take a very long time to reestablish themselves: young seedlings have difficulty in gaining a hold on the exposed crests of hills, for example. Downland pasture like this, cleared for grazing centuries ago, is becoming very uneconomical to farm using traditional methods.

Heathlands of purple flowering heather, once common in many parts of western Europe, are a habitat resulting from human activity. They need to be maintained by controlled burning if they are not to be invaded by trees such as the birches growing here. These heathlands in southern England support a rich wildlife, including many rare reptiles; they are disappearing rapidly as the land is used for urban development.

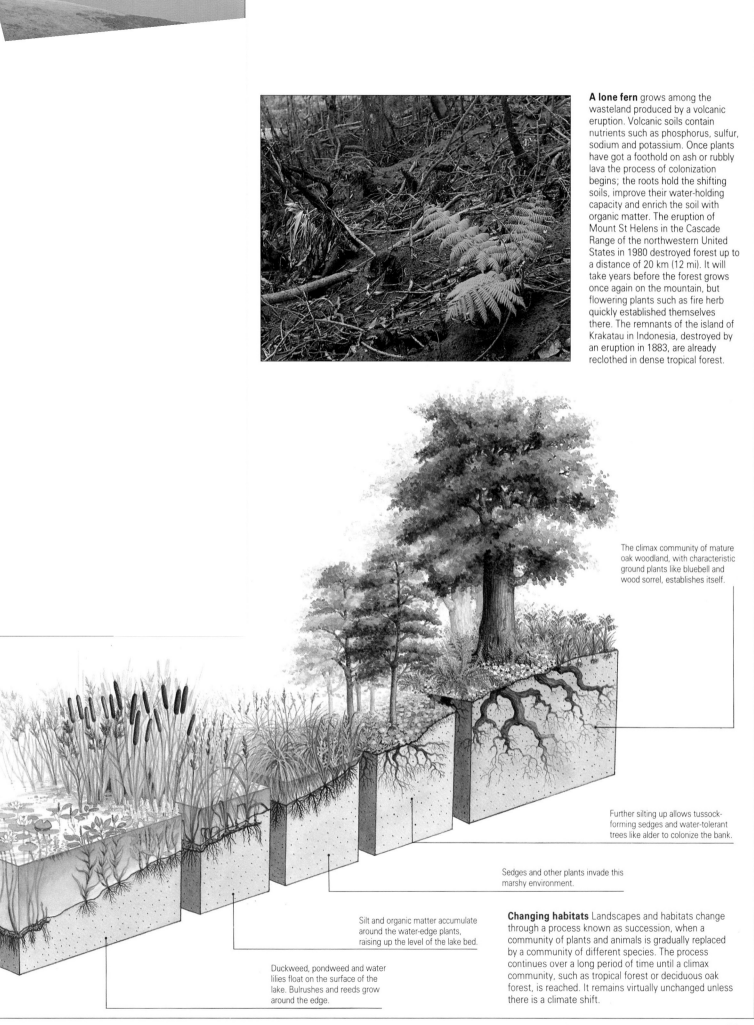

A lone fern grows among the wasteland produced by a volcanic eruption. Volcanic soils contain nutrients such as phosphorus, sulfur, sodium and potassium. Once plants have got a foothold on ash or rubbly lava the process of colonization begins; the roots hold the shifting soils, improve their water-holding capacity and enrich the soil with organic matter. The eruption of Mount St Helens in the Cascade Range of the northwestern United States in 1980 destroyed forest up to a distance of 20 km (12 mi). It will take years before the forest grows once again on the mountain, but flowering plants such as fire herb quickly established themselves there. The remnants of the island of Krakatau in Indonesia, destroyed by an eruption in 1883, are already reclothed in dense tropical forest.

The climax community of mature oak woodland, with characteristic ground plants like bluebell and wood sorrel, establishes itself.

Further silting up allows tussock-forming sedges and water-tolerant trees like alder to colonize the bank.

Sedges and other plants invade this marshy environment.

Silt and organic matter accumulate around the water-edge plants, raising up the level of the lake bed.

Duckweed, pondweed and water lilies float on the surface of the lake. Bulrushes and reeds grow around the edge.

Changing habitats Landscapes and habitats change through a process known as succession, when a community of plants and animals is gradually replaced by a community of different species. The process continues over a long period of time until a climax community, such as tropical forest or deciduous oak forest, is reached. It remains virtually unchanged unless there is a climate shift.

The Process of Destruction

WILDERNESS LANDSCAPES ARE NEVER unchanging. Over the course of centuries rocks are gradually worn down, rivers change course or dry up, plants invade lakes and ponds, and sand dunes slowly acquire a green cloak of vegetation. Sometimes the changes are more catastrophic, like those caused by the fires that swept Yellowstone National Park in the United States in 1988. Earthquakes and volcanic outpourings make permanent changes. Animals can change the wilderness, too. Elephants may reduce the vegetation to a wasteland, then move on to new pastures, leaving the habitat gradually to recover.

For thousands of years humans have been part of this process, clearing patches of forest to plant crops or pasture livestock – so-called slash-and-burn cultivation – and moving on when the soil became exhausted. Until recently these disturbances were local, and in many places the destruction was only temporary. At worst they left seminatural countryside, and even agricultural areas often supported large numbers of the original species of plants and animals. This destruction has greatly accelerated during the human population explosion of the last hundred years or so; in some places wilderness degradation has been irrevocable and permanent.

The human impact

The oldest causes of wilderness loss are settlement of the land and agriculture. Most of the natural forest disappeared from temperate regions centuries ago; now tropical forests are suffering the same fate. Some agricultural practices can be ecologically harmonious, but only at low population levels.

As the land has to support increasing numbers of people, the demands made on it are greater, and the soil is no longer left to recover. Clearance of unsuitable land leads to soil erosion and loss of protection for watersheds. Forests are also cleared or degraded by felling for timber and other forest products. Logging roads open up previously impenetrable country to settlers. In other areas alien species are planted, and may eventually replace the native species of many natural habitats such as heaths, peatlands and dunes.

The growth of cities and the spread of the industrial way of life of developed countries has created new pressures on wilderness areas. The cities spread out over wild places, and the people who live in them look to the surrounding countryside for recreation. The need to transport goods from the countryside to the cities and from city to city has spawned roads that open up more wild places to settlers and tourists.

The search for raw materials to support this way of life also opens up and destroys wilderness areas. The transport of oil, by pipeline and by supertanker, is a hazardous operation. Mines and their tailings disfigure and pollute the surrounding countryside and watercourses.

There is widespread poisoning of the

Extensive erosion in the Kilifi area of Kenya. Soil erosion has become a worldwide environmental problem, causing the loss of 1.1 m sq km (425,000 sq mi) of land each year. Although some of this is the result of natural processes, it is the increase in human population and activity that is mainly responsible.

Deforestation at Roraima, Brazil
Tropical rainforests are being destroyed at a rate of nearly half a hectare – the size of a football pitch – per second. The main causes are commercial logging and land clearance for cattle ranching. Rainforests are the habitat for nearly half of all the species on Earth.

land, air and water. Powerful, persistent pesticides seep into groundwater and work their way through the food chain. Acid rain resulting from the release of oxides of sulfur and nitrogen by industrial furnaces and motor vehicle exhausts destroys forests and renders lakes barren. Overenrichment (eutrophication) and poisoning of fresh waters results from the excessive use of nitrogen fertilizers. These processes know no boundaries.

On an even larger scale is the expected warming of the global climate because of the excessive release into the atmosphere of carbon dioxide and other "greenhouse" gases. This may bring about changes in temperature and weather patterns more rapid than anything plant and animal life has experienced before.

Increasing affluence in some parts of the world has brought about a boom in tourism. Coastlines are disappearing under development, ski resorts are springing up on vulnerable mountain slopes, and backpackers are trampling fragile upland vegetation. Tourism can play a significant part both in a country's economy and in financing conservation, but overexposure can ruin the very wilderness that tourists come to see.

Overgrazing of marginal land bordering deserts, as here on the edge of the Sahara in Tunisia, destroys the vegetation cover. This exposes the topsoil, which dries out and is easily blown or washed away. Poor irrigation, deforestation and intensive farming are also causes of soil erosion.

The Aims of Conservation

THE PRIME PURPOSE OF CONSERVATION IS sometimes thought to be simply the preservation of areas of spectacular scenery or rich wildlife, either for their own sake or for people to enjoy. In fact the conservation of natural ecosystems is far more important than this. Forests in upland water catchment areas help to maintain supplies of water for irrigation and drinking; savanna woodland prevents the invasion of desert; mangrove forests and coral reefs act as barriers against encroachment by the sea. The presence of extensive forests and large lakes or inland seas has a considerable effect on the climate of neighboring areas, influencing both the annual temperature range and the pattern of rainfall. Forests also play a vital part in maintaining the oxygen and carbon dioxide balance of the atmosphere around the world. These natural environments form the life-support system of the planet, and that makes their protection vitally important.

Preserving diversity

The enormous variety of species of both plants and animals provides a living laboratory whose resources have hardly begun to be tapped. Each year previously unknown species are being discovered that produce important drugs or other commerical products. For instance, childhood leukemia deaths have been dramatically reduced by a drug extracted from the rosy periwinkle of Madagascar. Study of the polar bear's tolerance of the cold has led to new ideas on carrying out surgery at low temperatures. New varieties of crops such as rice are being developed using genetic resources taken from wild populations. The human race cannot afford to lose this source of knowledge before all its riches have been tapped.

All species are unique and irreplaceable. The most effective way to preserve the diversity of living species is to preserve each of the main biomes in each of the main geographical regions, as there are different species not only in different biomes but also in the same biome in different parts of the world. An alternative approach is to focus on key species, because a reserve of the size needed by an animal near the top of a food chain, such as a tiger or an eagle, will automatically contain many other species in viable numbers.

Conservation is not simply about preserving wilderness areas. It is also about the wise and balanced use of resources, a use that will prevent them from becoming exhausted. It encompasses both the preservation of virgin and near natural wildernesses and the use of wilderness resources such as water and timber, the management of plant and animal stocks for food and sport, and the use of the areas for leisure pursuits. The needs of local people also have to be taken into consideration. It is no use trying to preserve a natural forest if the local population has no alternative source of firewood or other fuel.

These aspects of conservation management are not necessarily incompatible. However, as no wilderness area is immune from the effects of human activity elsewhere in the world, they can no longer be considered in isolation. There is an urgent need to extend the principles of conservation into every aspect of both local and regional planning.

Protecting the rural heritage The landscapes of western Europe, as here in the English Cotswolds, have been managed by humans for centuries.

Intensive farming techniques are now threatening even these landscapes and their associated wildlife. There are many calls to protect them.

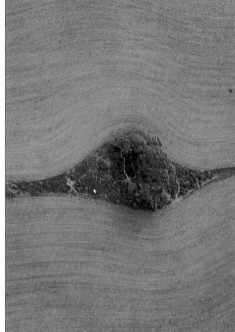

Pressure on the land An island of remnant mallee scrubland survives in a plowed field in southern Australia. The desire to increase crop yields on fertile land means there is little room left for wildlife unless the landscape is actively managed for conservation.

Rainforest in Colombia The ecological wealth of tropical rainforests and the alarming rate at which they are being destroyed has made them the object of international conservation programs and campaigns. Arguments for conserving these forests hinge on their great age, diversity of species, and their role in regulating the global climate.

KEYSTONE SPECIES

Some species play a key role in an ecosystem: remove one and a cascade of other species may decline into extinction. A study in Kalimantan, Indonesia, showed how a number of animals depend on figs. They include hornbills, parrots, bulbuls, fruit pigeons and many other birds, bearded pigs, civets, squirrels, fruit bats, macaques, gibbons and orangutans. In times of fruit shortage, life becomes difficult and the animals either have to travel in search of figs in other parts of the forest, or turn to less favored food; in this case a rare kind of coco plum attracted an influx of pigeons and hornbills.

If figs are permanently lost from the ecosystem, their removal will eliminate or deplete populations of animals. The loss of these animals will in turn prevent the seeds of figs and other plants being dispersed. Since figs are tree climbers, even low intensity logging drastically reduces their number because many grow on timber trees.

Protecting the Wilderness

AN ESSENTIAL PART OF ANY CONSERVATION program is a system of protected areas in which representative samples of the natural biomes are preserved. The Commission on National Parks and Protected Areas, established by the International Union for the Conservation of Nature and Natural Resources (IUCN), has instituted a scheme for identifying areas needing protection. It aims to preserve an adequate area, ideally 10 percent, of each biome in each geographical province. These needs have not yet been met, though all provinces have some area under protection. For example, by 1990 only about 4 percent of the tropical and subtropical humid forests were protected.

Such analyses are long overdue. Detail-ed study of vegetation maps of southern Africa, for example, shows that protection has been aimed at areas that support spectacular concentrations of large mammals, while other areas, which may contain greater biological diversity, are not so well conserved.

Since the first national park was established in 1872 at Yellowstone, in the United States, the number of protected areas has been increasing every year, particularly in the developing world. However, their distribution is patchy, and many are protected in name only. Some countries, such as Costa Rica, have protected large areas; but over half the countries in the tropics, where pressure on wilderness is greatest, have no properly established system of protected

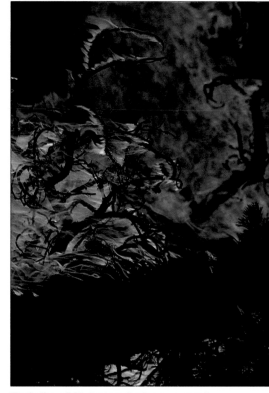

Fire in the wild helps to maintain some natural habitats. However, if allowed to get out of control, fire can cause enormous damage. Tropical forests may take centuries to recover from its ravages.

Successful conservation of the South American rainforest must take account of the indigenous Indian communities whose traditional lifestyles, based on sustainable cultivation and harvesting of forest resources, are in harmony with their environment.

GREENHOUSE CORRIDORS

Dramatic changes in climate are expected to result from the warming of the Earth's atmosphere because of the greenhouse effect. These would have a profound effect on habitat distribution and species diversity. If these changes take place, the major biomes will shift toward the poles. The greatest changes will be in the middle and higher latitudes. Much of the tundra will disappear, to be replaced by boreal forest, while the band of temperate deciduous forest will move north.

The species composition of biomes may also change, as not all plants and animals will be able to adapt to climatic change at the same rate. Some species, such as those of high alpine habitats, may become extinct as their habitat disappears altogether. New species will evolve to fill new ecological niches.

The consequence for the remaining wilderness strongholds could be disastrous, as their inhabitants migrate to unprotected areas. There is an urgent need to establish migration corridors of natural habitat between reserves so that species will be able to survive the coming changes. The United States already has plans for a network of "greenways" across the country. It is to be hoped that other countries will follow this example.

Slash and burn agriculture encroaches upon the wilderness. Crops are grown on land in the forest cleared by felling the trees and burning the vegetation cover. Once the soil is exhausted the cultivator moves on to a new area.

areas. The categories of reserves also vary greatly from one country to another, both in name and in significance. A national park in one country may receive strict protection, while in another it may support farming, industry and even mining within its borders.

The problem of reserve size

A reserve must cover at least the minimum area required both to maintain viable breeding populations of its various species, and to accommodate the migrations of nomadic species. This is becoming increasingly difficult to achieve as the world's landscapes are whittled away and fragmented. (An ordinary road driven through a rainforest is sufficient to prevent the movement of small birds and mammals.) Even areas where wildlife is protected may therefore slowly lose their species diversity.

An example of the problem is provided by Barro Colorado Island. The island was formed when the Panama Canal was excavated, and has remained a rainforest reserve with sloths, armadillos, howler monkeys, peccaries, tapirs, coatis and many other animals. But its predatory jaguars, pumas and harpy eagles have died out because the island is not large enough to support viable populations. As a result, several species of ground-nesting birds have also become extinct on the island because there are no longer any large predators to hold down the numbers of peccaries and coatis, which rob the birds' nests.

It is natural for changes in population to result in occasional local extinctions. Recolonization eventually takes place as species move in from surrounding areas. However, when populations are isolated from neighboring populations of the same species by farmland, roads or urban development, they cannot be restocked. This problem may be overcome by providing a series of smaller reserves linked by protected corridors or "stepping stone" reserves. This allows species to migrate between the major reserves, creating a larger genetic pool for interbreeding.

Managing the Reserves

IT IS NOT SUFFICIENT SIMPLY TO DESIGNATE protected areas. They also require careful management. Reserve managers need to be rather remarkable people. They must have the scientific knowledge to understand the complex interactions taking place in the ecosystems under their charge, and the management experience to know how to cope with them. Personnel management is also important. The reserve may have a local staff of porters, cooks, guards and guides, as well as visiting research staff of many different nationalitites.

The local people must also be involved with the reserve if it is to win their support. This means meetings with town and village officials, and an understanding of local traditions and problems. An education program may be desirable. Poaching may have to be combated using armed guards. The manager also has to finance the work of conservation. This involves selling the conservation ideal to business and to government, and managing effectively an often limited budget.

Conservation for the people

Poaching, encroachment for agriculture and the collecting of firewood pose some of the most serious threats to protected areas. Local people have traditionally used the land to provide the necessities of life, and their needs increase with rising expectations and growing populations. Clashes often result between local people and park management. These are in part due to the fact that many protected areas were set up as exclusive areas for wildlife, with little or no account being taken of local people's needs. Not surprisingly, they resent the fact that they are unable to touch an area containing a convenient source of meat, grazing and firewood.

A more enlightened approach to reserve management recognizes that local people should benefit from the presence of the reserve. This may take the form of a sustainable crop, as in the Royal Chitwan Park, Nepal, where people are allowed in once a year to cut grass for thatching; or it may be an indirect benefit such as a cash crop of meat from large game or income from tourism.

With a sensitive approach it is possible to manage a reserve and win the support of local people. The Parc des Volcans in Rwanda, home of the endangered mountain gorilla, provides a fine example. Rwanda is one of Africa's poorest and most densely populated countries. Pressure on the land is intense, yet surveys of public opinion have shown that while in 1980 51 percent of farmers wanted more of the park to be made over to agriculture, by 1984 only 18 percent still wanted this, an amazing swing due mainly to the efforts of the park officials and the Rwandan Office of Tourism and National Parks. The case for the gorillas is actively promoted by films and talks in villages and schools. Gorilla-centered tourism is one of Rwanda's largest sources of foreign currency, some of which finds its way back to the conservation program. The tourist industry is a major source of employment for many Rwandans.

The world's wilderness areas have many dedicated and talented managers working to ensure their future. Increasingly, knowledge and experience are being shared worldwide, and international aid and expertise are being applied to saving these last strongholds. There is a good prospect, therefore, that future generations will still have reservoirs of wilderness areas to study and to enjoy.

One important aspect of reserve management is the monitoring of species populations to record any changes in size or distribution. Even small variations may indicate major disturbance to the ecosystem.

An influx of elephants has caused major damage to the trees in this park. Human encroachment on their natural wide-ranging territory causes great concentrations of population.

A park warden burns encroaching vegetation in Kruger National Park. Careful management is needed to ensure the right balance of grassland is maintained to support the park's population of grazing animals.

INTERNATIONAL FINANCE FOR CONSERVATION

Many of the poorer countries of the world still lack properly constituted and managed systems of protection for their wilderness areas. Their priorities are development schemes, education, public health and employment. Yet sustainable management of living resources is vital to their longterm welfare.

International assistance and funding for protected areas can offer a solution. This is a developing field in which traditional conservation agencies such as the World Wide Fund for Nature (WWF), the International Union for the Conservation of Nature and Natural Resources (IUCN), the New York Zoological Society, the West German *Brot für die Welt*, and the United Nations agencies FAO and UNESCO have been joined by international banks, including the World Bank.

In 1986 the World Bank introduced a "wildlands policy" whereby it refuses to finance projects involving the conversion of wildlands (relatively unmodified natural habitats) unless allowance is made for wildland management or includes the protection of an alternative area.

In 1987 Conservation International, an organization based in the United States, established the idea of debt swap for hardpressed developing countries. In the first debt swap, Conservation International purchased $650,000 of Bolivia's national debt in return for the Bolivian government agreeing to designate 160,000 sq km (61,760 sq mi) of forest and savanna as the Beni Biosphere Reserve. The government also provided $250,000 in local currency to pay for staff to guard and manage the reserve. Debt swap offers a construcitve way of integrating economic development with the effective conservation of natural resources.

REGIONS OF THE WORLD

CANADA AND THE ARCTIC
Canada, Greenland

THE UNITED STATES
United States of America

CENTRAL AMERICA AND THE CARIBBEAN
Antigua and Barbuda, Bahamas, Barbados, Belize, Bermuda, Costa Rica, Cuba, Dominica, Dominican Republic, El Salvador, Grenada, Guatemala, Haiti, Honduras, Jamaica, Mexico, Nicaragua, Panama, St Kitts-Nevis, St Lucia, St Vincent and the Grenadines, Trinidad and Tobago

SOUTH AMERICA
Argentina, Bolivia, Brazil, Chile, Colombia, Ecuador, Guyana, Paraguay, Peru, Uruguay, Surinam, Venezuela

THE NORDIC COUNTRIES
Denmark, Finland, Iceland, Norway, Sweden

THE BRITISH ISLES
Ireland, United Kingdom

FRANCE AND ITS NEIGHBORS
Andorra, France, Monaco

THE LOW COUNTRIES
Belgium, Luxembourg, Netherlands

SPAIN AND PORTUGAL
Portugal, Spain

ITALY AND GREECE
Cyprus, Greece, Italy, Malta, San Marino, Vatican City

CENTRAL EUROPE
Austria, Liechtenstein, Switzerland, West Germany

EASTERN EUROPE
Albania, Bulgaria, Czechoslovakia, East Germany, Hungary, Poland, Romania, Yugoslavia

THE SOVIET UNION
Mongolia, Union of Soviet Socialist Republics

THE MIDDLE EAST
Afghanistan, Bahrain, Iran, Iraq, Israel, Jordan, Kuwait, Lebanon, Oman, Qatar, Saudi Arabia, Syria, Turkey, United Arab Emirates, Yemen

NORTHERN AFRICA
Algeria, Chad, Djibouti, Egypt, Ethiopia, Libya, Mali, Mauritania, Morocco, Niger, Somalia, Sudan, Tunisia

CENTRAL AFRICA
Benin, Burkina, Burundi, Cameroon, Cape Verde, Central African Republic, Congo, Equatorial Guinea, Gabon, Gambia, Ghana, Guinea, Guinea-Bissau, Ivory Coast, Kenya, Liberia, Nigeria, Rwanda, São Tomé and Príncipe, Senegal, Seychelles, Sierra Leone, Tanzania, Togo, Uganda, Zaire

SOUTHERN AFRICA
Angola, Botswana, Comoros, Lesotho, Madagascar, Malawi, Mauritius, Mozambique, Namibia, South Africa, Swaziland, Zambia, Zimbabwe

THE INDIAN SUBCONTINENT
Bangladesh, Bhutan, India, Maldives, Nepal, Pakistan, Sri Lanka

CHINA AND ITS NEIGHBORS
China, Taiwan

SOUTHEAST ASIA
Brunei, Burma, Cambodia, Indonesia, Laos, Malaysia, Philippines, Singapore, Thailand, Vietnam

JAPAN AND KOREA
Japan, North Korea, South Korea

AUSTRALASIA, OCEANIA AND ANTARCTICA
Antarctica, Australia, Fiji, Kiribati, Nauru, New Zealand, Papua New Guinea, Solomon Islands, Tonga, Tuvalu, Vanuatu, Western Samoa

North America

Central and South America

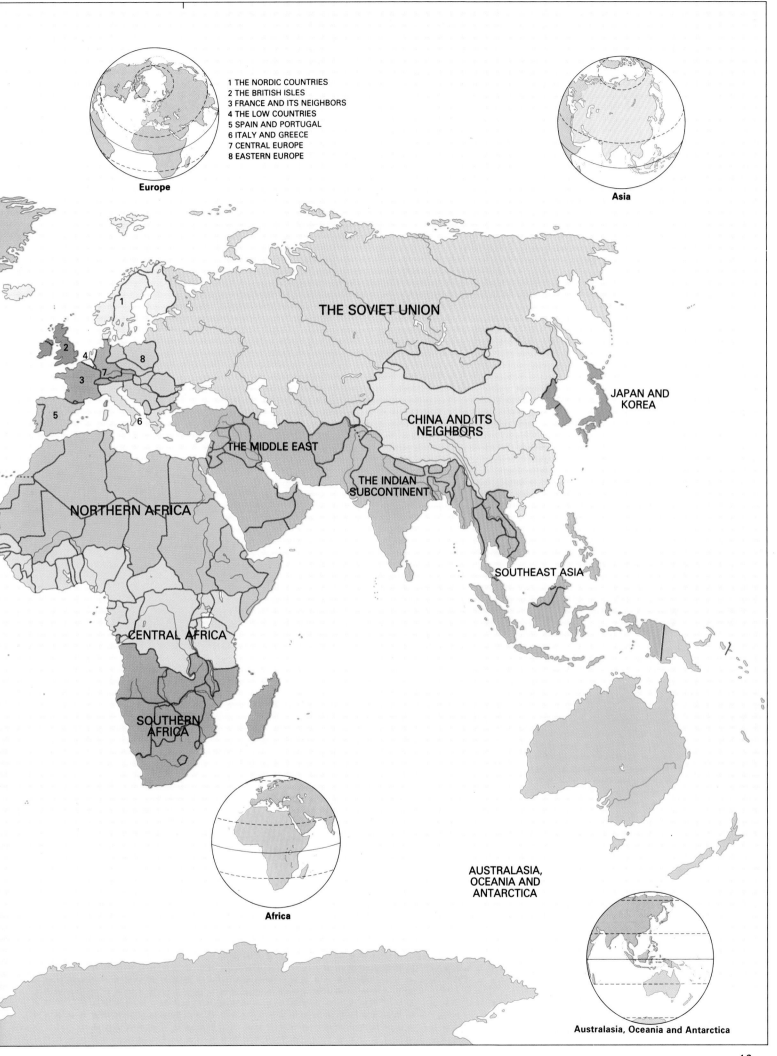

1 THE NORDIC COUNTRIES
2 THE BRITISH ISLES
3 FRANCE AND ITS NEIGHBORS
4 THE LOW COUNTRIES
5 SPAIN AND PORTUGAL
6 ITALY AND GREECE
7 CENTRAL EUROPE
8 EASTERN EUROPE

Europe

Asia

THE SOVIET UNION

JAPAN AND KOREA

CHINA AND ITS NEIGHBORS

THE MIDDLE EAST

THE INDIAN SUBCONTINENT

NORTHERN AFRICA

SOUTHEAST ASIA

CENTRAL AFRICA

SOUTHERN AFRICA

Africa

AUSTRALASIA, OCEANIA AND ANTARCTICA

Australasia, Oceania and Antarctica

WIDE OPEN SPACES

The vast tracts of Canada and Greenland extend from a narrow, urbanized zone bordering the United States to within a few hundred kilometers of the North Pole. They encompass a wide range of habitats from temperate rainforest to Arctic tundra, and from prairie grassland to dense coniferous forest. Most of their huge open spaces are sparsely inhabited, or not at all, and are likely to remain wilderness areas because of their harsh climate and inhospitable terrain. The region marks the furthermost limit for many American plants and animals that traveled northward at the end of the most recent ice age. It also shares many species with Europe and Asia that migrated across the Bering Strait over the land bridge formed by the advance of glaciers and changing sea levels many thousands of years ago.

COUNTRIES IN THE REGION	
Canada	
Major protected area	**Hectares**
Algonquin PP	765,345
Banff NP	664,106
Bruce Peninsula NP	27,000
Fathom Five NMP	147,000
Gros Morne NP WH	194,240
Jasper NP	1,087,800
Kejimkujik NP	38,151
Kluane NP WH	2,201,500
Kootenay NP	137,788
Laurentides/Charlevoix PP BR	966,300
Melville Bay NR (Greenland)	1,050,000
Mont St Hilaire BS BR	5,550
Mont Tremblant PP	124,800
Mount Revelstoke NP	26,263
Nahanni NP WH	476,560
Northeast Greenland NP BR	70,000,000
Northern Ellesmere Island NP	3,950,000
Northern Yukon NP	1,016,865
Pacific Rim NP	147,000
Point Pelee/Long Point PP BR	27,000
Polar Bear PP	2,410,000
Polar Bear Pass NP	81,000
Quetico PP	475,819
Riding Mountain NP BR	297,591
South Moresby NP	147,000
Waterton Lakes NP BR	52,597
Wood Buffalo NP WH	4,480,700
Yoho NP	131,313

BR=Biosphere Reserve; NMP=National Marine Park; NP=National Park; NR=Nature Reserve; PP=Provincial Park; WH=World Heritage site

Prairie wetlands, or sloughs, are valuable habitats for waterfowl and waders. The typical vegetation includes sedges, reeds, horsetails, bulrushes and floating duckweeds and water lilies. Most of the prairie grasslands have now vanished beneath the plow.

NORTHERN HABITATS

Canada contains large areas of relatively flat land that are dominated by mountains in the west; these form the northern end of the Cordillera belt, which extends southward through the United States to Central and South America. To the west of these mountains, the land is deluged in rain coming in from the Pacific. This supports a belt of temperate rainforest of huge, ancient evergreens.

To the east, in the rain shadow of the mountains, lie stretches of treeless grassland. This is the prairie zone, which experiences extremes of temperature in the annual alternation of airmasses flowing up from the subtropical south and down from the Arctic north. The prairie grassland ecosystem is maintained by drought, fire and grazing. Most of it has been destroyed by plowing and cultivation. Only a pitiful remnant is left among the vast fields of crops as a reminder of what was once a paradise of colorful flowers and rich animal life. The landscape is studded and pocked with dry sandhills and countless pools shaped by the retreating glaciers of the last ice age.

Forest zones
Forests cover a great swathe of Canada from the Yukon in the northwest to Newfoundland off the eastern coast. They form part of the taiga zone of coniferous forest that extends across northern Europe and Asia. The taiga is dominated by spruces, with pines, tamarack and fir, and intrusions of hemlock, deciduous aspen, beech and maple. Large areas consist of muskeg, a quaking, waterlogged peatland where tree cover is patchy. The

Biomes

- subtropical and temperate rainforest
- coniferous forest
- temperate broadleaf forest
- arctic desert and tundra
- temperate grassland
- mountain and highland system
- lake system

- ◆ major protected area
- ○ Biosphere Reserve
- ✕ World Heritage site

Map of biomes The Arctic tundra of the Canadian north gives way to coniferous taiga in the south. Prairie grasslands lie in the eastern rainshadow of the Rocky mountains; temperate rainforest grows in the west.

Virgin temperate rainforest is found on the islands off the Pacific coast of Canada. The mild climate and high rainfall combine to give a long growing season that allows this luxuriant vegetation to thrive. The forest is characterized by sitka spruce, western red cedar, western hemlock and Douglas fir.

forests are rich in animal species. Among the large mammals are moose, caribou, beavers, lynx and bears.

To the north of the forests, and merging with them in a ragged boundary, are the treeless expanses of the tundra, which stretches from the western shore of Hudson Bay to the tip of Labrador and the Arctic archipelago. The feature most striking to the human visitor is the summer plagues of mosquitoes and blackflies that breed in the meltwater ports. These support large populations of wildfowl and waders that migrate to the tundra to breed in summer. Caribou also visit the tundra for summer grazing. The nutrient-rich coastal waters support populations of seals, whales and seabirds, especially where currents maintain stretches of open water throughout the year.

In the southeast corner of Canada the climate becomes more moderate under the influence of the Great Lakes and the Atlantic. It was the first area to be settled by Europeans, who found its deciduous forests of oak, beech, sugar maple and basswood very similar to those of their home countries. Little of the original forest cover has survived the encroachment of agriculture, industry and urbanization in this densely populated area of the country.

A northern wilderness

Greenland has much in common with the Canadian Arctic. Only the coastal fringe is clear of permanent ice and supports a tundra vegetation with sparse populations of musk oxen, caribou, Arctic hares, lemmings, Arctic foxes and polar bears, together with visiting birds. The southwestern coast of Greenland, coming under the influence of the North Atlantic Drift, is relatively mild; stands of juniper, birch and rowan grow in sheltered valleys and sheep herding is practiced on a modest scale. The human population is only 53,000 and exploitation of mineral and oil resources in this wilderness area has barely started.

A NETWORK OF PARKS AND RESERVES

The preservation of Canada's wilderness started in November 1885 when 260 ha (643 acres) of country around Banff's Cave and Basin Hot Springs were set aside for public use. In the next decade more reserves were created: these were eventually to become the Yoho, Kootenay, Mount Revelstoke and Waterton Lakes National Parks. The creation of the Canadian Parks Branch in 1911 provided the impetus for many new national parks over the next 25 years, but changes in federal legislation concerning the transfer of land caused a lull after that. However, since the 1960s there has been renewed activity in park creation, especially in the empty west and north.

Today Canada has 34 national parks and national park reserves, comprising over 182,000 sq km (72,250 sq mi). The largest, Wood Buffalo National Park, covers an area greater than Switzerland. At the present time, nearly 2 percent of Canada's land area is within the national park network.

The primary aim of the national parks is to protect representative examples of the Canadian landscape. One major omission is a grasslands national park to protect the nation's most endangered ecosystem. The idea of a "magnificent park" for the grasslands was first suggested by the explorer George Catlin in 1832 when he predicted the demise of the prairie, along with the plains Indians and the bison herds that roamed the grasslands. Although serious discussion over the last 30 years has led to an agreement in principle between the federal and Saskatchewan provincial governments, the park has still not been established.

A second category of parks – the provincial parks – has the wider goal of catering for tourism and recreation. The first, the Algonquin Provincial Park, was established in Ontario in 1893. They now range from the 24,100 sq km (9,300 sq mi) Polar Bear Provincial Park on Hudson Bay to tiny urban sites scattered throughout the populated parts of Canada.

Other protected areas

There are a multiplicity of other protected areas in Canada, ranging from critical wildlife habitats, ecological reserves, refuges for migratory birds and wilderness areas to marine parks. Some areas are under provincial authority; others are federally administered. In July 1987 Ontario's Fathom Five Provincial Park off the Bruce Peninsula became the country's first marine park, and there is a growing system of Canadian Heritage Rivers.

Canada's protected areas have a significant place in the international framework of reserves. There are 17 wetlands considered to be of international importance (Ramsar sites), 4 Biosphere Reserves and several World Heritage sites. Waterton Lakes National Park is linked with the contiguous Glacier National Park on the other side of the border with the United States and is the world's first International Peace Park.

Components of the ecosystem
1 Heather family plants
2 Mosses and lichens
3 Marsh grass
4 Cotton grass
5 Peat
6 Permafrost
7 Caribou
8 Arctic hare
9 Ptarmigan
10 Arctic lemming
11 Snowy owl
12 Wolf
13 Arctic fox

Energy flow
⊏▷ primary producer/primary consumer
▭▷ primary/secondary consumer

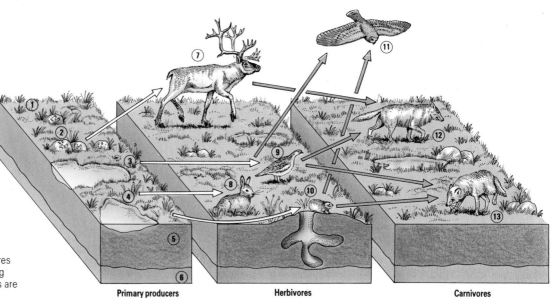

Primary producers   Herbivores   Carnivores

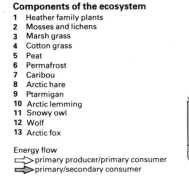

A tundra ecosystem Freezing temperatures slow down the decomposition and growing rates of plants. Lichens and cushion plants are vital components in the food chain.

Snow and ice cloak the land during the long winters in the Northwest Territories of Canada. The spring thaw brings flowering plants, such as this yellow poppy, along with swarms of insects. Migrating birds and mammals move north as the short summer begins.

Taiga forest on the mountains of the Kluane National Park in Yukon Territory. These coniferous forests of spruce, pine and fir provide a habitat for large mammals such as bears, moose, caribou and lynx.

Ecological reserves are a new and important form of protection. They are set aside either for ecological monitoring or to protect unique and endangered species and habitats. Many sites in Canada were identified by the International Biological Program in the 1960s. One, Polar Bear Pass on Bathurst Island, has become a national wildlife area; it is an "oasis" inhabited by 11 land and marine mammals, including polar bears, musk ox and caribou, and 42 bird species, one of the largest known concentrations of animals in the Arctic. Protection was opposed by the mining, oil and gas industries who feared it would lead to widespread restrictions on economic development in the Arctic. Similar controversy confronts other Arctic wildernesses.

The world's largest park

The population of Greenland is dependent on fishing and hunting, and conservation of natural resources is vital. Consequently egg collecting and other forms of harvesting are forbidden on certain cliffs and islands; in an area of Melville Bay on the west coast, where both hunting and motorboats have been banned, there is a reserve for polar bears and ringed seals. The most extensive protected area is the Greenland National Park: the largest in the world, and the emptiest. Most of it is polar desert, but it includes important wildlife habitats in the north.

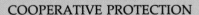

COOPERATIVE PROTECTION

The federal government of Canada is increasingly keen to gain the cooperation of private individuals and organizations in protecting isolated sites in urbanized and farmed areas. While such sites frequently contain reduced populations of species, they are often all that is left and their protection is consequently vital if Canada is to look to the future and prevent its natural heritage from being completely destroyed by human activity.

The Ontario Natural Heritage League (ONHL) was founded in 1982, and is now a coalition of 28 public and private organizations. Its guiding principle is cooperation; by securing support in both the private and public sectors and by encouraging private landowners to act as stewards of the countryside, the ONHL has attained goals that other, individual efforts have not.

Among its successes, the ONHL has helped to establish cooperative protection for more than 20 provincial ecological and geological sites. It was instrumental in obtaining final provincial government approval for a comprehensive land-use plan for the 300 km (186 mi) Niagara Escarpment. It is also helping to prepare a conservation strategy with the Nature Conservancy of Canada and the World Wide Fund for Nature, and gives support to the protection of the last remnants of the Carolinian forest in southwestern Ontario in the area bounded by lakes Huron, Erie and Ontario.

CONFLICT OF INTERESTS

There is no shortage of threats to protected areas in Canada. To people living in more crowded parts of the globe it may seem incredible that the Canadian wilderness is under pressure from such a sparse population, especially in the remote areas of the north and west. However, one threat comes from powerful forestry interests, especially in British Columbia and Ontario, which question proposals to preserve forest land. The widespread practice of clear-felling often destroys the environment, and has adverse effects that spread far beyond the cleared area through, for example, the disruption of rivers and streams.

Balancing resource use
Mineral extraction, with its associated pipelines, hydroelectric schemes and settlements, probably presents the greatest threat of all to the environment. The building of access roads endangers wildlife: outside Waterton Lakes National Park, for example, roads to natural gas

drilling rigs have enabled hunters to come within range of grizzly bears and bighorn sheep, whose winter migrations lead them out of the park.

Longstanding mining claims may have to be resolved before a protected area can be established. In British Columbia mining interests have succeeded in opening up some provincial parks for exploration and development, despite protests from local people and conservationists. Such experiences show that in Canada, as elsewhere in the world, legal declaration of a protected area is not always sufficient to ensure its preservation. Developments for hydroelectricity or oil in the Canadian Arctic affect huge areas of wilderness. With the degree of commercial investment involved, it is often hard to gain a hearing for the priceless value of wildland and wildlife conservation.

The increased demands of visitors for greater access to the wilderness areas are being felt in many areas. In the Rocky Mountains the Trans-Canada highway, which runs through Banff National Park, has been widened to four lanes. Visitors to parks pollute water supplies, trample

On the rocks in the Jasper National Park. The mountain goat is a rock climbing specialist; it lives on rock faces and in alpine meadows where it is safe from predators.

vegetation and start fires. The dilemma facing conservationists is that tourism is often cited as the means of bringing employment to a local community as an argument against allowing more damaging forms of development.

The threats to protected areas are not diminished by the attitude of federal and provincial governments. Not only can they hamper proposals for protected areas, but they can pass legislation to alter existing parks. Banff National Park's boundaries have changed ten times since 1885. Over half the area of Waterton Lakes National Park was removed by the government in 1921, so it no longer conserves an entire plant and animal community. More recently, however, Waterton has been designated a Biosphere Reserve which has led to greater co-operation over activities that may affect it.

Even where the federal government is willing to establish a national park, the provincial government may be reluctant

SOUTH MORESBY NATIONAL PARK

Rainforests are so linked in most people's minds with the tropics that it comes as a surprise to many to find that they exist on the western seaboard of North America. Giant, ancient evergreen trees thrive in the high rainfall of the Pacific coast and support a varied wildlife that has led to the islands of British Columbia being called the "Canadian Galapagos". But, just like the rainforests of the tropics, these temperate rainforests are coveted for their timber.

The story is familiar: new techniques have allowed logging operations to speed the destruction of the forests. It is estimated that at the present rate of felling the forests of British Columbia will be exhausted in the first decade of the next century.

In 1985, a group of Haida people, who have lived in the forest for generations, were arrested for blockading a logging road. The logging continued but the blockade helped to unite the opposition mounted by the Haida and a local group, the Islands Protection Society. As a result of their campaign, an agreement was made two years later between the federal and provincial governments to create the South Moresby National Park at the southern end of the Queen Charlotte Islands. Felling was stopped, though there were reports that the loggers had worked double shifts while negotiations were going on in order to remove as many trees as possible. The price exacted by the logging companies and provincial government in compensation was in excess of $100 million.

to cooperate in implementing it, and control of the land and natural resources lies with the provincial rather than federal authorities. Local communities may oppose protection of land out of fear that it will curb their activities.

Canadian rainforests suffer from the worldwide demand for timber. This interior hemlock forest, however, is protected as part of the South Moresby National Park. These isolated forests have given rise to endemic species such as the Columbian blacktail deer.

Indigenous land claims

Matters are most complicated in cases

Harp seal pups bask in the winter sun on the frozen Gulf of St Lawrence. The Gulf's populations of seals and whales are increasingly threatened by pollution from the chemicals that are dumped into the waters by the many industries in this highly-populated area.

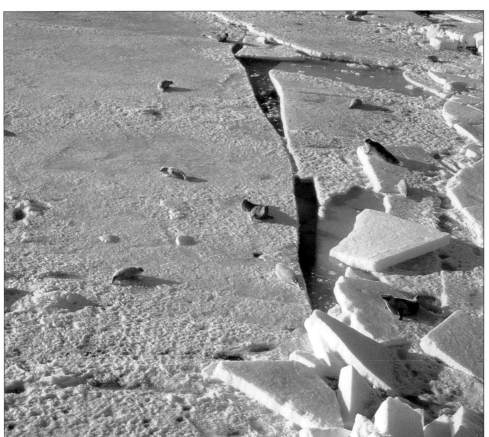

where indigenous peoples have land claims. These have assumed an increasing importance in Canada with the growth of oil and mineral extraction. Most claims are recognized by the federal government but not necessarily by the relevant provincial government. Since 1976 national parks in such areas are established as "national park reserves" pending final settlement of the land claim. The Inuit and other indigenous peoples regard the land not as wilderness but as homeland and habitat. Thus, while they may not want development of their land, they may also not want protective measures that would restrict their hunting.

However, land claims in northern and western Canada are proving to be a significant force for national park establishment. It is recognized that an area can be protected from resource development and opened for tourist use, while still being used by indigenous peoples for hunting and harvesting game. The Northern Yukon, Northern Ellesmere Island and South Moresby National Parks owe their existence at least in part to indigenous land claims.

The Northern Yukon

The Canadian Arctic is one of the empty parts of the world. Its hostile climate of dark, freezing winters and short, cool summers that barely thaw the soil has kept human settlement to a minimum. The wildlife that evolved to withstand and thrive in these conditions has provided food, clothing and shelter for the scattered nomadic bands of people who also adapted to this harsh but incredibly beautiful environment.

The success of these people is seen at the Engigstciak archeological site on the Firth river in the Northern Yukon National Park. It provides evidence of more than 5,000 years of occupation by many different people. Other archeological sites show the arrival of Arctic whalers, of trappers and hunters from the Hudson Bay Company and of the Royal Canadian Mounted Police. The last twenty years have seen the Arctic wilderness brought into the 20th century. Improved technology now enables the quest for oil and minerals to take place where once it would have been impossible, so that even the most remote wilderness stands in need of protection. One way is through the involvement of those people who have traditionally been part of this wilderness habitat.

The 10,168 sq km (3,925 sq mi) that make up the Northern Yukon National Park rise gently from the Beaufort Sea and reach a peak in the 1,800 m (5,900 ft) British Mountains close to the Alaska border. Near the southern edge the tundra is replaced by taiga, with open stands of stunted white spruce and balsam poplar marking the transition zone.

The park is home to one of North America's highest concentrations of large birds of prey and to large numbers of grizzly bears. The coastal plain of lagoons, spits, islands and river deltas makes an important corridor for millions of waterfowl, birds of prey, guillemots and shorebirds. But the park is best known as the

Fall in the north brings color to the alpine tundra in the Northern Yukon National Park. Above the treeline is tundra vegetation of low shrubs, grasses and lichens. The flush of growth in the spring and summer creates lush meadows that attract insects, birds and mammals. To cope with the harsh winters, many small mammals hibernate while the larger mammals and birds migrate down the mountains or to warmer habitats.

Remote wilderness The remote location and harsh climate of the Northern Yukon National Park means that it is wilderness territory. There are a few settlements of the indigenous peoples who have adapted to these severe conditions over the centuries. They have been instrumental in ensuring that the area remains protected and have negotiated exclusive rights to hunting game.

On the southern edges of the park taiga takes over from Arctic tundra. The freezing winters, short growing season and poor acidic soils produce a vegetation of coniferous trees, with deciduous trees such as birch, poplar and aspen in clearings. It forms the largest ecological zone in Canada, stretching from the Yukon to Newfoundland.

home of a 150,000-strong caribou herd. Their summer calving grounds lie along the coastal tundra and foothills; the animals move north in summer to graze on the tundra, and then travel southward to spend the winter feeding on caribou moss in the more protected forest.

Local involvement

The park is closely involved with the local Inuvialuit people. It came into being through agreement between the Inuit Committee for Original People's Entitlement (COPE) and the federal government, and was formally designated in 1984 as part of the settlement of the Inuvialuit land claim. The agreement gives the Inuvialuit certain rights in the park. These include exclusive rights to harvest game, preferential rights to employment and economic opportunities within the park, and the right to be involved in planning and management.

Wilderness of ice

Although much of the Arctic Ocean is permanently frozen it supports a surprisingly rich food chain. Light fliltering through the ice supplies energy to a film of algae on the undersurface. Shrimp-like crustaceans graze the algae, and are food for fish and migratory whales. The fish are eaten by marine mammals such as seals and walruses, and by a host of migratory birds that breed on the tundra.

In places, the water is kept ice-free by ocean currents. These oases of open water, or polynias, are important feeding grounds for marine mammals and seabirds. They also attract unwelcome human activity in the form of oil and gas extraction along the coast of North America. There are proposals to cut a channel through the ice to take tankers from the oilfields to the ports around Hudson Bay. In the event of a spill, oil would spread for hundreds of kilometers under the ice, destroying the algae – the basis of the Arctic food chain – and blocking the oxygen supply to the animals living in the waters below.

Polar bears on the frozen Arctic. In winter the sea ice is so extensive that it is possible for these animals to walk all the way from Greenland to the Svalbard Islands off the coast of Norway – a distance of almost 1,000 km (620 mi).

A LAND OF PLENTY

HABITAT PATTERNS · THE GLORIOUS DIVERSITY · MANAGING THE WILDERNESS · THE NATIONAL PARK
SERVICE · WILDERNESS INVADED · ENDANGERED WETLANDS

Embraced by two great oceans, the continental United States is extraordinarily varied, with high mountain ranges such as the Rockies and the Appalachians, deep valleys, rolling hills, spacious plains, many lakes, streams and wetlands, mighty rivers such as the Mississippi and Colorado, volcanic oceanic islands and a saltwater coastline some 154,000 km (96,000 mi) long. Its many different climates and soils support a rich and very diverse wildlife. In recent geological times, ice sheets have advanced and retreated over this vast area, and land bridges to Europe and Asia have formed and been submerged again, allowing frequent migrations and the mingling of species of many different origins. The United States contains some of the world's largest pristine wilderness areas, and some of its most dramatic scenery.

COUNTRIES IN THE REGION	
United States of America	

Major protected area	Hectares
Aleutian Islands BR	1,100,943
Badlands NP	195,000
Bering Land Bridge NPr	1,127,515
Big Bend NP BR	283,250
Big Cypress NPr	228,000
Champlain-Adirondack BR	3,990,000
Denali NP BR	2,441,295
Dinosaur NM	83,636
Everglades NP BR WH	566,000
Glacier NP BR	410,202
Glacier Bay–Admiralty Island NM BR	1,510,015
Grand Canyon NP WH	272,596
Grand Teton NP	124,140
Great Smoky Mountains NP WH BR	209,900
Haleakala NP BR (Hawaii)	11,600
Hawaii Volcanoes NP BR WH (Hawaii)	92,800
Isle Royale NP BR	215,740
Katmai NP	1,655,870
Mesa Verde NP	20,830
Mojave and Colorado Deserts SP NM BR	1,297,264
Mount Rainier NP	97,550
Olympic NP WH BR	363,379
Redwood NP WH	44,200
Rocky Mountains NP WH	106,710
Sequoia NP	163,115
Wrangell–St Elias NP NPr	5,339,400
Yellowstone NP BR WH	898,350
Yosemite NP WH	307,900

BR=Biosphere Reserve; NM=National Monument;
NP=National Park; NPr=National Preserve; SP=State Park;
WH=World Heritage site

HABITAT PATTERNS

The distribution of the plant and animal communities (biomes) in the United States is influenced by the north–south alignment of the great mountain ranges of the west, the Rocky Mountains, and the different climatic conditions surrounding them. The prairie grasslands of the Great Plains lie in the rainshadow of the Rockies and gradually give way to the temperature forest lands of the east.

Both the soils and the climate of the eastern United States favor broadleaf as well as coniferous trees. The deciduous forests of New England are famous for their brilliant colors in the fall. Here the white pines grow tall and straight, while in the warmer land to the southeast, yellow pines predominate; there are also mixed forests containing sycamore, elm, hickory, pecan, ash, birch and dogwood

In the Rocky Mountains is situated the world's first international peace park, established in 1932, embracing Waterton Lakes National Park in southern Alberta, Canada and Glacier National Park in the northwest of Montana in the United States.

interspersed with laurels and azaleas. The forests once covered so much of the land that, it is said, a squirrel could travel from the Atlantic coast to the Mississippi river without touching the ground. Hardwood forest once covered the entire state of Ohio. Moss-draped cypress trees stand in the swampy creeks (bayous) of the southeast coastal plain, while the shores are lined with mangroves.

Vast flat plains lie between the forested Appalachian Mountains in the east and the ranges of the Rockies in the west. The continental climate of the Great Plains means that the summers are long, hot and dry, and the winters severe. Even the Great Lakes to the northeast become choked with ice. In the plains these conditions encourage the growth of many species of native grasses.

To the north, in the states of Alaska and Washington, temperate rainforests grow along the coast. Lichens and mosses drape from the tall trees, and the forest floor is a rich larder of seeds and berries. The far northwest is coniferous forested land, with stands of spruce, fir and cedar.

Mountain and desert

The Rocky Mountains are forested with aspen below the treeline, while above it,

Bison graze the summer pastures at Yellowstone National Park. In the winter bison, wapiti (elk) and moose gather around the hot springs, where clumps of grass remain free of snow.

the alpine meadows, rich in wildflowers in the spring, give way to tundra vegetation and snowfields. The parallel range of the Sierra Nevada to the west is snow-covered in winter, and bears firs, cedars and hardy pines – lodgepole pine, yellow pine, tall sugar pine with its spiral of branches and long sticky cones, tough white-bark pine at the treeline and, on the eastern slopes, the pinyon pine. On the western slopes stand the great redwoods, with trunks so immense that early loggers held dances on their stumps.

Deserts extend across much of the southwest of the country. They support gray-green cacti, brilliant yellow rabbit brush and drought-tolerant herbs such as sage. Tall cottonwoods line the banks of rivers, whose waters have incised deep canyons in the rock, exposing millions of years of the Earth's history.

On the northwest coast sea otters live among dense forests of kelp (seaweed).

Thousands of kilometers to the southwest, the volcanic Hawaiian islands of the warm Pacific Ocean are fringed with coral reefs. Many plants and animals unique to the islands (endemic species) have evolved in this isolated archipelago.

The wilderness reduced

The people who first inhabited the region many thousands of years ago found a land rich in animals. Fish filled the rivers; deer and antelopes were plentiful; bison (buffaloes), wolves and the great plains grizzly bears roamed the prairies of tall grass; beavers dammed the streams; black and brown bears, foxes, mountain lions, lynx and smaller mammals inhabited the brushlands; and the forests contained many species of birds. The abundance of wildlife remained virtually undisturbed until after the arrival of the Europeans, nearly 500 years ago. For many of the early pioneers the land seemed boundless and its resources limitless. The broadleaf forests of the east, like those of Europe, were quickly cleared for agriculture and for fuel to drive the young country's growing industries. The rush to exploit the resources of the vast open spaces west of the Mississippi river in the 19th century brought large areas of the prairie grasslands under the plow and destroyed the way of life of the indigenous Amerindians, who had lived in harmony with nature.

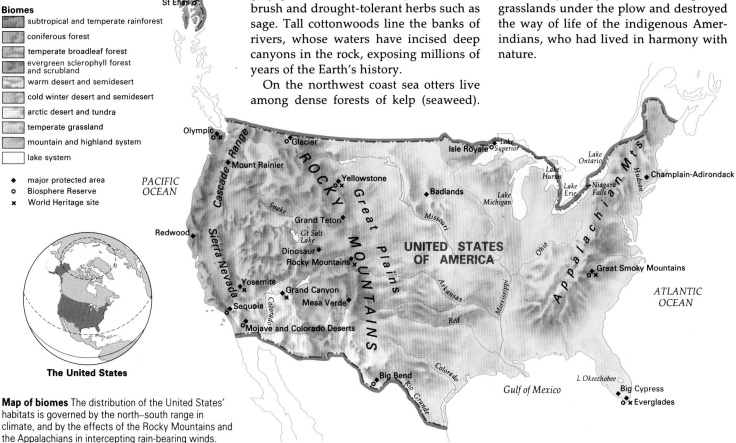

Biomes

- subtropical and temperate rainforest
- coniferous forest
- temperate broadleaf forest
- evergreen sclerophyll forest and scrubland
- warm desert and semidesert
- cold winter desert and semidesert
- arctic desert and tundra
- temperate grassland
- mountain and highland system
- lake system

◆ major protected area
○ Biosphere Reserve
✕ World Heritage site

The United States

Map of biomes The distribution of the United States' habitats is governed by the north–south range in climate, and by the effects of the Rocky Mountains and the Appalachians in intercepting rain-bearing winds.

THE GLORIOUS DIVERSITY

The great diversity of the plants, animals and habitats in the United States owes much to the continent's underlying structure. The high, young mountain ranges of the Rocky Mountains and the Sierra Nevada and the older mountains of the Appalachians all run from north to south. Consequently, during the recent ice age there were no barriers to the movement of ice, and icesheets swept south unhindered. When the ice retreated some 10,000 years ago, species from the warmer south migrated northward, while species that preferred colder conditions found refuge on the higher mountains. The alignment of the mountains also allows birds to migrate freely from north to south. Three major corridors – up the Pacific coast and along the Mississippi valley and the east coast – are used by birds traveling between the tropics and the tundra of Canada and Alaska.

During periods of drier climate, an extensive land bridge linked Alaska with what is now the Soviet Union. Plant species of Asian origin – azaleas and rhododendrons – can still be found in the Appalachian Mountains. The Appalachians alone contain more species of trees than the whole of Europe. The land bridge of Central America links the two great continents to the north and south, and has also helped species to spread. The opossum, for example, a marsupial that survived for millions of years in South America, is also common in the United States. This diversity was further enhanced by the native Indian peoples, who opened up small sections of forest for hunting and cultivation, creating new

habitats. Even on the apparently monotonous prairies, some 80 species of grass may be found.

Riches of the west coast

The greatest diversity is found along the west coast in California and Oregon, where the cold Californian ocean current gives rise to a Mediterranean-type climate. Here the coastal forests contain some 18 species of conifers. Farther to the south a scrublike vegetation of drought-tolerant, hard-leaved (sclerophyll) plants flourishes – the chaparral. This area experienced not only the fluctuating climate of the ice age, but also the great volcanic and tectonic upheavals that shaped the young mountain ranges of the Rockies

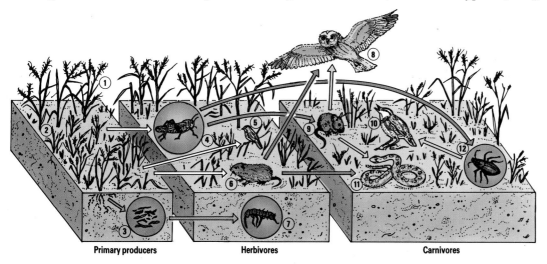

Primary producers — Herbivores — Carnivores

Components of the ecosystem
1 Big bluestem
2 Little bluestem
3 Detritus
4 Grasshopper
5 Grasshopper sparrow
6 Vole
7 Springtail
8 Short-eared owl
9 Deer mouse
10 Eastern meadowlark
11 Rattlesnake
12 Carabid beetle

Energy flow
⇨ primary producer/primary consumer
⇨ primary/secondary consumer
⇨ secondary/tertiary consumer
⇨ dead material/consumer
⇨ death

A prairie ecosystem has rich soil and grassland. It supports many small animals, but large grazing mammals are now extinct here.

there are also literally millions of lakes. Southeast Alaska contains many immense snow fields and glaciers, along with densely forested islands.

Alaska is home to by far the greatest number of free-roaming animals in the United States – caribou, moose, musk oxen, bears (black and grizzly), wolves, wolverines, foxes, lynx, mountain sheep and goats, beavers, ermine, weasels, walruses, seals, whales and a great multitude of migratory birds.

Until recently almost all the state was held as "vacant, unappropriated public land". Since the late 1950s, however, this vast land has been progressively subdivided, with ownership passing to the state, the indigenous people and – as national parks, national wildlife refuges and designated Wild and Scenic Rivers – to all the people of the United States.

In Alaska, for the first time in the history of the National Park System, the boundaries of the new national parks were deliberately drawn to take in whole plant and animal communites. Alaska's 16 national parks incorporate entire river systems and volcanoes. Here and farther south international reserves have been established. Glacier National Park in Washington, for example, joins with Canada's Waterton Lakes National Park. Together with the Wrangell–Saint Elias National Park and Preserve, which joins Canada's Kluane National Park and Game Sanctuary, it forms a parkland extending over nearly 770,000 sq km (300,000 sq mi). These provide expansive preserves that enable the animals to roam freely as they have always done, unhindered by local or national boundaries.

FRAGILE ISLAND ECOSYSTEMS

The isolated islands of the Hawaiian archipelago, which lie in the Pacific Ocean some 3,200 km (2,000 mi) south of the nearest mainland, contain some unique habitats. Covered in lush vegetation with forests of feathery ironwood, red-blossomed ohia trees and graceful coconut palms, these islands are eroded in places into deep and vividly colored canyons and valleys. Their sandy beaches and rocky coasts are fringed by creamy white surf, while offshore the ocean waters above the coral reefs are a clear blue-green.

Many new species of plants and animals evolved on these isolated islands, whose location offers some protection even today. European settlers have, however, altered many of the native habitats, and the careless introduction of alien species into the fragile island ecosystems has played havoc with some native species.

Pigs provide a typical example of the effect that introduced species can have. Those that have gone wild now live in the forests. By rooting in the vegetation of the forest floor, they have opened it up to the weather, and mosquitoes now breed in the pools. The mosquitoes carry a lethal kind of malaria that has decimated local bird populations. Furthermore, the lack of plant cover leads to soil erosion; the exposed soil is washed by heavy rains down to the coast, where it smothers and kills the reef-building corals.

Saguaro and cholla cacti flourish in the Sonoran Desert which lies inland to the north of the Gulf of California in the southwestern United States. Their wide-spreading shallow root systems intercept the infrequent rainfall before it sinks far into the ground. Rows of pale spines reflect the intense solar radiation and ward off grazing animals, for whom the succulent cacti are the main source of water. Filled with water-storing tissue, cacti can survive prolonged droughts.

and the Sierra Nevada. Many ancient species, such as redwoods, survive here, and many new species have also evolved. There are well over 2,000 endemic species.

A cold peninsula

Alaska – one of the planet's largest peninsulas, and the United States' most extensive wilderness area – covers 1,518,000 sq km (586,000 sq mi). Its tidal coastline is nearly 72,000 km (45,000 mi) long. The landscape of this northern state includes rolling hills, tundra sculpted into humps and hillocks by permafrost below the surface, inland sand dunes, and meadows full of flowers in summer. The fast-flowing streams and the multichanneled (braided) rivers drain into rich deltas, and

Lava flow in Hawaii Lava buries old habitats and creates new ones, which are quickly colonized by plants. The isolation of small areas by lava has led to the evolution of many new plant and animal species.

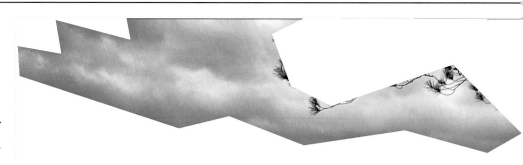

MANAGING THE WILDERNESS

Despite, or perhaps because of, the early pioneers' careless treatment of the land and its resources, the concept of conservation began to take hold during the latter part of the 19th century. There was particular concern for the country's fisheries and wildfowl. The first national park (Yellowstone, declared in 1872) and the first forest reserves were established by the end of the 19th century, and many more federal (state and county) reserves were to follow.

Almost 18 percent of the land in the United States remains in the hands of the federal government (not including land held by the military), and since the turn of the century four different federal land agencies have been created to administer it. Each agency is charged with specific management tasks.

The United States Forest Service was formed in 1905 under the Department of Agriculture. It is in charge of 770,000 sq km (297,300 sq mi) of land in 44 of the 50 states, including national forests, national grasslands and land utilization projects. Forest Service lands also include mountains, glaciers, lakes, streams, rangeland and even rainforests. The Forest Service is a "multiple use" agency, managing its lands for timber, minerals, forage, wildlife habitats, recreation and wilderness.

The Fish and Wildlife Service came into being in 1940 under the Department of the Interior. It manages 370,000 sq km (142,860 sq mi) of land in national wildlife refuges located in all the states. It has charge of the country's major wetlands as well as other wildlife habitats. This too is a multiple use agency, with programs for the management and protection of the country's fisheries and its free-roaming mammals, its wildfowl and other migratory species, and its endangered species. It also provides for hunting, fishing, trapping and outdoor recreation in many of its refuges.

Established in 1946 as part of the Department of the Interior, the Bureau of Land Management (BLM) has traditionally been in charge of the "vacant and unappropriated public land" in the United States. It now manages some 1.1 million sq km (424,700 sq mi) of land. With the exception of classified wilderness areas, all BLM lands are open to grazing, hardrock mining and mineral exploration, oil and gas development, logging, hunting, fishing, recreation – in fact, every kind of multiple use.

A fourth agency, the National Park Service (established in 1915 under the Department of the Interior), is charged with preserving its lands (some 320,000 sq km/123,550 sq mi) in perpetuity for the use and enjoyment of the people.

Wilderness preservation

Strict wilderness protection was unknown until the 1920s and 1930s, when the Forest Service became the first federal agency to classify some of its lands for strict wilderness preservation. Agency regulations, however, were not binding, and as pressure for development continued to mount throughout the 1940s, the supporters of wilderness protection joined forces to press for the legalization of wilderness preservation. When the National Park Service was challenged over dam-building in Dinosaur National Monument, Colorado, and the Forest Service began declassifying its protected lands, public support for the wilderness movement gained ground. The result was the Wilderness Bill of 1964, which established the National Wilderness Preservation System.

Wilderness was defined as "an area of federal land retaining its primeval character and influence without permanent improvements or human habitation ... (with) outstanding opportunities for solitude or primitive and unconfined type of recreation ... (and with) ecological, geological or other features of scientific, educational, scenic or historical value." No motor vehicles, motorized equipment or other form of mechanical transport would be allowed, nor would any structure or installation. However, established grazing and mining were allowed to continue. Some 400,00 sq km (154,400 sq mi) of protected federal land were immediately classified as wilderness, and the Act provided for the ongoing review and enlargement of the Wilderness Preservation System on lands administered by all four agencies.

The philosophies of the four agencies differ. The concept of preserving lands "unimpaired" fits well into National Park Service policy, while the Fish and Wildlife Service puts management for wildlife productivity first. The Forest Service – the first agency to classify its lands for wilderness values – prides itself on the strictness of its wilderness management. Primarily a multiple-use agency, the Bureau of Land Management has been slow to classify any of its land as wilderness and has resisted setting aside any large areas under this classification.

By 1990, the total area of land in the National Wilderness Preservation System covered almost 3,600,000 sq km (1,400,000 sq mi), much of it in Alaska.

WILD AND SCENIC RIVERS

In 1968 the Congress of the United States decided that the established national policy of dam construction on certain "appropriate" rivers needed to be complemented by a policy to preserve other rivers. To this end, it passed the Wild and Scenic Rivers Act to protect "selected rivers of the Nation which, with their immediate environments, possess outstandingly remarkable scenic, recreational, geologic, fish and wildlife, historic, cultural or other similar values".

These rivers are to remain forever free-flowing "for the benefit and enjoyment of present and future generations". The act provides for a 0.4 km (0.25 mi) zone of protection on either side of the river. Wild and Scenic Rivers are established by Congress or by the secretary of the interior at the request of a state or states.

By 1990, 75 rivers or sections of rivers had been designated within the Wild and Scenic Rivers System. They range from small streams such as the Loxahatchee in Florida (12 km/7.5 mi) to grand swift-flowing rivers like the Smith in California (630 km/394 mi). In all, 12,334 km (7,709 mi) of rivers and river stretches are protected, 11,129 km (6,956 mi) of which are administered by the federal government and 1,205 km (753 mi) by 29 states.

The deciduous forest of the Adirondack Mountains in New York State. The Adirondacks are the first mountains to intercept clouds blowing from the industrialized American midwest, and as a result their forests and lakes are being ravaged by acid rain.

Mount St Helens, in Washington State, is now covered by a mantle of vegetation less than ten years after a major eruption laid waste its slopes.

The Okefenokee Swamp in America's Deep South, where tall bald cypresses provide a rich habitat for birds, reptiles, amphibians and a host of epiphytic and climbing plants. Cypress swamps depend for their survival on periodic fires as well as water.

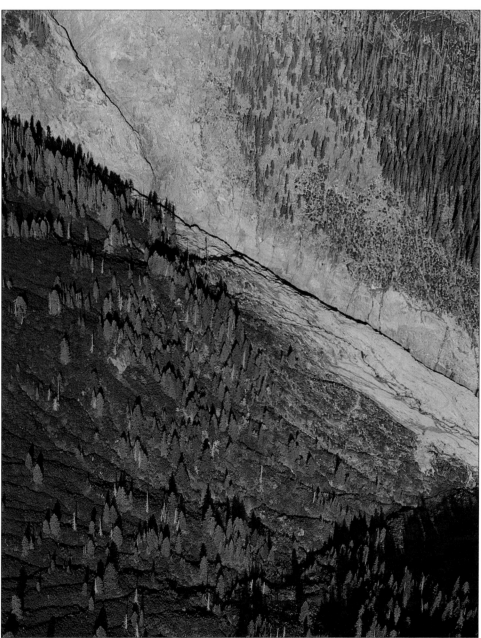

Yellowstone National Park

The first national park in the world, and still the United States' largest park (covering a total of 898,350 ha/2,218,920 acres), Yellowstone National Park was established in 1872. Part of Yellowstone is an enormous dish-shaped crater or caldera – roughly 48 by 72 km (30 by 45 mi) – formed as a result of a volcanic eruption of tremendous force that took place perhaps 600,000 years ago. Shaped by glaciers as well as by lava flows, this caldera marks one of the hottest "hot spots" known on Earth: there are some 10,000 geysers and hot springs in the park. Petrified forests stacked one on top of another provide evidence of repeated volcanic activity. Scientists consider the area to be geologically unique.

pelicans. It is also one of the last refuges of the grizzly bear.

These animals range widely outside the park boundaries, especially in winter. Most of the area surrounding the park has been preserved as National Forest lands, fish and wildlife refuges and the adjacent Grand Teton National Park. Although there are growing pressures to log, mine and develop the lands outside the park, there is at the same time growing awareness of the value of preserving ecological units intact, especially when they are as remarkable as Yellowstone.

Eighty percent of the area is covered in forest, mainly of lodgepole pine. Sagebrush lines some of the drier valleys, and in summer the mountain meadows are

Regenerating forests Yellowstone's forests are still recovering after the devastating fire of 1988. Such fires open up the forests, allowing diverse ground species of plants to flourish. The tender new growth of trees and grasses is welcomed by the park's grazing mammals.

Morning Glory Pool owes its brilliant colors to the algae and bacteria that thrive in its warm, mineral-rich waters. Thousands of brine flies breed in these pools, supporting many predatory spiders.

A rich wilderness

Most of the Yellowstone ecosystem has been very little disturbed. Although the park boundaries encompass a relatively modest area within the much larger caldera, Yellowstone boasts more than 1,000 species of flowering plants and 13 species of trees, and supports a herd of wapiti deer numbering more than 25,000, as well as bison, bighorn sheep, black bear, mule deer, pronghorn antelopes, moose, bald eagles, trumpeter swans and

filled with wildflowers. The lakes are among the highest in the world, and are well stocked with fish. Even the brilliantly colored hot pools contain their own miniature ecosystems, and are teeming with microscopic life.

Protecting the wilderness

Designated a Biosphere Reserve in 1972, Yellowstone National Park was declared a World Heritage site in 1978. The park is visited by more than 3 million people a

year. This invasion from the cities brings city problems with it – litter, theft and vandalism. Park officers have to be trained in law enforcement. Automobiles are restricted to about 5 percent of the park; a 228-km (142-mile) loop road contains the traffic, which is often bumper-to-bumper in summer. There are well over 2,000 camp and trailer sites, and garbage has to be collected regularly to minimize the risk of bears coming into contact with tourists.

Fire is the other major management problem. Such forests catch fire naturally from time to time, ignited by lightning. Indeed, the cones of lodgepole pine need fire to release their seeds. After the burn, the newly admitted light stimulates the growth of plants on the forest floor, encouraging their diversity and providing fodder for grazing mammals. If too long a period passes without fire, a dense layer of dead underbrush accumulates, and the ensuing fire may then rage out of control. A series of devastating fires swept through Yellowstone in 1988. Controlled burning may in future be used to prevent such large-scale fires taking hold.

The colorful Hot Pots trail is part of 1,600 km (1,000 mi) of trails that penetrate Yellowstone's backcountry. Determined hikers can still find havens of wilderness and wildlife untouched by tourists less adventurous in their exploration of the park.

THE NATIONAL PARK SERVICE

Following the establishment of Yellowstone National Park in 1872, Congress designated several more national parks under the aegis of the Department of the Interior; these were to be managed by the United States Army. In 1906 Congress gave the president the right to create "national monuments" by proclamation. As the number of parks and monuments increased, so did public pressure for a separate national park system. In 1916 Congress passed the act that established the National Park Service within the Department of the Interior. Under this legislation, Congress would still establish the parks and the president would proclaim the national monuments.

The newly formed agency was charged with managing all the national parklands, with the specific aims of conserving the scenery, natural and historic objects and the wildlife, and also of providing for the enjoyment of them in such a way that they would remain unimpaired for future generations – a daunting task, as the two aims are often in direct conflict.

A lake in Yosemite National Park, which is visited by millions of tourists each year. There are plans to designate up to 89 percent of Yosemite as protected wilderness, restricting leisure developments to peripheral areas.

LOVING A PARK TO DEATH

Yosemite National Park epitomizes one of the great challenges facing the managers of protected areas in the United States: how to keep people from loving their parks and wildernesses to death.

Established in 1890, the park contains not only the "Incomparable Valley", with its glacier-sculpted cliffs and cascading waterfalls, but also some of the loveliest wilderness backcountry in the Sierra Nevada, with its groves of giant sequoias, among the mightiest trees on Earth. More than 3 million people visit the park each year.

The valley has always been the principal gateway to the park, and over the years a small city has grown up within it. Thousands of people crowd into its limited living space – 5 sq km (2 sq mi) – with as many as 18,000 people sleeping in its camping grounds and hotels. Traffic jams clog its narrow roads; on peak weekends, automobiles have to be turned away. Accommodation includes a church and hospital, and rangers do double duty as policemen. The number of visitors continues to grow.

Use of Yosemite's backcountry is also increasing. Wilderness trails are being eroded by heavy foot and horse traffic; camping areas are beaten down; ski lifts scar the landscape; noisy parties of river rafters disturb the silence; the garbage people leave behind must be disposed of. Permits are now required for recreational use, and it may eventually be necessary to ban automobiles and campers altogether from the congested Yosemite Valley.

As the prime federal land protection agency, the National Park Service received a growing number of national parks and monuments to care for, as well as numerous "historic objects", historical and aboriginal sites. As time went by, the National Park System was expanded so that it would also include national lakeshores, seashores and recreation areas such as urban parklands.

The National Park System has grown from a few parks to more than 340 units located in 49 of the 50 states, the District of Columbia, Puerto Rico, the Marianas and the Virgin Islands. It is even charged with taking care of the United States Capitol, the White House, and other important sites in Washington DC. The National Park Service is responsible for approximately 320,000 sq km (123,550

rams. It has to negotiate with and control the organizations providing food, gas, groceries, gifts, lodging of all types, and other amenities. Yet the Park Service must also maintain its backcountry and wilderness for the wellbeing of its wildlife as well as for the pleasure of its human visitors. More than 300 million people now enjoy the national parklands every year. Consequently the Park Service has had to impose tight regulations, including permit systems, in some of the more popular parks: advance reservations for overnight camping and wilderness use are now essential.

It was said when the National Park Service was still young that it was the best idea the United States had ever had. With all its problems, it has enriched the nation's culture and has offered inspiration for the establishment of national parks in more than a hundred other countries all over the world.

Strong support

The difficulties of the National Park Service are to some extent eased by the growing strength of the support it receives from the public. Wildlife and conservation groups are numerous, some of them very large. The Sierra Club, in particular, has been a champion of the national parks. With the National Audubon Society and New York's Appalachian Club, it was one of the earliest conservation organizations in the United States, founded in 1892. It spearheaded many early conservation efforts, campaigning for the national park status of the Yosemite, Sequoia and Mount Rainier wildernesses. Notable successes include preventing dam construction in Dinosaur National Monument and Grand Canyon National Park.

Of growing influence is the National Wildlife Federation (NWF), to which many conservation societies are affiliated. As well as running educational programs of all kinds, it assists in raising finance for conservation projects both in the United States and overseas and, where appropriate, it may challenge or take legal action against government, federal, private or international organizations that pollute or commit other environmentally unfriendly acts. American conservationists continue to carry the message worldwide, as they have done since they established the world's first national park more than a hundred years ago.

The Yosemite Valley, dominated by giant domes of granite and high waterfalls. The towering pines, firs and incense cedars conceal campgrounds, lodges and roads that cater for over 14,000 vehicle loads of visitors a day during the summer.

American wilderness areas are a paradise for backpackers, but as numbers increase the conflict between recreation and preservation intensifies.

sq. mi), and the remarkable diversity of its units is reflected in their numerous different categories. The Park Service employs some 21,000 people, including historians, landscape architects, engineers, wildlife biologists, botanists, marine biologists, naturalists, park rangers and park planners. More than 42,600 volunteers contribute over 1,000 work years – worth about $17,000,000 – to the Park Service each year.

A difficult task

The intricate role of the National Park Service is complicated by the conflict between its two founding aims. It is obliged to provide access to its parklands, and must therefore provide developed campgrounds, sanitary facilities, visitor centers and educational centers and prog-

Prairies plowed

Tractors and heavy machinery have damaged the structure of the soft prairie soils, and the newly planted cereal crops do not bind the soil as firmly as the native grasses. This has led to extensive loss of soil during periods of drought. Soil also washes into the rivers, choking river life.

Much of the natural prairie grassland has fallen under the plow, though scientists are now attempting to recreate prairies. They have introduced native species into areas that once supported prairie, with some measure of success.

The prairie potholes, deep natural pools, are important wildfowl hatcheries – about half of the United States' wild ducks hatch out in potholes. But some 15,000–20,000 potholes are being drained each year, resulting in a dramatic reduction in the numbers of wildfowl.

Aliens in the wilderness

The European settlers brought with them not only their crops and livestock, but also many other non-native species, some deliberately, some by accident. Japanese honeysuckle and kudzu (a leguminous vine) are running riot in the forests of the Appalachians, while the rich eastern chestnut forests have been decimated by an accidentally imported fungal disease to which they have no natural resistance. Wild boars that have interbred with domestic pigs are overrunning the forests of the Great Smoky Mountains in the Appalachians on the east coast.

When new species are introduced with

WILDERNESS INVADED

The Europeans who found and settled the United States less than four hundred years ago brought with them a land ethic very different from that of the indigenous people. They saw the land as personal private property, and an economic asset that could be put to work for them. "Wilderness" was often seen as a challenge, an unwholesome environment that required changing, a wild place that must be tamed. They cut down the forests and used the timber for building and fuelwood; they plowed the grasslands and planted them with wheat and corn; they took the fish, then harnessed the streams and the rivers for power; they decimated the bison and replaced the deer and antelopes with cattle.

As increasing numbers of people came to the country, so the natural landscape was altered more and more, especially with the development of sophisticated machinery during the 19th and 20th centuries, and the need to provide food for the rapidly growing population.

Desert soils near the Grand Canyon reflect the nature of the underlying rocks. Competition for water leads to a more or less even spacing of the desert plants.

Migrating snow geese rest and feed at Gray Lodge State Refuge, California. Both natural and artificial lakes are often maintained with the support of wildfowlers.

Ancient temperate rainforests in the Olympic National Park in Washington State, dominated by Douglas firs and ponderosa pines more than 250 years old. Like rainforests the world over, they are threatened by excessive logging.

good intentions but inadequate knowledge, the results may be very different from what was hoped. Earthworms were introduced by the settlers, who believed them to be essential for good soil fertility. But the prairie soils had evolved in the absence of worms, which when present undermined their structure and accelerated erosion. Another example is provided by opossum shrimps, which were introduced to feed the salmon. However, in Glacier National Park the great salmon migration runs, which until only a few years ago attracted the greatest known gatherings of bald eagles – well

over six hundred at a time – have been greatly reduced. This has been a direct result of the activities of the shrimps, which compete with the tiny salmon fry for their plankton food, and are multiplying rapidly at the expense of the salmon.

The insatiable demand for timber

The rate at which American forests are being felled is causing alarm. Ninety percent of the ancient forests of the Pacific northwest have already been felled, and there are only 810,000 ha (2 million acres) left. The present rate of exploitation is not sustainable, yet it continues. The Forest Service claims that conservation of these forests will devastate the local economy, but already timber mills are closing as the larger trees disappear and the competition from overseas increases. Access

roads needed by the loggers open up new areas of forest to other forms of exploitation and to poachers. The Forest Service is even building more roads than it needs for its present timber sales in an attempt to open up the forest before stricter wilderness protection measures come into force.

A similar strategy is affecting the Alaskan forests. Lands were returned to self-managing indigenous groups, who were given a mandate to make a profit, a concept that is alien to their view of natural resources. Consequently, many such groups are felling the forests at speed to beat a change in the law to limit the tax loss schemes that help to finance even unprofitable timber extraction. Natural habitats are being destroyed, salmon streams choked and land eroded.

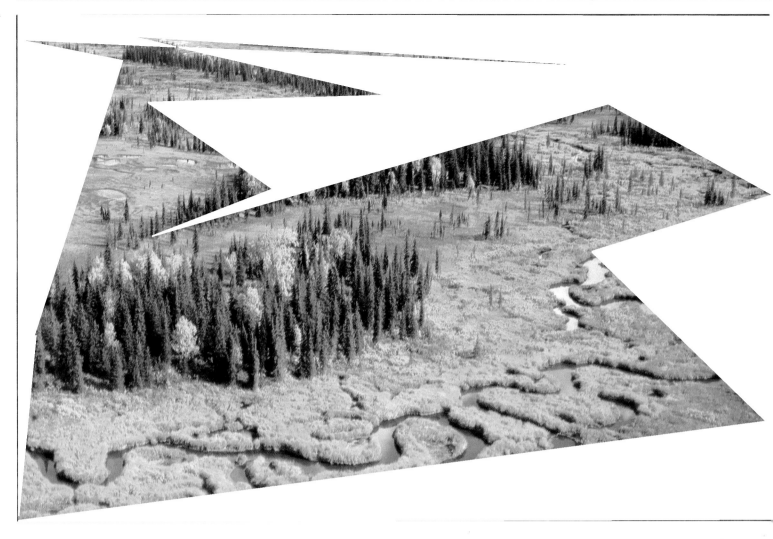

ENDANGERED WETLANDS

The direct effects of human activity are not the only threats to the American wilderness. When wetlands are drained, rivers dammed or land irrigated, the effects on the water system can be felt over a wide area.

Wetlands help to control the seasonal floods by acting as a sponge for the excess water. They are also important agents in recharging underground water supplies, trapping water long enough for it to sink into the ground. River diversions and dams destroy wetland and meadow habitats downstream. The sporadic use of water for hydroelectric power from dams on the Colorado and Green rivers has led to fluctuations in river level of as much as 4 m (13 ft) in the Grand Canyon, washing away riverside vegetation, placing several species of fish in jeopardy, and disrupting bird nesting and roosting sites.

Drainage continues apace. In 1988 the wild duck population of the United States fell to its second lowest level in 30 years due to loss of wetlands. The United States took a major step forward in the protection of wetlands in its 1985 Farm Bill, with the introduction of the "swampbuster" provisions: federal crop subsidies may not be given to farmers who drain wetlands in order to plant crops.

In Florida another unique solution has helped to extend wildlife refuges. The government gave the Aerojet General Corporation 11,350 ha (28,000 acres) of land in Nevada, which it needed for testing rocket engines, in return for some 1,850 ha (4,500 acres) of wetlands in Florida and additional money, which will be used to buy more land for wilderness protection.

Other attempts to protect wetlands have gone sadly astray. Flood prevention measures on the Mississippi river and the spread of cities such as New Orleans over parts of the river delta have led to the river being channeled in concrete, preventing it from flooding and distributing its vital silt over the delta. As a result, the growth of the delta no longer balances the incursions of the sea, and the coast is rapidly being eroded.

Threats from afar

Pollution does not respect state boundaries. Acid rain, spawned by the factories of the Midwest, is killing millions of trees in the Appalachian Mountains, while the lakes of the Adirondack Mountains in New York State are becoming too acidic to support fish. Pollution can also

The Alaskan tundra, interspersed with boreal forest, or taiga. Slight undulations in the glacial drift are sufficient to create islands of birches and conifers, bogs and riverside swamps, in a rich mosaic of subarctic habitats. Herds of caribou wander across these plains in search of grazing, followed by predatory wolves and Arctic foxes.

The effect of pollution The stark outline of a dead tree in the Californian desert stands silhouetted against the smog that killed it.

enters wetlands from farm waste, which contains pesticides and herbicides as well as high levels of nitrates and phosphates, and from industrial effluents.

Extraction industries can seriously pollute wilderness areas. The most dramatic example in recent years was the polluting of Prince William Sound in Alaska by the oil spill from the grounded tanker *Exxon Valdez* in 1989. Eleven million gallons of oil were discharged, killing seabirds, bald eagles, seals and sea otters, and devastating shellfish beds and other marine life. Through the food chain, the oil also affected animals that feed on the migrating salmon – bears, whales and sealions – as well as seriously damaging the salmon industry. Yet there is still pressure to prospect for oil in Arctic refuges. Millions of gallons of oil a day pass through the 1,400-km (880-mile) long Alaskan pipeline, which passes under 350 rivers and through areas of earthquake activity. In these cold climates, where rates of decomposition are slow, the land takes many years to recover from such disasters.

How safe is the wilderness?

Until recently, ecological considerations played little part in drawing the boundaries of the national parks and wilderness areas in the United States. Landscape factors were given priority, and the boundaries were traced on maps in an office thousands of kilometers away. As a result, many of the older-established national parks and wilderness areas are only fragments of major ecosystems, so their wildlife is often dependent upon the lands outside as well as inside the park or wilderness boundaries.

Even where the surrounding land is administered by the United States Forest Service or the Bureau of Land Management, the ecosystem may not be safe. These agencies are dedicated to multiple

Mono Lake, California, where pinnacles of tufa have been exposed as the lake's water level drops. Starved of water by Los Angeles' thirsty population, the lake's millions of invertebrates and the birds that feed on them face the threat of extinction.

use, and allow extraction industries on land not protected by wilderness legislation. National parks and wilderness areas may be left as vulnerable natural islands in a sea of developed land.

At the same time the pressure from people visiting protected areas is growing. Park managers are well aware of the threats, but funding for adequate management is not a federal priority. Planners are attempting to mitigate the impact on irreplaceable resources, gain more public understanding of parks and wilderness areas, and limit facilities and use.

MONO LAKE – DYING FROM THIRST

High in the Sierra Nevada lies Mono Lake, a unique salt lake fed by springs that rise through beds of saline sands, and drained by the Owens river, which enters the Pacific Ocean at Los Angeles.

Despite its salinity, Mono Lake teems with life. Microscopic algae thrive on the minerals. They, in turn, are preyed on by millions of brine shrimps belonging to a species found nowhere else in the world. There may be as many as 50,000 brine shrimps per cubic meter of water. Brine fly larvae feed on other algae that coat the submerged tufa; each square meter of shore supports some 40,000 adult brine flies.

All this food attracts more than a million birds each year. For example, about one-fifth of the world population of California gulls nest on the islands in the lake, and huge numbers of Wilson's and northern phalaropes pause here on

their way south in late summer, as do some 80 percent of the world's eared grebes.

This whole ecosystem is in grave danger of extinction as more and more water is being drawn from Mono Lake to supply the population of Los Angeles. The lake's volume has been halved, and its salinity has doubled to 9.5 percent. The islands on which the birds nest can now be reached by coyotes, who plunder the nests. There are plans to take still more water to satisfy the city's needs. Yet it has been estimated that to save Mono Lake, Los Angeles need only water its lawns in the evening instead of during the heat of the day, and sweep instead of hose dead leaves from the pavements. This is a small price to pay to save a unique living laboratory and a vital refueling stop for migratory birds.

Everglades National Park

The mangroves of the Everglades, in the peninsula of Florida in the southeastern United States, provide resting and nesting sites for millions of birds, including the rare wood stork, the snail kite, the reddish egret and the great egret (seen here).

A marshy backwater in the Everglades, the largest subtropical wilderness in the mainland United States. These are both brackish and freshwater areas, and support a variety of animals. They contain alligators, turtles and a rich underwater life.

Florida Indians called the great wet area that forms the southern part of the state *pa-hay-okee* – Grassy Waters – an apt name for the 233,100-ha (576,000-acre) swamp at the heart of which is the Everglades National Park. This watery ecosystem begins about 440 km (250 mi) north of the tip of Florida in the Kissimmee river, which flows into Lake Okeechobee. The lake spills out into a vast, shallow body of water 80 km (50 mi) wide and 15 cm (6 in) deep that creeps slowly south and southwestward to the sea. Its waters seep into the underlying porous limestone to charge a complex system of aquifers – the water supply of southern Florida. Over much of it grow thick grasses and sedges as much as 3 to 4 m (10 to 13 ft) high. This massive "river of grass", with its associated "hammocks" (small tree-covered islands rich in orchids and ferns), dwarf cypress and pine forests, and wandering channels and marshes, is unique.

A wetland refuge
The Everglades are the last stronghold of the wood stork, the snail-eating kite, the reddish egret and the southern race of the bald eagle. There are at least 12 species of turtle in the park, including the giant loggerhead turtle, which nests on its beaches. The American alligator is king of the swamp, but there are other predators here too – raccoons, otters and the rare Florida panther – as well as fish-eating birds: anhingas (relatives of cormorants), pelicans, spoonbills, ibises and a multitude of species of herons and egrets. On the park's seaward edge, red mangrove swamps invade the coast of Florida and the Gulf of Mexico, including the park's Ten Thousand Islands. The mangrove swamps are an important nursery for fish and crustaceans, which feed the Everglades' huge population of birds. The mangroves also serve as a vital barrier against the erosive power of the sea.

Everglades National Park was established in 1947, and in 1976 Congress classified a large proportion of the park's 566,000 ha (1,400,000 acres) as protected wilderness. Shortly afterward, 228,000 ha (563,400 acres) of the adjacent land northwest of the park was designated the Big Cypress National Preserve.

Problems on the periphery
Despite this protection, the Everglades National Park is typical of parks that are incomplete parts of larger ecosystems: what happens outside its boundaries is

Cypress swamps, with their dense growth of epiphytic plants and moist mossy floors, are home to many amphibians, reptiles and poisonous snakes. They are rich hunting grounds for the rare Florida panther, the Everglades mink and the ubiquitous raccoon.

critical to its survival. The surrounding area has been altered on a massive scale. Its wetlands have been drained and it is crisscrossed by a network of canals, dikes and raised roadbeds. During the 1960s the Kissimmee river was shortened and channeled, and Lake Okeechobee was encircled by a huge dike to prevent its annual overflow. Hundreds of thousands of drained hectares were put under intensive cultivation or were built over. Lake Okeechobee has become a giant septic tank for runoff wastes and pesticides, yet it also serves as a holding pond for the water that is now carefully rationed to the Everglades. As supplies of fresh water dwindle, demand for it increases (some 900 people move to Florida daily). There is extensive oil exploration in Big Cypress National Preserve. The United States Air Force wants to use the air space 30–150 m (100–500 ft) above the park as a training ground for fighter planes. The noise will undoubtedly disturb the thousands of birds that nest here.

With all its problems, the Everglades continues to support 300 species of birds, 600 kinds of mammals and 45 endemic plants. Designated a Biosphere Reserve in 1976, and in 1979 a World Heritage site, it remains the United States' largest subtropical wilderness.

New life from the flames

In several regions of the world a scrub vegetation characterized by lowgrowing, woody perennial plants with small, leathery leaves is found wherever long, hot periods of summer drought are followed by mild, wet winters. This semiarid biome occurs in Europe's Mediterranean coastal areas (maquis and garigue), southern Australia (mallee and mulga), Chile (matorral) and the southern tip of South Africa (fynbos). In California and other parts of the southwestern United States the vegetation is called chaparral.

The plants have developed defensive strategies against insect and animal predators. Many are thorny, while others rely on chemical defenses: their leaves contain unpalatable aromatic oils. Competition between the plants for the limited resources is fierce. Many shrubs produce poisons that inhibit the growth of their neighbors or prevent their seeds from germinating.

The plants are adapted to withstand fire – a common hazard in the chaparral, ignited by lightning – by putting out new leaves and shoots from fire-resistant trunks or underground stems. The outer parts of the plants are highly flammable: the oils in their leaves burn fiercely, and so does the litter of small, dry leaves. The fires destroy the poisons left in the soil by the shrubs, allowing seeds lying dormant in the ground to germinate.

A semidesert habitat in Utah supports scanty chaparral vegetation – stunted bushes and small trees with tiny drought-resistant leathery leaves.

BETWEEN TWO CONTINENTS

THE RICHES OF NATURE · APPROACHES TO CONSERVATION · FORESTS UNDER THREAT

In their habitats and wildlife Central America and the Caribbean are among the richest areas on Earth. The mountains, plateaus, plains and islands encompass many different types of tropical forest, hot and cold deserts, mangrove swamps and coral reefs. The region is home to plants and animals from both North and South America. Life on the two continents sometimes developed in isolation, but when the Central American land bridge surfaced during periods of low sea level, as happened some 3.5 million years ago, the region provided a unique mixing ground for many species. The isolation of populations of plants and animals on the islands of the Caribbean led to the evolution of hundreds of unique species; adaptation to the great variety of landscapes and habitats of the mainland gave rise to many more.

COUNTRIES IN THE REGION

Antigua and Barbuda, Bahamas, Barbados, Belize, Costa Rica, Cuba, Dominica, Dominican Republic, El Salvador, Grenada, Guatemala, Haiti, Honduras, Jamaica, Mexico, Nicaragua, Panama, St Kitts-Nevis, St Lucia, St Vincent and the Grenadines, Trinidad and Tobago

Major protected area	Hectares
Armando Bermúdez NP	76,600
Asa Wright R	None given
Barro Colorado NM	5,400
Bonaire MR	2,600
Corcovado NP	41,790
Darién NP BR WH	597,000
Gandocá-Manzanillo WR	70,000
Guanacaste NP	70,000
Guanica SFR BR	4,006
Half-Moon Caye NM	4,144
Inagua NP	74,330
Kuna Yala R	60,000
La Amistad R BR WH	584,590
La Michilia BR	42,000
La Tigra NP	7,570
Luquillo NP BR	11,340
Mapimé BR	103,000
Martinique RNaP	70,150
Montes Azules BR	331,200
Nonsuch Island (Bermuda)	None given
Río Plátano BR WH	500,000
Sian Ka'an BR WH	528,147
Sierra del Rosario BR	10,000
Sierra Maestra NP	500,000
Tikal NP WH	59,570

MR=Marine Reserve; NM=National Monument; NP=National Park; RNaP= Regional Nature Park; R=Reserve; WR=Wildlife Refuge; WH=World Heritage site

THE RICHES OF NATURE

The landscape of Central America is so varied that it is possible to travel from mountainous terrain, tropical dry forest and swamp forest to rich coastal plains in the space of a single day. From the coast the land rises rapidly to high mountains. Volcanic peaks that soar to heights of 3,660 m (12,000 ft), tower above sub-alpine meadows and the deep emerald green forests, inhabited by a variety of animals, including monkeys, sloths, tapirs, toucans and snakes. Birds such as the rare and beautiful quetzal and distinctive scarlet macaw are still found here. Jaguars and pumas stalk their prey and brightly ringed coral snakes feed on birds and rodents.

The islands of the Caribbean, known as the Antilles or West Indies, also contain numerous ecosystems, from coral reefs to tropical upland forest, many with unique plant and animal species. The coral reefs provide habitats for marine species such as conchs, lobsters, sea anemones, and many kinds of reef fish; the mangrove forests and sea grass meadows support manatees and other plant-eaters, and also act as nurseries for the young of many reef- and ocean-dwelling fish. Sea turtles and shrimps migrate seasonally from one habitat to another.

The ecosystems on the islands themselves are just as diverse as those of the coast. There is a succession of forest types as the land rises from the coast to the mountains in the center of the larger islands. The evergreen woodland that fringes the coast gives way to dry lowland forest, then to moist rainforest, misty cloud forest, and finally to mountain thickets and "elfin forest", with strange, twisted dwarf trees. The proper management of the forests is vital to the prosperity of the island communities. They protect the watershed, ensuring that there is fresh water during the dry season, and also provide fuelwood and construction material; they shelter the islands from storms and hurricanes, and protect the coastal systems by checking the pace of soil erosion.

Exploiting the land

Tragically, this extraordinary region is falling victim to human disturbance. It is a process that has been going on for centuries – environmental destruction is rooted deep in the history of the region. Before the Europeans arrived in Central America, the Aztec and Maya peoples, with populations that were in some areas greater than those of today, destroyed large tracts of forest. However, the agriculture of these indigenous peoples did not degrade the land: they cultivated Indian corn (maize), beans, squash and root crops, and supplemented these by hunting and fishing. The land was able to sustain their agriculture without becoming exhausted.

Systematic exploitation began in the 16th century, when the lure of vast mineral resources attracted the Spanish Conquistadors. The region's gold and

Riding the storm – a coral island off the coast of Belize. Coral reefs form a living barrier that protects the island against storm damage. Corals grow in warm, shallow seas where photosynthesis is high, providing energy for a large and complex food chain.

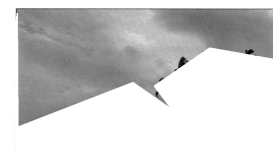

silver deposits were soon depleted, and the Spanish turned their attention to natural resources such as timber for export to the Old World. They began to cultivate native crops such as coffee, sugar and bananas for export, and later raised cattle on a large scale.

Fueled by rapid population growth and economic need, the overexploitation of natural resources that began with the Conquistadors has reached dangerous proportions. Less than 40 percent of the forest cover remains in Central America, and 3 percent of the remaining unprotected forest is being destroyed by landless peasants every year. The coastal mangroves are being cleared to make way for urban development. The timber is logged to provide wood chips, charcoal and poles for construction work.

Severe reef deterioration is characteristic of almost all the Caribbean islands. These are increasingly popular resorts for tourists, attracted by the warm climate, white sandy beaches, the reefs and the exotic marine life, but in catering for them heavy demands are made on the islands' resources. These are just a few of the threats that need to be overcome if the world is not to lose forever the beauty and genetic diversity of one of its biologically richest regions.

Cloud forest in Costa Rica At heights of between 900 and 1,500 m (2,500 and 5,000 ft) the forest is dim, damp and cold. There may be fewer than ten cloudless days a year, and trees are stunted and overgrown with lichens, mosses and other plants.

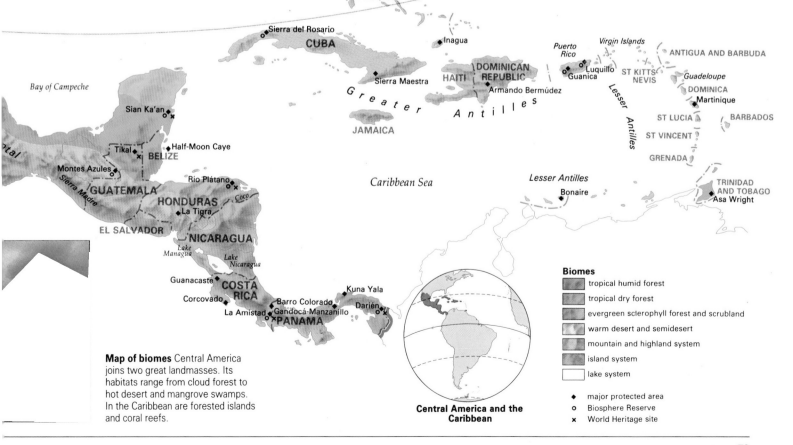

Map of biomes Central America joins two great landmasses. Its habitats range from cloud forest to hot desert and mangrove swamps. In the Caribbean are forested islands and coral reefs.

Central America and the Caribbean

Biomes
- tropical humid forest
- tropical dry forest
- evergreen sclerophyll forest and scrubland
- warm desert and semidesert
- mountain and highland system
- island system
- lake system

◆ major protected area
○ Biosphere Reserve
✕ World Heritage site

APPROACHES TO CONSERVATION

In response to the worsening situation, the Central American and Caribbean countries have made considerable progress in conservation. Puerto Rico, for example, has a system of forest reserves covering 32,400 ha (80,060 acres), 4 percent of its territory. Caribbean governments have for the most part concentrated on marine reserves, which they see as beneficial to the tourist industry upon which many of the islands depend. In 1985 Caribbean tourism generated $5 billion, and countries such as the Bahamas, half of whose gross national product is generated by tourism, are aware that their appeal is based almost exclusively on climate, environment and marine life.

Since 1970 well over a hundred marine reserves have been created throughout the Caribbean, though only a quarter of them are as yet reported to be properly managed. The management of underwater areas includes preventing overexploitation, promoting nature-based tourism such as photography, underwater marine trails and scuba-diving, and protecting the marine environment from damage by urban development, dredgers, pleasure boats, souvenir hunters, inexperienced divers and pollution. Despite the problems there have been many success stories, such as the marine reserve on the island of Bonaire in the Lesser Antilles, where tourism and conservation coexist to mutual advantage.

Unfortunately, each country's efforts have taken place in isolation, depriving conservation of an important regional focus, though the Caribbean Conservation Association and the Eastern Caribbean Natural Management Program have taken the initial steps in this direction. More disturbing is the fact that governments value the reserves more for tourist development than for conservation. This increases the likelihood of habitat destruction. The paradox is that in catering for the tourist industry, they are destroying the very environmental assets on which the industry depends.

Except for local efforts, systematic regional habitat conservation in Central America began only in 1974. Since then, each country has established protected areas. Costa Rica leads the field, with 10 percent of its total land area designated as

A typical peasant smallholding Many small farmers make a living on the edges of the forest. Their system of growing mixed crops, as well as raising a few cattle, makes few demands on the land, which is left fallow for periods so that secondary forest growth develops. These farmers also protect the water sources in the high forest.

Lush forest plants have sprung up where a stream runs through a lowland rainforest reserve. It creates an opening in the upper tree canopy so that light penetrates to the ground, stimulating growth. The understorey in the depths of the forest is much more open, but here the forest is impenetrable.

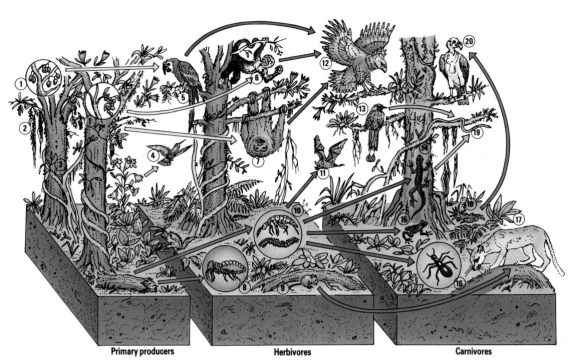

Components of the ecosystem
1 Fruits and seeds
2 Epiphytic plants
3 Liana
4 Hummingbird
5 Macaw
6 Squirrel monkey
7 Three-toed sloth
8 Termite
9 Mouse
10 Herbivorous insects
11 Bat
12 Harpy eagle
13 Insectivorous bird
14 Basilisk lizard
15 Poison arrow frog
16 Ants
17 Ocelot
18 Dead animal
19 Blunt-headed treesnake
20 King vulture

Energy flow
primary producer/primary consumer
primary/secondary consumer
secondary/tertiary consumer
dead material/consumer

A tropical forest ecosystem
showing the interaction of organisms between the soil, forest floor, understorey and tree layers.

Primary producers   **Herbivores**   **Carnivores**

The desert peninsula of Baja California, which runs parallel to the northwestern coast of Mexico, has proved inhospitable to human settlement. This has been a major factor in its survival as a wilderness. Plants and animals have adapted to these dry and arid conditions. Cacti, typical plants of New World deserts, have developed their characteristically swollen stems and reduced leaves in order to ensure that water loss through transpiration is minimal.

wildlands, the majority of which are well managed; no stretch of forest over 30,000 ha (74,000 acres) remains unprotected. Since 1970, when conservation laws were first developed, the number of legally designated protected areas in Central America as a whole has grown eightfold to approximately 350. However, the quality of management in many of these areas lags behind that found in Costa Rica.

Managing the land

Lack of government funding and local protection remain major obstacles to effective management. One proposed solution is to entrust the management of an area to local people whose traditional use of the land does not overexploit its resources. In Mexico, for example, the Lacandón Indians cultivate forest farm plots (milpas). After clearing or burning the forest they plant root crops and a mixture of others such as corn, beans and squash. In a typical Lacandón field up to 80 crops are grown over a period of five to seven years. The land is then left fallow, and produces secondary forest growth that supports game animals.

Another solution is to reduce pressures on a reserve's resources by establishing projects that emphasize sustainable production. At the Gandocá–Manzanillo

Wildlife Refuge in Costa Rica, farmers within and around the refuge receive help to diversify their crops. Over a hundred different species are cultivated. The cacao tree, of great commercial value for its seeds (cocoa beans), is the most widely grown. It is suited to the area; like the natural forest it does not expose the soil to erosion, and the crop is both easy to cultivate and profitable. Farmers here are participating in cooperative marketing schemes, and are working to secure rights of land ownership to help stabilize the peasant community.

In the absence of sufficient government finance, nongovernmental organizations have multiplied. They raise funds both

THE NONSUCH ISLAND PROJECT

The reef system and mangrove zones of Bermuda are the most northerly in the Atlantic. They form a crescent-shaped chain of approximately 150 coral-fringed limestone islands. Bermuda faces considerable environmental problems. Its population of some 60,000 crowded into just 54 sq km (21 sq mi) puts increasing pressure on the land. The reefs are overfished to cater for the 600,000 annual visitors. The presence of so many tourists exacerbates the problems of sewage and garbage disposal, which pollutes the reefs and kills the reef-building coral polyps.

Bermuda has one of the most ambitious conservation experiments in the Caribbean. Since the Nonsuch Island project was set up in 1963 more than 8,000 seedlings of rare native shrubs and trees have been planted. Once-

endangered animals such as the endemic skink and Bermuda cicada are now thriving. The dwindling population of green turtles, which for centuries have returned to breed every 15 to 30 years, has been restocked.

Most attention has focused on the campaign to restore the island's only endemic bird, the cahow (a kind of petrel), which once lived here in its thousands. Hunted for food by settlers, these ground-nesting birds were also easy prey for introduced cats and rats. They were thought to be extinct, but in 1951 one breeding pair was discovered. Under careful management their numbers have slowly built up, and by the 1980s 40 nesting pairs had been identified. The Nonsuch Island project is so successful that it has become a model for island conservation.

Regreening the forest The agriculture practiced by the Maya before European conquerors destroyed their civilization did not degrade the land though it sustained large populations. Trees grow densely around the ruins of a Maya settlement at Palenque in Mexico, showing how far the forest has been able to regenerate.

internally, from membership dues and government grants, and externally from such organizations as the International Union for the Conservation of Nature and Natural Resources (IUCN), US-AID, the World Wide Fund for Nature (WWF) and Nature Conservancy International. It is not uncommon to find international funds being channeled through these private organizations, which have assumed responsibility for the management of some conservation areas.

A future for conservation?

Habitat conservation in Central America and the Caribbean has come a long way in the last 25 years, though many areas are still struggling to find a workable balance between exploitation and preservation. Despite the current economic, social and political problems, it has been predicted that by the year 2000 as much as 8 million ha (20 million acres), or 15 percent of the region, will be grouped within as many as 500 protected units; that protected corridors will join remote areas; and that species that migrate seasonally between different altitudes will be protected across the various vegetation zones.

FORESTS UNDER THREAT

The achievements of the conservation movement are threatened everywhere in Central America. In areas of military conflict forests and crops have been destroyed by scorched-earth tactics; the population has been terrorized and dislocated, making it difficult to stabilize the way the land is used. Elsewhere, forests are cleared to make way for roads to consolidate government control over distant provinces, and this encourages destructive colonization by providing access to virgin areas.

Pressures on the land

Central America has a small landmass and a large – and rapidly growing – population, with relatively little industry. It is dependent on agriculture. Overwhelmed with foreign debts, governments try to improve their economic position both by industrial development and by the exploitation of natural resources – oil, gold and other minerals, timber, and water for hydroelectric power – as well as by producing cash crops for export. All these practices destroy the forest.

Half the population farms for its livelihood, but 80 percent of the farmers do not possess enough land to support their families. Most of the agricultural land is owned by a wealthy minority (in Guatemala 80 percent of the farmland is held by 2 percent of the population) that continues to expand large cattle ranches; already 65 percent of the richest agricultural land is under pasture.

The number of landless peasants has tripled since 1960 as the population has grown. This puts increasing pressure on remote, unprotected forested land for which ownership has not been established. Traditional peasant agriculture often cannot be sustained by the land, which is hilly and stony, with thin soils that do not retain plant nutrients and are easily eroded. After only a few years, when soil productivity declines, the peasants move

Deforestation in Haiti Once richly forested with pines and hardwoods, less than 7 percent of the forest now remains. Its disappearance has been hastened by slash-and-burn agricultural techniques. Once the vegetation is cleared, overcultivation exhausts the soil, leading to severe erosion and dereliction of the land.

REBUILDING THE FOREST

One of the most threatened types of forest in Central America is dry tropical forest. This unique forest has been reduced to only 2 percent of its former size, much of it in Costa Rica. The small patches of dry forest survive amid large expanses of degraded agricultural land and cattle ranches. A new national park is being created in Guanacaste, Costa Rica, by buying up this land and restoring the original forest.

Centered on three existing reserves, Guanacaste National Park should eventually cover as much as 7,000 ha (17,300 acres). It will sustain breeding populations of 150 mammal species (including rare tapirs and spotted cats), 300 bird species, over 6,000 species of butterflies and moths, 6,000 plant species and about 15,000 other species. The park includes migration routes from the dry forest to moist forest refuges used during the dry season by birds, mammals and millions of insects.

The abandoned agricultural land is dominated by African pasture grasses, which prevent forest seeds regenerating and are prone to severe fires. Under the management plan designed to bring back the trees, controlled grazing by livestock will help to keep down the grasses, and firebreaks protect the regenerating forest. Windblown seeds of forest trees reach the area unaided; large trees and hedgerows planted in the pastures attract birds, which bring the heavier seeds of forest fruit trees.

Efforts are also being made to ensure that the local people benefit by offering them employment as guards, managers and research assistants, by providing educational programs, and by encouraging tourism. The complete recovery of the forest is expected to take about 300 years, but significant progress will be made long before that.

been given international conservation status as the region's first Biosphere Reserve and a World Heritage site. It contains every endangered species of mammal in Central America. Despite its enormous size – over 200,000 ha (nearly 500,000 acres) – the entire reserve is in danger of devastation.

Invading peasant farmers, loggers and Miskito Indians who have fled across the border from war-torn Nicaragua practice shifting cultivation, logging and cattle ranching, activities that inflict severe erosion in the area. The wildlife is increasingly threatened both by habitat destruction and by hunting: animals command a high price on the international black market. Fishing with dynamite is responsible for the marked absence of river life. The government has neither the means nor the political will to stop the invading cultivators and establish real protection for the reserve.

Problems and solutions

The general economic crisis in the region limits efforts to conserve designated areas of forest. Parks and reserves are forced to operate on minimum budgets and are understaffed. The lack of financial support has taken its toll. Low salaries, minimal operating revenue, appalling living conditions in many parks and reserves and declining morale have all made it more difficult to tackle the growing problem of encroachment effectively. A further consideration is the resistance of the local population: to a land-hungry peasant the creation of a park in an area suitable for even short-term agriculture seems absurd. The associated problems can be equally serious. At La Tigra, Honduras' pilot national park, squatters are felling trees and destroying the watershed that supplies the capital, Tegucigalpa, with half its water.

The problem is not confined to the forests. In Belize the barrier reef, second only to Australia's Great Barrier Reef in size and complexity, remains largely unprotected despite its importance as a fishing resource and tourist attraction. The situation is similar throughout the Caribbean, where tourism and local population pressures are degrading the island ecosystems.

The lesson of recent years is that conservation must prove itself economically productive if it is to survive. Nature-based tourism, emphasizing the appeal of unexploited natural landscapes and encouraging activities such as river rafting, is being promoted. Conservationists have also recognized the importance of acquiring trained staff, of long-term financing and of gaining the support of the local population. Land-use practices and technology suited to the region are being developed to help bring prosperity to local communities.

on again. They lack the legal title that would allow them to take advantage of government improvement schemes, so they sell their holdings to ranchers, who thus acquire new pasture cheaply. Every year 1 million ha (2,470,000 acres) of forest are cleared in this way.

Honduras, with its weak economy and sizable landless population, typifies the problems facing conservationists in Central America. The Río Plátano area has

Kuna Yala Reserve

Throughout Central America the traditional way of life of indigenous peoples is threatened by the migration of landless peasants attracted by the prospect of unexploited cultivable land. Population pressure, together with economic incentives to develop the natural resources, has prompted successive governments to construct roads, which has increased the threat to Indian culture. In certain cases the indigenous people have resisted this process. In Panama the Kuna people have taken conservation into their own hands. They are applying traditional agricultural methods in their homeland in the territory (*comarca*) of the Kuna Yala Reserve, which occupies the archipelago of San Blas and the neighboring northeast coast of Panama.

Kuna Yala is inhabited by 30,000 Kuna Indians, who live on some of the 350 forested coral islands lying close to the narrow coastal plain that extends into the steep San Blas mountain range. For their livelihood the Indians fish, harvest coconuts, raise pigs and chickens and farm the coastal plains, in addition to hunting and gathering in the forest. The reserve was established in 1938 after a successful war against the government. Since then the tribe has deliberately isolated itself from Panamanian society, but by early 1970 its independence was threatened by the construction of an all-weather road that brought peasant farmers, practicing their slash-and-burn agriculture, to the limits of the reserve.

At first the Kunas supported the construction of the road, believing that it would bring prosperity, but they soon became alarmed at the threat posed to their land and way of life. To demonstrate their rights to the area, and to conserve the forest, they initiated the Kuna wildlands project. The project aims to promote scientific research and natural history, to generate revenue by encouraging low-key tourism, and to preserve the integrity of the Kuna Yala Reserve by asserting ownership of the region. It began by demarcating the *comarca*'s boundary, installing patrol trails and stations and building a headquarters at Nusagandi. The forest guards patrol the critical boundary areas, and to date there has been no encroachment on the reserve.

The success of the project is a unique example of an indigenous people planning, implementing and managing a conservation area. The Kunas' harmonious relationship with the environment, and in particular with the forest, which serves as a supermarket, pharmacy and lumberyard, forms the basis of their prosperity and independence. Their home in the reserve is only 150 km (100 mi) from Panama City, and some of them have migrated to it, attracted by the prospect of employment on United States' military bases. The value of Western education has been recognized – there are primary schools on all the major islands, and the best students are encouraged to continue their education, both in Panama City and by traveling abroad.

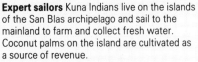

Expert sailors Kuna Indians live on the islands of the San Blas archipelago and sail to the mainland to farm and collect fresh water. Coconut palms on the island are cultivated as a source of revenue.

Plantains being harvested Farming takes place along the banks of rivers. Two or three families farm each plot as a cooperative, growing fruit, corn and coffee for themselves and cacao as a commercial crop.

In harmony with their surroundings The way of life of the Kuna Indians preserves the natural processes of the forest and islands. Rivers provide the only means of transport inland and as a valuable source of fresh water, they are particularly respected.

The Kunas are a confident people, aware that they can provide a model for their fellow Indians, and proud of the fact. They accept technological advances such as electric generators and modern transportation methods, but reject materialism as a way of life. Private ownership is still an alien concept, and decisions that affect the future of the people are made collectively, according to tradition. The Kunas are succeeding in protecting their traditional lands, taking advantage of the modern world yet continuing to live in harmony with the unchanged wilderness of the reserve.

RICHES OF THE RAINFOREST

CONTINENTAL DIVERSITY · WILDLIFE PROTECTION · PLANNING FOR THE FUTURE

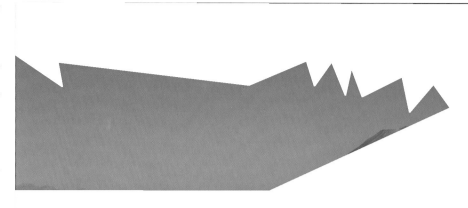

South America shelters the richest plant and animal life found anywhere on Earth. This tremendous diversity is the result of the climatic influence of the Andean mountain chain, which runs almost the length of the continent along its western rim and determines the pattern of its plant and animal communities, as well as of the geological past. With the shifting of the Earth's tectonic plates, South America gradually became separated from the ancient, drifting supercontinent of Gondwanaland and its animal and plant species evolved in isolation, until the creation of the Central American land bridge about 3.5 million years ago enabled the species of North America to mix with those of the south. Its great diversity of habitats range from barren, ice-covered mountain peaks to lush, species-rich rainforest.

COUNTRIES IN THE REGION

Argentina, Bolivia, Brazil, Chile, Colombia, Ecuador, Guyana, Paraguay, Peru, Surinam, Uruguay, Venezuela

Major protected area	Hectares
Amazonia (Tapajós) NP	1,000,000
Banados del Este NP RS BR	200,000
Brasília NP	28,000
Canaima NP	3,000,000
Cara-Cara FBR	70,000
Chiribiquete NP	1,000,000
Cotacachi-Cayapas ER BR	45,000
Defensores del Chaco NP	780,000
Fray Jorge NP BR	9,960
Galapagos NP WH BR	691,200
Iguaçu NP	1,100,000
Isiboro Sécure NP BR	1,000,000
Kaieteur NP	11,655
La Neblina NP	1,360,000
Lauca NP BR	9,960
Los Alerces NP	263,000
Los Glaciares NP WH	445,900
Manu NP BR	1,532,800
Monte Pascoal NP	22,500
Nahuel Huapi NP	428,100
Noroeste BR	226,300
Pacaya Samiria NR	1,387,500
Sanguay NP WH	370,000
Sierra Nevada de Santa Marta NP BR	383,000
Torres del Paine NP BR	16,300
Ulla Ulla NFR	200,000
Vicente Pérez Rosales NP	220,000
Wia-Wia NR	36,000

BR=Biosphere Reserve; ER=Ecological reserve; FBR=Federal Biological Reserve; NFR=National Floral Reserve; NP=National Park; NR=Nature Reserve; RS=Ramsar site; WH=World Heritage site

CONTINENTAL DIVERSITY

One physical feature jumps out at even the most casual observer of a map of South America: the Andes. This still-growing mountain chain exerts tremendous influence on the climates and soils of the region, and consequently upon the distribution of plants and animals. Because it acts as a barrier to the moist Pacific weather systems, most of the continental rainfall is supplied by systems originally in the Atlantic. The driest part of South America is the cold fog desert flanking the Pacific coast of Peru. In some places, rain may fall no more than once a century; only acacias, mimosa and a few scattered cacti manage to survive. In stark contrast are the lush rainforests of the Amazon basin. This vast jungle, with its network of rivers and swamps, contains 10 percent of all the world's species – at least one million. Its value as a world resource has only just begun to be appreciated, yet some 10 percent of the rainforest has already been felled or burned.

Along the east coast are pockets of Atlantic rainforest. These, too, are rich in species, many of them unique to the area (endemic species). Atlantic rainforest is under even greater threat than the forests of the Amazon Basin – little more than 2 percent survives today, and even this remnant is in danger of being destroyed.

The subantarctic beech forests to be found in southern Chile (of which only tiny fragments remain) are all that survive of an ancient vegetation that is also found in parts of New Zealand and Australia. Once all belonged to the ancient super-continent of Gondwanaland. The forests are dense and evergreen in the north, and more open and deciduous in the south, where the beeches are mingled with larches and araucaria pines (monkey-puzzle trees). They are lively with parakeets and hummingbirds; the endemic species include a primitive marsupial (pouched mammal) called a rincolesta.

Seas of grass

Between the great snow-capped volcanic ranges of the Andes lie the *altiplanos*, a series of high plateaus broken by hills and valleys, lakes and marshes, where

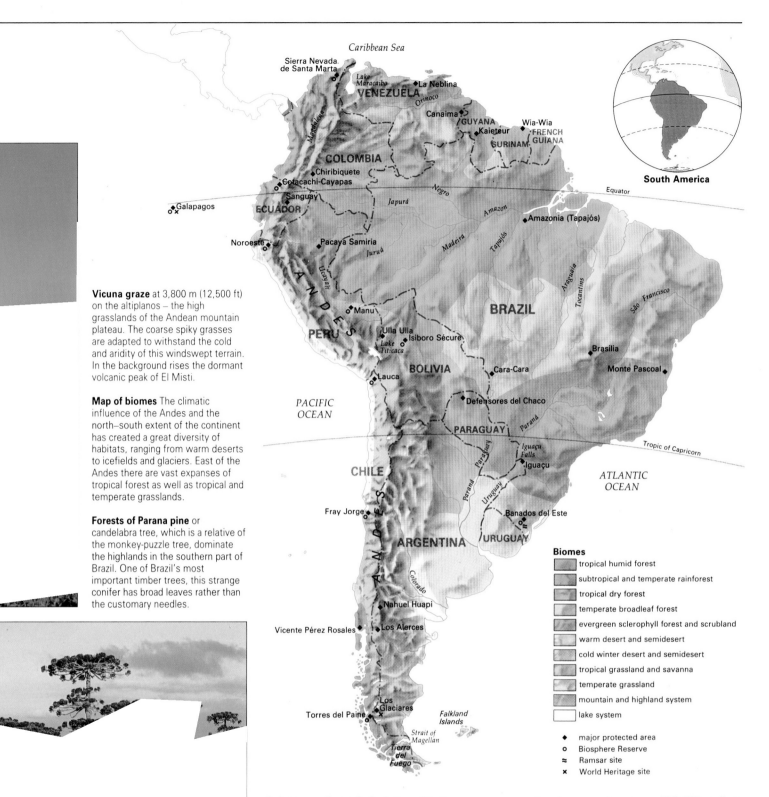

Vicuna graze at 3,800 m (12,500 ft) on the altiplanos – the high grasslands of the Andean mountain plateau. The coarse spiky grasses are adapted to withstand the cold and aridity of this windswept terrain. In the background rises the dormant volcanic peak of El Misti.

Map of biomes The climatic influence of the Andes and the north–south extent of the continent has created a great diversity of habitats, ranging from warm deserts to icefields and glaciers. East of the Andes there are vast expanses of tropical forest as well as tropical and temperate grasslands.

Forests of Parana pine or candelabra tree, which is a relative of the monkey-puzzle tree, dominate the highlands in the southern part of Brazil. One of Brazil's most important timber trees, this strange conifer has broad leaves rather than the customary needles.

Biomes
- tropical humid forest
- subtropical and temperate rainforest
- tropical dry forest
- temperate broadleaf forest
- evergreen sclerophyll forest and scrubland
- warm desert and semidesert
- cold winter desert and semidesert
- tropical grassland and savanna
- temperate grassland
- mountain and highland system
- lake system

- ◆ major protected area
- ○ Biosphere Reserve
- ≈ Ramsar site
- ✕ World Heritage site

alpine meadows fade into cold steppe on the higher slopes. These support mainly tough grasses, ground-hugging rosette plants, shrubs and cacti. Members of the South American llama family, guanacos and vicunas, together with deer and increasing numbers of domestic livestock, graze these windswept grasslands. They share the habitat with rodents, mainly chinchillas and guinea pigs, and predators such as pumas and foxes. Carrion-eating condors soar high above, scanning the ground for carcasses. Farther south, the Andes become barren wastes of permanent snow and ice.

To the east of the Andes, and in its rainshadow, is the *pampas*, a huge area of grassland measuring some 770,000 sq km (300,000 sq mi). Few animals live here other than rheas (large flightless birds), grazers like guanacos and vicunas, and small, burrowing rodents such as maras and viscachas.

The grasslands of the Brazilian plateau to the north, the *campos cerrados*, are poor in species, but the marshy area around the Paraguay river – the Pantanal – provides a habitat for lungfish and the rare maned wolf. By contrast, the subtropical *llanos* grasslands of Venezuela contain many species. However, large parts of this habitat have been destroyed by overgrazing, a threat that is increasing in other South American grasslands.

Island ecosystems

Off the southeastern coast of South America are the Falklands, South Georgia and the South Sandwich Islands. Hilly, barren and swept by Atlantic gales, they are covered with peat, low dense shrubs and tussock grass. No native land mammals live on them, but their shores and the surrounding waters support porpoises, dolphins, elephant seals, penguins and many species of seabirds, including the black-browed albatross and the red-backed buzzard.

On the opposite side and at the opposite end of the continent the rugged volcanic islands of the Galapagos lie nearly 1,000 km (600 mi) from the mainland, too far to be colonized by mammals. This meant that other species moved in to occupy their niche in the ecosystem by evolving special adaptations to exploit the islands' different habitats. The islands therefore contain an extraordinarily large number of endemic species, including giant tortoises, seaweed-feeding marine iguanas and flightless cormorants. There are several species of finches, each of which has its own particular diet and a specially modified beak to deal with it.

Rockhopper penguins on the Falkland Islands. The rich waters of the South Atlantic provide food for large colonies of seabirds and sea mammals such as southern fur seals. Now abundant, they were once hunted near to extinction by fur traders.

WILDLIFE PROTECTION

Slowly but surely the concept of protecting South America's outstanding natural areas has taken hold. The continent's first protected area was Argentina's Nahuel Huapi National Park, established in 1922 as the Parque del Sur. Chile followed soon after with Vicente Pérez Rosales National Park, a trans-Andean continuation of Nahuel Huapi. Then Ecuador created the Galapagos National Park in 1934.

Conservation speeded up in the 1960s and 1970s. Between 1972 and 1984 the number of parks and reserves doubled from 126 to 253, increasing the area of land under protection from 180,000 sq km (69,500 sq mi) to 450,000 sq km (173,000 sq mi). In the Amazon areas of Brazil, Bolivia, Ecuador and Venezuela alone, the creation of new parks and reserves accounted for more than 1 million sq km (386,000 sq mi).

Even more significant is the quality of these new areas. Many of them take in ecosystems that were previously under-represented, demonstrating a readiness to use ecological criteria in selecting sites. Most South American countries are seeking to establish protected areas that will adequately represent each of the major habitat zones, as well as protect unique or endangered landscapes and species. They are also realizing that increased tourism can give an added economic value to well-protected areas.

Multiple-use reserves

There has been a recent move in the management of parks toward establishing multiple-use systems – or Biosphere Reserves – which allow for the full integration of indigenous peoples and their methods of food production. Such reserves contain a highly protected core zone, surrounded by buffer and transition zones. Local people live within these areas, which are also used for experimental research and species-monitoring.

In a reserve on the Colombian border, the Awa Indians are being trained in reserve management and organization, as well as learning how to deal with government and conservation agencies. The Indians in the northeastern part of the Amazon basin are being encouraged to raise pig-like peccaries and pacas (a type of rodent), activities which do not disrupt the surrounding environment. In Peru's

Manu Biosphere Reserve the local economy has been given a boost by the development of fish farming and the encouragement of smallscale tourism.

By 1986 there were 19 Biosphere Reserves in South America, but not all are managed in accordance with the internationally recognized criteria for such reserves, and many exist on paper only. Bad planning, economic constraints, corruption and the pressure of prospectors, speculators and settlers make it difficult to ensure the proper protection of these areas. Multiple-use reserves place a heavy burden on administrators, whose duty it is to ensure the survival of the intact ecosystem. It is also their responsibility to secure the cooperation of the local people and to safeguard their livelihoods. Achieving the correct balance is an enormous challenge and a difficult task. However, there have been some notable successes, of which the Manu Biosphere Reserve is an outstanding example. Its strictly protected core zone alone covers some 1.5 million ha (3.75 million acres). It is home to many large mammal predators and grazers, and boasts some 1,000 species of birds, 13 species of monkeys and more than 15,000 plants species.

Action by the people

Until recently there had been relatively slow growth in popular conservation movements in South America, but education programs are expanding. Ecuador's Nature Foundation, a nongovernmental organization created in 1978, shows what can be achieved. After producing a report on the country's many environmental problems, it prompted the formation of an association of environmental journalists, and produced a range of imaginative radio programs, films and television shows, as well as environmental education programs for the use of schools and government officials.

Perhaps the Foundation's greatest contribution has been to organize a debt swap in which part of Ecuador's outstanding international debt was purchased at a discount in return for increased government financing of conservation projects, especially the national park system. If the other organizations that are springing up elsewhere in South America follow this example, it may help to reduce the pressures and financial constraints that at present assail the hard-pressed reserves in this region.

HARVESTING NATURAL RESOURCES

The Pacaya Samiria National Reserve lies on the upper waters of the Amazon in central Peru, and covers more than 13,800 sq km (5,325 sq mi). Its management exemplifies the regional policy of combining protection of valuable natural resources with the controlled harvesting of some species in order to supply the living requirements of local people. Almost 50,000 people live around the reserve's borders.

The Pacaya and Samiria rivers were originally set aside in the 1940s as fishery reserves to preserve the paiche, a huge catfish found in the waters of the Amazon. In 1982 the reserve was enlarged to its present size, and its conservation program extended to all its plants and animals. The reserve contains a wide range of wildlife, including many species not found in Peru's better-known park, Manu. In addition to the paiche, peccaries, deer, monkeys and pacas are hunted by the local people to a limited extent.

A management plan established that many species could be sustainably harvested, both for food purposes and for scientific research, without depleting their numbers below the level necessary for survival. The aim was to improve the situation for endangered species and to maintain or increase already abundant wildlife populations, particularly paiche, caymans, some monkeys, and capybaras and pacas. A series of research stations is planned throughout the reserve to monitor this work. The local people will be allowed to harvest the reserve's resources for their own needs, and it is hoped that the wildlife program will eventually supply them with a steady income and involve them directly in the management and benefits of what is perhaps the world's largest tropical multiple-use conservation project.

A river ecosystem Both algae and decomposing plants that fall to the river bed are important primary producers at the bottom of the food chain.

Swampy ground high in the Andes at Sangay National Park, Ecuador, provides a habitat for a rare marsupial frog. Alpine cushion plants stud the drier slopes.

Components of the ecosystem
1   Plant plankton (algae)
2   Amazonian water lily
3   Plant detritus
4   Animal plankton
5   Terecay turtle
6   Capybara
7   Neon tetra
8   Leporinus fish
9   Amazonian dolphin
10 Piranha fish
11 Giant arapaima
12 Cayman

Energy flow
primary producer/primary consumer
primary/secondary consumer
secondary/tertiary consumer
dead material/consumer

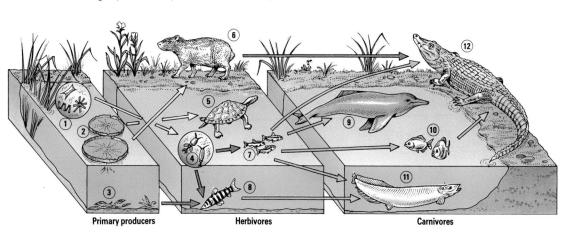

Primary producers   Herbivores   Carnivores

PLANNING FOR THE FUTURE

Although great strides have been made in recent years to increase the number of protected areas in South America, many difficulties remain. Most are the result of the overwhelming economic and social problems facing all the countries of South America. There is a rapidly rising population – numbers will have reached about 400 million by the year 2000 – but the real problem is not overpopulation so much as unequal land distribution. In non-Amazonian Brazil, for example, less than 5 percent of the landowners own 81 percent of the country's farmland, and 70 percent of the population has no land at all. The situation is roughly similar in most other countries.

To alleviate landlessness, most governments have sought to colonize previously unsettled lands, with little regard to their suitability. Most of these areas are in the tropical rainforests. The settlers are inevitably forced to practice unsustainable forms of shifting cultivation because of the thin, infertile forest soils, while the driving of roads into previously impenetrable forest allows access for hunting, mining and other activities that destroy these fragile ecosystems. Even the national parks are not exempt: Manu National Park has had its forest areas periodically threatened with a road that would cut the park in half in order to facilitate colonization.

A recent innovation may help to stem this invasion of the wilderness. This is the concept of conservation-for-debt: part of a country's debt to a Western bank is sold at a discount to a charity or conservation agency, which then extracts an agreement from the debtor country to put the full value of the debt purchase into a conservation project. Such schemes are rare, and it is difficult to ensure that the promised work is carried out.

Menace in the forest

The greatest cause of deforestation is the clearance of land for largescale agriculture and cattle ranching. Second to this comes logging for commercial purposes. The loggers remove only about 3 percent of each area they work, but in the process they damage more than 50 percent of the remaining vegetation. The building of logging roads into the forest opens it up to further exploitation.

A SPECIES-RICH HABITAT AT RISK

The Amazon Basin is among the most biologically rich areas on Earth. Drained by the 6,570-km (4,080-mi) long Amazon river system, its vast expanse of seemingly uniform tropical rainforest contains at least three major ecological zones and over 1 million species of plant, animal and insect life. According to one reckoning, a single patch of rainforest measuring 10 sq km (4 sq mi) in extent may contain as many as 1,500 species of flowering plants, 750 different tree species, 125 mammals, 400 birds, 100 reptiles, 60 amphibians and 150 butterflies.

The consequences of the eventual destruction of the Amazon rainforest are incalculable. Species whose value we scarcely yet know will simply disappear. Only a tiny percentage of the Amazon's incredible abundance of species is fully understood by science, but we know that many plants are rich in secondary compounds that have potential use as medicines, pesticides and herbicides. Many of today's principal food crops originated from wild ancestors growing in these forests. Even when secondary forest growth regenerates on cleared areas of land, the diversity of the forest species can never be recovered.

Of even greater significance are the possible global consequences of continued deforestation. Most scientists agree that the loss of forest cover contributes to a buildup of carbon dioxide in the atmosphere. This absorbs infrared radiation reflected from the Earth's surface, causing a gradual increase in the Earth's overall temperature – the "greenhouse effect".

Much forest has been cleared to convert the wood to charcoal for iron smelting and other industrial processes, and to allow mineral extraction to take place. The wasteheaps of excavated soil are washed away by the heavy tropical rains and silt up the rivers. Gold prospecting can be even more devastating: the mercury used to extract the gold runs into rivers, killing aquatic life and poisoning the water, often with fatal consequences for both animals and humans living further downstream.

The removal of vegetation cover from watersheds allows more water to reach the rivers and streams. This causes them to flow faster and to become more turbulent. Whereas fish were able previously to browse on the nutritious floodplains during periods of high water, now they are unable to hold their own against the increased current. Seasonal rain is lost in floods, and many rivers dry up altogether for the rest of the year.

Industrial devastation Mining in Brazil has reduced the rainforest to muddy pools poisoned by heavy metals. The environmental consequences of such widespread destruction are enormous, and spread far beyond the original site of activity. The waste soil produced silts up rivers, and their waters are contaminated with pollutants.

Water hyacinths form a thick blanket across a marsh. Unchecked, they can prove a great nuisance by choking waterways, but they are kept naturally under control by manatees and other browsing aquatic animals. The floodplains of South America's rivers are rich in plants and animals.

Savanna grassland in Brasília National Park, Brazil. Puma and jaguar once hunted these plains, but the arrival of human predators has forced them to retreat into the forests. These once-extensive grasslands, too, are in retreat as ranchers plow them up and re-seed them with European grasses to feed their livestock.

Islands on the continent

Originally most of South America's parks and reserves were created in the middle of large tracts of unspoilt wilderness. But as a land-hungry population encroaches on these areas, they are left as islands in a sea of altered and degraded land.

The survival of most species depends on there being an adequate breeding population – usually a minimum of about 500 individuals – to ensure sufficient genetic variation. Some mammals, and even some rainforest trees, require many thousands of square kilometers to support the feeding and reproductive needs of that number of individuals. In addition, many animals, especially birds, need to live within huge areas because they migrate seasonally to different feeding or breeding sites. The hard fact is that most parks and reserves are too small to protect every species.

The creation of buffer zones, as in multiple-use reserves, is one solution to this problem. Another is legislation that rewards landowners for conserving large parts of their land in a natural state. This could help to preserve corridors of vegetation linking protected areas, thus helping to conserve genetic variation by allowing animals and seeds to travel between them. The future outlook for protected areas in South America is promising in spite of the immense problems that still have to be confronted.

Chiribiquete

Chiribiquete National Park, created in 1989, covers 1 million ha (2.5 million acres). It is the latest stage in a land transfer process that is designed to make Colombia's Indians the protectors of the country's tropical forests, which rank among some of the richest in the world.

Traditionally the Indians have lived in harmony with the forest, taking no more than they need from it and resting the land between cultivations to allow the forest to regenerate. The concept of harvesting a surplus in order to generate a profit is quite alien to them.

The mainstay of their traditional way of life is the *chagra* – a garden in the forest. The Indians clear a patch of forest and grow crops such as yuccas, peppers, avocados, papaya, mangoes, lemons and other fruits. After a couple of years the soil starts to lose its fertility and the Indians move on to clear a new *chagra*, returning from time to time to harvest the fruit and to hunt the game it attracts. Wood and vines for building and for making canoes, as well as game, fish and plants, are collected from the surrounding forest, rivers and lakes. It has been estimated that each member of a traditional Indian family uses some 1,000 ha (2,500 acres)

Natural renewal The rainforest is always changing – when a tree falls and decays it leaves a clearing that is then gradually recolonized by plants and animals. This natural process is believed to encourage diversity by allowing new combinations of species to develop. Chiribiquete contains very many endemic species.

Imitating nature The Indians' *chagras* mimic the natural process of forest regeneration; they fire the fallen wood after felling but do not burn or cultivate the clearing's perimeters. The sites for their plots are selected very carefully; they are well-drained and avoid the river banks.

A vast resource The forest provides all the Indians' needs. A family group of about 10 people may roam over as much as 10,000 ha (25,000 acres) in their hunting and foraging expeditions. Canoes constructed from forest trees are used to travel along rivers – their only form of transport.

of forest. The exploitation of forest resources is controlled by the local headman or guardian, who has a natural understanding of the environment and decides how many animals and plants of each species may be taken at any one time; he also imposes food and sexual restrictions on the other members of the group as he judges necessary.

The *chagras* are managed in such a way that the forest is able to regenerate fairly quickly; they mimic natural forest clearings created by fallen trees, though they are considerably larger. The Indians' practice of removing all the trees and burning them results in the death of some species of pioneer seedlings, but their fruit trees attract birds and mammals which bring in new seeds. A natural clearing takes only 40 years to return to mature forest, but an Indian *chagra* may require up to 200 years for complete recovery. This is a vast improvement on the time it is estimated that sites bulldozed by settlers will take to regenerate.

In recognition of the Indians' efficient management of the land, the Colombian government is returning over 1.8 million ha (4.5 million acres) of the Amazon to their ownership, thereby creating the largest single territory in the world legally in the hands of ethnic peoples. The plan involves considerable financial outlay in order to compensate landowners and peasant farmers who have been forced to move away. However, Colombia's rainforest areas have suffered relatively little invasion by colonists.

The decision to make the Indians entitlement to the land irreversible has generated some controversy: there are those who wish to retain the option of developing this land in the future, while others point out that the forest will be protected only if the Indians continue to live in their traditional way. A provision in the land settlement goes some way toward meeting this objection by prohibiting the sale of the land to outsiders.

Colombia's far-sighted policies are in stark contrast to the situation in neighboring countries such as Brazil, where uncontrolled exploitation and colonization of the tropical forests are rampant. Other countries are watching developments in Chiribiquete and other reserves – already Bolivia has sought Colombian advice on creating Indian reserves. If the idea catches on, it could revolutionize conservation in the Amazon.

Life at the top

On the cold, rocky mountain heights few plants can survive. Those that do are mostly cushion plants, typical of high-altitude habitats the world over. Their low growth reduces exposure to chilling winds and allows them to shelter under an insulating blanket of snow in winter. Tufts of spiny grasses and cushions of small-leaved herbs dot the landscape, often sheltering in the lee of protective rocks.

Few animals can survive on these sparse rations and in these cold temperatures, but on the high plains of the Andes the plants are grazed by herds of guanaco and by Darwin's rheas, large flightless brown birds that stand some 3 m (9 ft) high. Smaller herbivores include mice and chinchillas, well-protected against the bitter climate by their long silky fur. Pumas still stalk the guanaco, while the smaller animals are hunted by smaller predators – Geoffroy's cats, and red and gray (South American) foxes. The lakes attract many rare birds, including Coscoroba and black-necked swans and colorful Chilean flamingoes. The sky is the undisputed domain of the Andean condor, the world's largest vulture.

A bleak windswept landscape in the Torres del Paine National Park in Argentina, an area of outstanding natural beauty, with jagged mountain peaks, cold blue lakes, waterfalls, rivers and grassy plains.

LAND OF FJORD AND FOREST

WILDERNESS OF THE NORTH · A WELL-PROTECTED REGION · THREATS FROM AFAR

The Nordic countries contain some of the largest wilderness areas in Europe. Extending far beyond the Arctic Circle, yet warmed on its western shores by the North Atlantic Drift, the region lies at the western limit of the taiga, the great, unbroken coniferous forest that stretches to Siberia and is home to the wolf, wolverine, brown bear, lynx and moose. The sparkling lakes and bogs of Finland and northern Sweden are the summer breeding homes of millions of waterfowl and waders. Norway's rugged coast supports a profusion of underwater life. Away from mainland Europe volcanic Iceland has its own unique landscapes; huge numbers of seabirds nest on the cliffs of the Faeroe Islands; and the polar desert of Svalbard in the Arctic Ocean is inhabited by one of the world's largest polar bear populations.

COUNTRIES IN THE REGION	
Denmark, Finland, Iceland, Norway, Sweden	

Major protected area	Hectares
Abisko NP BR	7,700
Børgefjell NP	108,700
Breidafjördur NR	270,000
Dovrefjell NP	26,500
Femundsmarka LPA NP	38,600
Hardangervidda NP	342,800
Lemmenjoki NP	280,000
Muddus NP	49,300
Mývatn Laxá NR RS	44,000
Northeast Svalbard NR BR (Svalbard islands)	1,903,000
Øvre Anarjåkka NP	139,000
Øvre Dividal NP	74,100
Padjelanta NP	198,400
Pallas-Ounastunturi NP	50,000
Rogen NR	41,166
Rømø Island NaP	12,100
Rondane NP	57,200
Sarek NP	197,000
Sjaunja BS RS	290,000
Skaftafell NP	50,000
Skagen NR	4,300
Stora Sjöfallet NP	127,800
Thingvellir NP	4,200
Urho Kekkonen NP	253,000

BR=Biosphere Reserve; BS=Bird Sanctuary; LPA=Landscape Protected Area; NaP=Nature Park; NP=National Park; NR=Nature Reserve; RS=Ramsar site

Forest and ice Snowcovered mountains rise above the forest and lakes of Norway's Rondane National Park, which lies within the subarctic taiga zone of coniferous forest. On the northern fringes of the taiga, and at altitudes where the trees become scattered and more open, lichens predominate among the ground cover, and are grazed by herds of wild reindeer.

An old beech tree is covered in mosses, ferns and a bracket fungus, providing food and shelter for insects. Northern Scandinavia still contains large areas of virtually untouched forest. Not all this boreal forest is dark and evergreen. In the valleys, and where drainage and fertility allows, there are significant deciduous woodlands. The one shown here is of beech and rowan. Sunlight penetrates between the young saplings, allowing ground cover to develop.

WILDERNESS OF THE NORTH

Northern Scandinavia (Norway, Sweden and Finland), which is largely unpopulated, still contains significant areas of unspoilt wilderness. Above 1,000–1,200 m (some 3,300–4,000 ft) there are glaciers and patches of permanent snow, rocky peaks and block screes, a hostile land almost bare of vegetation. Below the exposed peaks lies a treeless, windswept landscape of heaths and low-growing willows, the mountain tundra, which leads down to a treeline of scattered birch trees, outliers of the birch forest that occupies most alpine valleys in Scandinavia and all mountain slopes up to 500–800 m (1,600–2,600 ft) above sea level.

At lower levels the birch forest gradually merges into the boreal forest or taiga, an enormous and rather uniform area of dark coniferous forest, sometimes hilly, sometimes flat and interspersed with bogs and lakes. Scots pine dominates on coarse soils and bedrock, and spruce on richer soils. Mixed in with the conifers are deciduous trees such as birch and alder. Despite widespread logging there are still significant areas of protected virgin forest in parts of northern Sweden and Finland, especially in the mountains.

Lakes and bogs are widespread on waterlogged ground, their abundant insect life attracting millions of migrant birds in summer. Here are some of the last nesting grounds of the red-throated loon or diver, whooper swan and bean goose. Peregrines breed on treeless bogs; ospreys fish in the lakes, nesting in nearby trees.

The southern forests

The southern limit of the taiga, around 60°N, is also the northern limit of oak. On the lower slopes of the mountains, in sheltered valleys and in the warmer south the coniferous forest gradually merges into deciduous temperate forest. A mixed forest of coniferous and deciduous trees is found throughout much of southern Sweden and in small areas of southern Finland and Norway. With their great variety of nuts and seeds, these mixed forests support a rich birdlife, as well as numerous species of small mammals, deer, foxes and badgers.

True deciduous forest, characterized by beech, oak, elm and ash, covers southernmost Sweden and a small area in southern

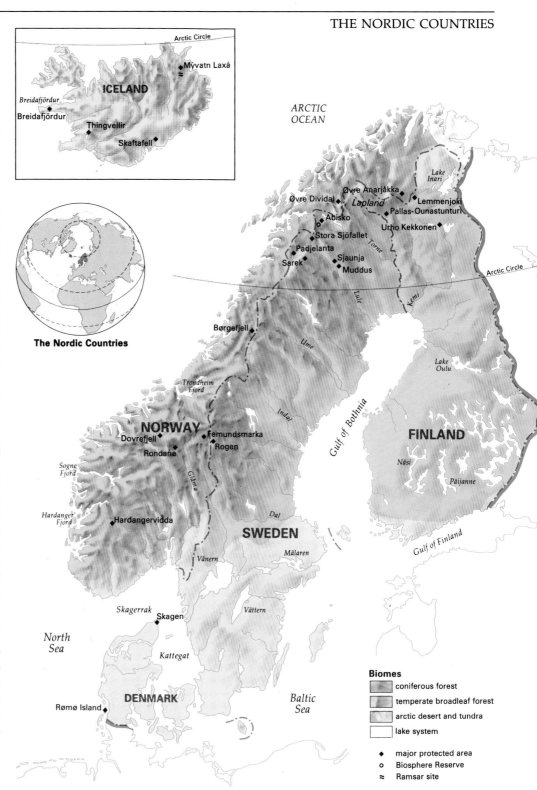

Map of biomes Temperate broadleaf forest merges into the taiga coniferous zone within the Nordic region. Its numerous lake systems are threatened by acid rain. Iceland has sparse tree cover and peat bogs, which are covered in snow and ice in winter.

Norway, as well as 10 percent of Denmark. Although rich in species, very little of this forest is completely untouched by human activity. Altogether, 30 percent of Norway, 55 percent of Sweden and 65 percent of Finland is covered in forest of one type or another.

In the south, the deep deposits of material deposited by melting ice sheets at the end of the last ice age have formed a rich soil, and a gentler agricultural landscape dominates. Denmark's sandy coast, with its mudflats and lagoons, attracts waders, ducks and geese.

Islands of the far north

The northern islands are quite different from the rest of the Nordic countries. Their isolation and unique past have given each its own character.

Iceland's landscapes are a unique blend of polar snow and ice, extensive peat bogs and volcanic deserts. There is continuous vegetation cover over only a quarter of the island. The only natural forest in this windswept land is birch. Foxes were the only land mammals here until humans introduced domestic species, including reindeer, and accidentally brought in rats and mice; there are no amphibians or reptiles. But the richness and variety of the birdlife more than compensates for the lack of other vertebrates.

The treeless, windswept Faeroe Islands in the north Atlantic have no indigenous mammals, but the seabirds nesting on the cliffs are so numerous as to provide a source of income from food and feathers for the local people.

A lakeland paradise for birds Lake Mývatn in Iceland is named after the mosquitoes that swarm over the surrounding countryside in the summer. Biting insects are a scourge throughout this region, but provide an important source of food for birds and fish. Mývatn is famous for the rare species of ducks it supports.

A WELL-PROTECTED REGION

The Nordic countries were among the first to practice modern conservation, and today they boast one of the most sophisticated conservation systems in the world. Conservation considerations are closely integrated with land use planning, and "green" issues have come to dominate the decision-making political scene.

Some of the earliest conservation legislation was passed in Sweden in 1909 in the form of a conservation law that established the first nine national parks in Europe. Denmark followed suit with a conservation law in 1917, Finland in 1923 and Iceland in 1928. Today it is a general principle in these countries that all species of wild birds and mammals not scheduled as game are protected.

Since this early start, protected areas have increased in number. The largest and most scenic are usually the national parks, but their numbers are not a reliable indication of the degree to which nature is protected. Nature reserves are more numerous than national parks: Sweden, for example, has some 1,000 nature reserves but only 20 or so national parks. Denmark has no national parks at all, but a quarter of the country is covered by nature parks in which town and country planning controls safeguard habitats: building, digging and planting are forbidden within 300 m (1,000 ft) of forest fringes and 100 m (330 ft) of ancient monuments.

Game reserves, where both hunting and public access are regulated, are widespread in Scandinavia, often affording considerable protection to wild species. The nature reserves protect specific sites of aesthetic, scientific and recreational value, or the breeding sites of birds and mammals. Sometimes specific habitats are protected as well.

A further category of natural monuments protects smaller features, such as waterfalls, clumps of trees and, in Iceland, volcanoes and hot springs. Special reserves may also be created to protect ecosystems under particular pressures, such as Finland's fragile peatlands and

Components of the ecosystem
1 Pine, spruce, fir, larch trees
2 Fallen cones, seeds, needles
3 Aphids, beetles
4 Red squirrel
5 Nutcracker
6 Bark beetle
7 Northern red-backed vole
8 Red ant in nest of pine needles
9 Pine grosbeak
10 Siberian tit
11 Western capercaillie
12 Goshawk
13 Pine marten
14 Gray owl
15 Dead vole
16 Red fox

Energy flow
⇨ primary producer/primary consumer
⇨ primary/secondary consumer
⇨ secondary/tertiary consumer
⇨ dead material/consumer

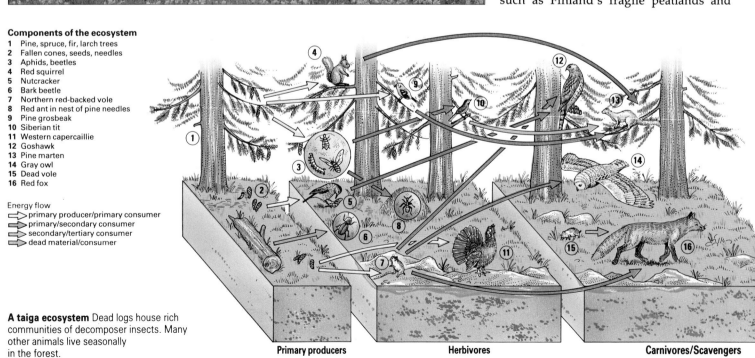

A taiga ecosystem Dead logs house rich communities of decomposer insects. Many other animals live seasonally in the forest.

Primary producers   Herbivores   Carnivores/Scavengers

600 types of vegetation. It takes into account such factors as the nature of the bedrock, the soil, plant and animal communities, current land use and the history of modification by human activity. Devising such a system has involved many new surveys. These will provide an invaluable environmental data base.

This information will be used not only to identify sites that are worthy of protection, but also to assess the ability of the local environment and natural resources to withstand various types of exploitation, such as industry and urbanization, agriculture, forestry and recreation. The data will also aid planners in finding suitable sites for reservoirs and the regulation of water flow.

A place for the people

In some places there is a conflict between the interests of wildlife and habitats and those of local people and visitors. Many people living in the country are suspicious of regulations against hunting, snowmobiles, overflying and fishing, and the nomadic reindeer-herding Lapps still retain the right to practice their traditional way of life in some conservation areas. Tourists can damage vegetation as, for example, in Sweden's Sarek National Park, where the bogs are easily destroyed by trampling feet. Conservationists must strike a balance between protecting the environment from harm, and providing opportunities for people to enjoy the wilderness alone and in peace.

bogs. Sweden's Nature Resources Act specifically protects several important and unspoiled river systems against hydroelectric power developments, and also safeguards 12 very large, roadless and uninhabited mountain districts against commercial exploitation.

In the Nordic countries the right of public access to the land is almost universal, but transport may be regulated. In fragile environments the use of snow scooters and other motorized vehicles is often prohibited within national parks and reserves, and sea and river traffic may be restricted near breeding birds.

The Nordic countries' plethora of reserves is well justified, for much of the region's wildlife heritage is of international importance. Many wetland areas have been designated as Ramsar sites under the international convention of 1971 – 38 in Denmark (11 of these in Greenland), 20 in Sweden, 14 in Norway, 11 in Finland and 1 in Iceland. There are also a number of Biosphere Reserves, set up to protect significant examples of particular habitat types. These include the tundra and polar desert areas of the Svalbard archipelago, and the Abisko Biosphere Reserve in the area of Lake Torne in Sweden. Although all the Nordic countries except Iceland are party to the World Heritage agreement, no natural wilderness site has yet been designated in the region.

Protecting the remote wilderness Sjaunja Nature reserve in the north of Sweden has been added to the region's extensive network of protected areas because the shores of its lakes make it an important feeding and breeding site for many species of northern birds.

A unique blueprint for conservation

The Nordic Council, to which all the countries of the region belong, is pioneering a system for assessing the importance of various sites for conservation. This complex system of classification recognizes 76 geographical regions and some

MÝVATN AND LAXÁ NATURE CONSERVATION AREA

Public outcry greeted the proposal to turn Iceland's Lake Mývatn into a reservoir for hydroelectric power. As a result of this strong opposition, a special protected area, covering at least 44,000 ha (108,600 acres), was established in 1974 by act of parliament.

Lake Mývatn is one of the island's largest lakes. It is very irregular in shape, quite shallow and extremely rich in birdlife. Vast numbers of midges and blackflies breed here, providing an important source of food for fish and birds. All 15 species of Icelandic ducks breed here, between 8,000 and 9,000 pairs, a number that includes the entire European population of the Barrow's goldeneye.

The lake is studded with volcanic islands and surrounded by hot springs and old craters. When the explosion crater of Hverfjall was created, lava

dammed the lake. Its outlet river, Laxá, still follows the course of an old lava flow. Steam erupting through molten volcanic rock has created a fascinating scenery of sculpted lava.

The main conservation aims for Lake Mývatn are to prevent its exploitation for hydroelectric power, and to control building and environmental disruption in the surrounding area, which has been settled for centuries. There are several farmsteads, and one small village, Reykjahlid, has grown up to service a diatomite factory. Diatomite, used as a filter material, is made from the skeletal remains of minute algae (diatoms). The method of extracting the raw material is a cause for concern among conservationists, as it involves dredging the muddy bottom layers of the lake, upon whose productivity Lake Mývatn's wildlife depends.

THREATS FROM AFAR

The most serious threat to the plant and animal life of the Scandinavian wilderness is the airborne acidification of land and water. Toxic chemicals, discharged from industries and cars, disperses over great distances. In Nordic countries such pollutants, which increase the acidity of the rain, snow, sleet and mist, mostly originate in other western European countries, and the heavily industrialized Soviet Union in the east.

The major contributors to acid rain are oxides of sulfur and nitrogen, which dissolve in rainwater to form weak acids. In some remote places, whether protected or unprotected, high acidity and contamination by poisonous heavy metals have been found. The great wilderness area of the Rogen Nature Reserve in Sweden and the neighboring Norwegian Femundsmarka National Park are so badly polluted that lime has had to be spread in the lakes to counteract the acidity. In southern and central Sweden the fisheries of some 18,000 lakes are now affected, as the increased acidity leads to high egg and fry mortality.

Other sources of water pollution include effluent from factories and farms, and pesticide residues. During the last 80 years the concentrations of phosphates and nitrogen compounds in the Baltic Sea have increased tenfold, affecting the breeding success of seals, sea eagles and salmon. The fish of many lakes are poisoned by mercury, cadmium and other heavy metals; this has contributed to the almost total disappearance of the otter from southern Sweden.

The mountain forest battle
The decision of Sweden's state-owned forestry company in 1980 to do away with its voluntary logging boundary in the mountains of Swedish Lapland surprised many people. These areas of high altitude were supposed to be unsuitable for reforestation, but the company claimed that new methods had now made this economic. Conservationists became very worried, and the authorities began to compile an inventory of virgin forests – forests untouched by human hand. It revealed that Sweden had the largest area of virgin coniferous forest in Europe.

Voluntary organizations then mobilized public opinion, and the conflict was taken

SVALBARD – A POLAR WILDERNESS

On the edge of the Arctic Ocean lies the Svalbard island group, covering some 62,000 sq km (24,000 sq mi) of polar desert, bare rock and ice, and windswept tundra. In winter the islands are linked by sea ice. Thousands of waterfowl and waders breed here in spring, and large numbers of seabirds nest on the cliffs. Ringed and bearded seals, walrus and belugas (white whales) hunt offshore; arctic foxes and herds of native reindeer live on the islands. Svalbard is most famous because one of the world's largest denning areas for polar bears is to be found here.

These polar habitats are very fragile. Only a thin surface layer of soil thaws in summer, and the waterlogged vegetation is easily damaged by the trampling of feet and by vehicles; in such low temperatures regeneration is very slow. Trash rots extremely slowly, and may take decades to disappear.

Svalbard has large deposits of coal. The islands belong to Norway, but are controlled by an international treaty between Norway and the Soviet Union. Three national parks, 2 nature reserves and 15 bird sanctuaries were established in 1973, covering 45 percent of the land area. The Northeast Svalbard Nature Reserve has since been designated a Biosphere Reserve.

Throughout the islands vehicles are forbidden on any thawed ground, with or without vegetation cover, and the dumping of waste into the sea is not allowed. Harvesting from the sea bed and diving are forbidden. Aircraft and boats need permission to approach the reserves, to avoid disturbing nesting birds. The location of hostels, hotels and mining and exploration installations is strictly controlled. Any disturbance to wildlife is illegal.

Svalbard is a welcome example of international cooperation that enables natural resources to be exploited while at the same time protecting an unspoiled wilderness.

A devastated landscape Logging is an important industry in Nordic countries but large areas of cleared forest are almost useless for wildlife.

Reindeer graze in a birch forest The international nature of environmental problems was highlighted when herds of reindeer in Scandinavia were affected by fallout following the explosion at the Soviet nuclear power plant at Chernobyl in 1986.

Threatened river life Disruption of the water flow by hydroelectric power schemes has had a serious impact; it is one of the most damaging environmental problems facing the northern wilderness.

to the Swedish government, which decided to designate 55 of these areas as nature reserves. This was one of the greatest victories conservation has ever had in Sweden, but some of the old mountain forests that are not classified as virgin forests remain at risk. These forests are important components of the scenery and of the local ecosystem, and Sweden's Nature Protection Association is pressing for a new boundary based on ecological considerations, above which no forestry should be allowed.

Protecting the natural resources
A major task for conservation authorities in Scandinavia is to stop resource exploitation in wilderness areas, especially forestry, mining and hydroelectric power installations. Powerful interest groups can sometimes overcome legal protection. Only four of Sweden's major river systems have escaped damming for electricity generation, and these remain at risk despite so-called protection. The in-

crease in hydroelectric power results from the decommissioning of nuclear power plants following public pressure: the shortfall in energy production must be made up. In Norway there has been a long battle over the spectacular Svartisen area lying on the Arctic Circle. It was proposed for national park status as early as 1936, but its potential for hydroelectric power has prohibited conservation, and the outcome is still uncertain.

Logging is of vital importance to the economies of the Scandinavian countries. But the creation of large felled areas have spoiled the scenery and habitats in many areas, especially in Sweden and Finland. Despite legislation that requires consideration of the environment, old wood is removed and large areas totally cleared, depriving local wildlife of valuable cover and nesting areas.

The larger mines in Scandinavia are old and present little current danger to the wilderness, but so long as prospecting continues, there remains a potential

threat. This exploitation has many secondary effects, such as the construction of new roads. Roads open up the wilderness to poachers and tourists. Developers follow, to build leisure complexes and other facilities for tourists.

Such construction has led to several conflicts in the past. In the 1970s conservationists fought in vain against a decision to construct a road through Abisko National Park in the far north of Sweden, and other protected areas are still threatened by plans to run roads across some of their very wildest parts.

The very attractiveness of these northern wildernesses causes problems. The growing numbers of tourists are causing increasing damage to the fragile habitats of bogs and wetlands. More and more people build summer homes, which now number hundreds of thousands, especially around more accessible and scenic lake shores. These have transformed the surroundings of reserves, leading to human disturbance of the wildlife.

The Lapland National Parks

Two of the first national parks declared in Sweden in 1909 were Sarek and Stora Sjöfallet. They share a common border, and in 1963 a further park – Padjelanta – was added. Altogether the three parks cover 523,200 ha (1,292,300 acres) of mountain habitats – the largest area of protected wilderness in Europe.

Sarek National Park has the most alpine landscape of the three, with hundreds of glaciers and sharp peaks (the highest 2,090 m/6,854 ft above sea level), big rivers and deltas; in the east are deep valleys with birch forests that shelter wolverines, brown bears, lynx and moose. Above the birch, dwarf willow scrub gradually gives way to alpine tundra, which in summer is speckled with the large white flowers of the mountain avens. Golden eagles and gyrfalcons hunt the slopes, whooper swans and many different species of duck breed in the wetter areas, and long-tailed skuas nest on the stony tundra. Sarek is a true wilderness. Access is difficult, and there are no marked trails or mountain huts. Even so the park receives about 2,000 visitors a year.

By contrast, there is easy access to Stora Sjöfallet via the road that runs through its center, and about 12,000 visitors find their way here each year. The road was constructed for the hydroelectric power installation that ruined the park's main attraction, an impressive waterfall. A line of lakes above the waterfall became over-dammed, and the fall now forms an ugly waterless scar for much of the year. However, the surrounding mountains remain pristine wilderness, fully deserving their national park status.

Padjelanta, the largest of the three parks, has a quite different kind of mountain scenery. It is more open, and consists of low, undulating ridges, some high, isolated massifs and large, wide lakes. The area has many rare flowers and a rich birdlife. Remnants of primeval forests of Scots pine and Norway spruce are home to bears, lynx, wolverines and moose. As in Stora Sjöfallet, there are some marked trails and huts for walkers and cross-country skiers, who number about 2,000 a year. There is no strong tradition of scientific research and education in the Swedish parks, but in Padjelanta research into plants, animals, geology, glaciology and hydrology has been going on since the beginning of this century.

The three parks are managed by a central authority, the National Swedish Environmental Protection Board. This is situated in Stockholm, but has a local office. Because of the remote situation of the parks, few people visit them for much of the year, so there are no serious management problems, except for the politically sensitive issue of the Lapps.

Lichens growing on rocks in Stora Sjöfallet National Park. Lichens flourish in clean air, so they are good indicators of the presence of airborne pollutants. The greatest environmental threat to these Lapaland parks comes from hydroelectric power schemes.

A moose in Sarek National Park Moose take leaves from trees and shrubs and also plants from the forest floor. The species is found through much of northern Europe and Asia as well as North America, evidence of the land route that once existed between them.

Remote Sarek National Park has mountain scenery unrivaled in Sweden. The lower slopes are clad with birch and willow; above lies the open moorland. This fragile wilderness is threatened by tourism and by some of the practices of the Lapps, who wish to herd reindeer here as they have done for centuries.

The Lapps have several thousand reindeer in the parks during the summer months, but in winter the reindeer migrate to the forests of the eastern lowland. The Lapps have retained the rights to commercial hunting of ptarmigan and willow grouse. This is questioned by conservationists, who are also worried about their use of motorbikes to round up their animals: the fragile tundra vegetation is easily damaged by vehicles. There are also claims that the Lapps illegally kill predators such as wolverines and eagles, which may threaten their animals. The conflict between the parks and the interests of the Lapps needs to be resolved to safeguard many of Scandinavia's protected mountain areas. The Lapps have farmed their reindeer here for centuries, and feelings can run high.

In the 1970s there were fears of an increase in tourism, especially in one of the most interesting areas, Sarek's Rapa river valley. Fortunately, the number of visitors there has decreased in recent years. However, the problem may recur again in the future, because the fragile environments of these parks can carry only low numbers of tourists.

Lakes of the northern forests

Dotted like jewels in the broad belt of boreal coniferous forest, or taiga, that circles the northern hemisphere are numerous lakes. These may be extensive bodies of water, like the Great Lakes of North America, or deep and ancient, like Lake Baikal in Siberia, which contains one-fifth of the world's fresh water. Others are shallow, formed in depressions in waterlogged ground where meltwater from the winter snows is prevented from draining away by permafrost lying below the surface.

The ecosystems of lakes are highly complex. Planktonic plants and animals, and detritus from lakeside plants, feed a variety of mollusks, shrimps and worms as well as insect larvae. The northern lakes and bogs are breeding places for millions of midges and mosquitoes. All these lifeforms are eaten by different species of fish, which in turn support other animal predators. The reed beds fringing the lakes provide nesting sites for the large populations of water fowl that migrate to these rich, northern feeding grounds in summer.

A reed-filled shallow lake surrounded by coniferous forest in northern Finland.

A MICROCOSM OF DIVERSITY

THE ANCIENT WILDWOOD · PROTECTING THE HERITAGE · PRESSURES ON THE LAND

Before the advent of human settlement, temperate deciduous forest covered most of the British Isles, though the far northern areas probably extended into the boreal forest, or taiga, zone. Little natural vegetation remains today, except in isolated localities such as mountain peaks and offshore islands. There are secondary oak and beech woodlands in the south, and remnants of the ancient Caledonian pine forests in the north. Grassland covers much of the uplands. In the wetter areas are some of Europe's most important peatlands; on poorer soils heaths take over. The extensive agricultural landscape of fields, hedgerows and small patches of woodland has its own gentle beauty, and contains habitats rich in wildlife. This great diversity of landscapes is enclosed within some 10,000 km (6,215 mi) of spectacular and varied coastline.

COUNTRIES IN THE REGION

Ireland, United Kingdom

Major protected area	Hectares
Argyll FP	24,000
Beinn Eighe NNR BR	4,800
Brecon Beacons NP	134,400
Burren NP	proposed
Caelaverock NNR BR RS	5,400
Cairngorms NNR NSA RS	23,500
Connemara NP	2,000
Dartmoor NP	95,400
Exmoor NP	68,600
Forest of Dean FP	10,900
Killarney NP BR	8,500
Lake District NP	228,000
Lundy Island MNR	50
Moor House–Upper Teesdale NNRs BR	39,400
New Forest FP	37,600
North Bull Island/Foreshore BR	1,320
North Norfolk Coast NR BR RS	5,500
North York Moors NP	138,000
Peak District NP	142,285
Pembrokeshire Coast NP	58,500
Rhum NNR BR	10,700
St Kilda NNR BR WH	850
Snowdonia NP	218,800
The Broads NP	28,800
Yorkshire Dales NP	176,000

BR=Biosphere Reserve; FP=Forest Park; MNR=Marine Nature Reserve; NNR=National Nature Reserve; NP=National Park; NR=Nature Reserve; NSA=National Scenic Area; RS=Ramsar site; WH=World Heritage site

THE ANCIENT WILDWOOD

Very little primeval forest survives in the heavily populated British Isles. Ireland is the least wooded country in Europe, with only about 50,000 ha (123,000 acres) of ancient "wildwood" left. In Britain the remaining native deciduous forests survive mainly in the southeast, persisting farther north only where valleys are particularly sheltered.

The type of woodland found depends upon many factors, from soil type to vegetation history. In the damp western areas of Ireland and Britain durmast (sessile) oakwoods are common, while beech occurs in the drier parts of southern England. In the central and eastern Scottish Highlands the Caledonian pine forests (dominated by Scots pine) mark the transition to the conifer forest of the subarctic taiga zone.

Britain's uplands lie mainly to the north and west, in the arctic-alpine zone. Here

An oakwood in the west of Britain This modified remnant of the old wildwood has survived on a steep hillside despite a long history of coppicing and grazing. Such oakwoods are richly carpeted with flowers, and mosses and ferns grow densely on the trees. Woodland birds such as pied flycatchers, redstarts and wood warblers are found here.

the plants and animals are adapted to extreme cold and high levels of ultraviolet light. These habitats are largely absent from Ireland. The humid oceanic areas of northern and western Britain, together with much of western and central Ireland, are renowned for yet another ancient and natural formation – peatlands. In drier areas the peatlands are replaced by coarse grasslands.

Where the soils are too poor to support grassland the original forests have been replaced by moors and heaths, secondary plant communities that are dominated by heather-like (ericaceous) shrubs. These uplands are clothed in bare, windswept heather, cowberry and crowberry moorlands, often without a tree in sight;

The British Isles

Biomes

- temperate broadleaf forest
- mountain and highland system
- lake system

- ◆ major protected area
- ○ Biosphere Reserve
- ≈ Ramsar site
- × World Heritage site

Map of biomes The British Isles lie within the zone of temperate broadleaf forest, while conifers grow in the northern highlands. Centuries of human activity have destroyed much of the natural vegetation.

Coasts and wetlands

The British Isles possess some of the richest and most varied coastlines in Europe. In general the western coasts are steep and rocky, offering nesting cliffs for thousands of seabirds. The more sheltered eastern shores are mostly low and sandy, with wide, muddy estuaries that attract large numbers of waders and waterfowl. There is tremendous variation, from sand dunes and reed-fringed lagoons to shingle banks, from precipitous cliffs to saltmarsh, from Scottish sea lochs to the drowned river valleys of the extreme southwest.

The many small offshore islands form natural sanctuaries for wildlife. The Isles of Scilly, off the southwestern tip of Britain, are warmed by the Gulf Stream and support many plants not found farther north. They are a popular staging post for migrant birds, and also quite regularly attract stray birds, and even monarch butterflies, that have somehow crossed the Atlantic by accident.

The lochs of Ireland and Scotland and the myriad lakes and pools of Britain's Lake District are areas of great scenic beauty that attract both tourists and wildlife. There are also several sizeable rivers, such as Ireland's river Shannon and Britain's Severn and Thames, many of them rich in trout, salmon and other fish.

Near the coast lie important wetlands, such as the Somerset Levels in the southwest and the Cambridgeshire Fens in the east, areas of lowlying peatland interspersed with reedbeds. These are rich in rare plants, insects and birdlife.

Thousands of seabirds such as fulmars nest on steep and rocky cliffs along the coastlines of Britain and Ireland. Many of these sites are regarded as internationally important habitats.

farther south the heaths of Ireland and southwest England support gorse and the occasional stunted hawthorn tree.

Today more than half of England, Wales and Ireland is covered in permanent grassland. Reclaimed from the woodlands more than 5,000 years ago and since maintained by humans and their animals, these rich seminatural habitats vary with the soils and underlying rocks. The rolling chalk downlands of the southeast and the limestone grasslands of central England have plants that are adapted to lime-rich soils, while the granite uplands of southwest England support pastures on acidic soils.

PROTECTING THE HERITAGE

Conservation started early in the British Isles. Privately financed conservation projects were started in the 19th century, and eventually led to the growth of voluntary organizations such as the National Trust. The first government legislation was not passed until 1949, when the National Parks Act led to the establishment of ten national parks in the 1950s. The Act also provided for the selection of Areas of Outstanding Natural Beauty (AONBs). Today the Countryside Commission, a government agency, is responsible for both designations.

Most of these national parks are still privately owned. They contain within their boundaries towns, roads and other long-established developments; mining, industry and urban development are permitted within certain limits. As a result, they are insufficiently protected to qualify as national parks under the definition set out by the International Union for the Conservation of Nature and Natural Resources (IUCN); they are categorized as protected landscapes only.

The Republic of Ireland possesses three national parks, each established under a separate Act and currently the responsibility of the National Parks and Monuments Branch of the Office of Public Works. The Irish national parks are state-owned, but financial restraints limit size. Killarney – the largest park in Ireland – covers less than 5 percent of the area of Britain's most extensive national park, the Lake District in northwest England.

In 1990 there were still no national parks in Northern Ireland, but there was pressure to designate the Mourne Mountains as one. There are eight AONBs. In the absence of national parks in Scotland the Scottish Countryside Commission has instead proposed some 40 "national scenic areas", considered to be hybrids between national parks and AONBs, and enjoying a similar protection against unsightly development.

Attempts at stricter protection

In Great Britain the Nature Conservancy Council plays a valuable part in conservation. The Act of 1949, extended and reinforced by the Wildlife and Country-side Act 1981 and the Countryside (Scotland) Act 1981, enabled the Council to designate the National Nature Reserves (NNRs). It was also given a statutory duty to notify landowners if they were in possession of areas of particular wildlife

An estuary ecosystem has two main sources of energy – seaweeds and algae, and decomposing shoreline or saltmarsh plants.

Components of the ecosystem
1 Sea aster
2 Glasswort
3 Townsend's cord-grass
4 Bladder wrack
5 Detritus
6 Periwinkle
7 Mussel
8 Lugworm
9 Spire shells
10 Cockle
11 Shrimp
12 Oystercatcher
13 Ringed plover
14 Crab
15 Redshank

Energy flow
⇒ primary producer/primary consumer
⇒ primary/secondary consumer
⇒ dead material/decomposer
⇒ death

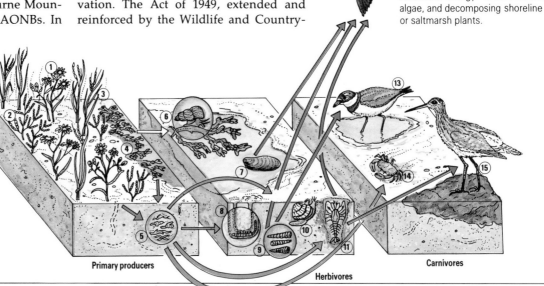

Primary producers

Herbivores

Carnivores

A lowland heath in southern Britain is one of the last fragments of a habitat once common throughout northwest Europe. Encroaching development is eating into these heathlands, home to some of Britain's rare species of reptiles such as sand lizards, smooth snakes and natterjack toads.

A moorland valley In the past these northern habitats provided refuges for rare animals such as golden eagles, but they are subject to increasing disturbance from walkers, climbers and other visitors, and by their use as catchment areas for urban water supplies.

THE BURREN

The Burren is an expanse of limestone cliffs and terraces on the west coast of Ireland. At present about 300 ha (750 acres) at Mullagh More have been acquired by the state, but a much larger area is destined to become the Republic of Ireland's fourth national park.

The summit exhibits the bare slabs and sheltered crevices characteristic of limestone pavement, while the marginal areas support dense hazel scrub. The climate is mild and moist, but although 1,700 mm (80 in) of rain falls each year, it rapidly disappears underground. The Caher is the Burren's only surface river, and there are many turloughs, or dry lakes, to the east.

The unique combination of geological, climatic and historical factors has produced a mixture of arctic-alpine and Mediterranean vegetation that is unrivaled elsewhere in Europe, together with plants of both limestone and acidic habitats. Many species have disjointed, or "relict", distributions: once more widely distributed, their range has been affected by climatic or other changes and they now survive as scattered isolated populations.

The Burren is home to 26 of Ireland's 33 butterfly species, and is the only Irish locality where the pearl-bordered fritillary is found. The pine marten has one of its last Irish strongholds here, and lesser horseshoe bats hang among the stalactites in the spectacular limestone galleries.

The limestone summit of Mullagh More The stark but beautiful landscape of the Burren shelters a rich variety of plants and animals, including many of Ireland's rarest species.

value, called Sites of Special Scientific Interest (SSSIs), though little power was given to enforce their protection. Most NNRs now also have SSSI status.

These Acts included provisions for protecting the remaining areas of limestone pavement, and for declaring marine nature reserves, of which only one – Lundy Island, off the southwest coast of England – has yet been established. More marine reserves are planned.

Similar powers were given to the Department of the Environment's Northern Ireland section in 1985. Little advantage has been taken of these opportunities: two years later only 15 Areas of Special Scientific Interest (the equivalent of SSSIs) had been notified (0.2 percent of the total land surface), compared with a total of 4,729 SSSIs in Great Britain (more than 6 percent).

In the Republic of Ireland the Wildlife Act of 1976 established a national network of nature reserves, of which 39 had been declared by the late 1980s, managed by the Forest and Wildlife Service of the Department of Energy, and mostly owned by the state. About 0.5 percent of the land area is currently protected; nearly three-quarters of the protected areas lie within the national parks.

Protection by the people

Literally hundreds of voluntary organizations have also contributed greatly to environmental protection in the British Isles. In Northern Ireland, for example, the Ulster Trust for Nature Conservation owns or manages as many as 15 nature reserves, exceeding 405 ha (1,000 acres) in extent. Britain's National Trust and its equivalent in the Republic of Ireland, An Taisce, have also acquired large areas of great scenic beauty and wildlife interest. The Irish Wildbird Conservancy is responsible for some 46 bird sanctuaries, while in Britain the Royal Society for the Protection of Birds (RSPB) purchased a large reserve in the Scottish Highlands in 1987, thought to be the most expensive ever acquired by a voluntary organization in Europe. The county-based Trusts for Nature Conservation, coordinated by the Royal Society for Nature Conservation, between them owned or managed about 1,400 British nature reserves totaling nearly 45,000 ha (about 111,200 acres); they continue to add to these.

This appreciation of wilderness has brought its own problems. The pressure of visitors has led to erosion of fragile upland turf and coastal sand dunes. The demand for improved access and the desire to build second homes in scenic areas has led to an increase in the numbers of roads in these areas, with accompanying disturbance to wildlife.

PRESSURES ON THE LAND

There are two main threats to the remaining wild areas of the British Isles. The pressure on space for both dwellings and recreation is extreme in many areas, especially in the southeast, where the density of population is very high and relatively little wilderness survives. These threats are generally local, such as road improvement schemes or the damage to sites of wildlife importance caused by motorcycles and cross-country vehicles – some 130 such sites were damaged in 1987, more than half of them SSSIs.

Farther north, larger expanses of wild country persist, but where there are fewer people objections are less forceful, and large-scale exploitation causes immense damage to irreplaceable natural resources. The stripping of Ireland's internationally important peatlands to provide fuel for turf-burning power stations, and the march of pine plantations across the blanket bogs of northern England and parts of Scotland are examples of the worst type. Land drainage also destroys wetland ecosystems that are valuable for wildlife: the Blackwater catchment drainage scheme, for example, aims to "improve" over 3,200 ha (7,900 acres) of farmland to the southwest of Ireland's Lough Neagh by dredging, widening and deepening some 600 km (370 mi) of existing watercourses. Unique areas of peatlands in the Somerset Levels have been cut to supply peat for gardens, or drained by farmers with the aid of grants from the European Community. The resultant lowering of the water table has damaged even untouched nature reserves in the vicinity.

The threat to the coast

The coastline of the British Isles is particularly at risk. Scotland's Morrich More, one of the most important sand dune systems in Europe, is earmarked for the construction of a huge pipeline assembly yard for the North Sea oil industry. There is a constant threat of oil spillages in the vicinity of oil refineries. On the westernmost peninsula of Wales oil tankers pass along a coast renowned for its seabird colonies. There is a similar problem at Sullam Voe in the Shetland Islands, and oil from a pipeline leak in 1989 contaminated tracts of saltmarsh in the Mersey estuary in northwestern England.

Estuary wetlands are also threatened by plans to erect tidal power barrages, and – in the case of the Taff/Ely estuary in southern Wales – by a road embankment that will span the river mouth to create an amenity lake inland. These developments are sure to disturb wintering and feeding grounds of international importance for waterfowl and waders. Elsewhere dock extensions and associated channel deepening will cause similar disturbance to estuarial wildlife. On the south coast of Ireland, Cork harbor suffers from pollution and industrial development, as does the Shannon estuary on the Atlantic. Land reclamation poses additional problems here, while domestic refuse dumping has almost destroyed the wildlife and scenic value of other estuaries.

Saving the countryside

Britain does, however, have some of the most stringent wildlife and landscape legislation in Europe, backed up by a strong voluntary movement: it is estimated that over 3 million people subscribe to non-statutory conservation organizations. There has been a recent shift from the protection of areas of known scientific interest to the preservation of the wider countryside, and to the maintenance of the traditional farming practices that have created a visually attractive landscape of great wildlife value.

With this in mind the Agriculture Act of 1986 makes provision for the selection of Environmentally Sensitive Areas (ESAs) within which landowners are encouraged to farm the land using traditional methods, any loss of profit being offset by financial contributions from the government. About 20 ESAs had been designated by 1990; the Nature Conservancy Council hopes eventually to extend the scheme to cover the whole country.

In the Republic of Ireland the prevail-

Wicken Fen – a tiny remnant of eastern Britain's vanished fenlands. Once these wetlands were rich habitats for birds and fish. Most have now been drained and converted to arable farmland.

A northern forest, at Abernethy in Scotland. The ancient Caledonian pine forest is home to birds that are rare in Britain, including capercaillies, crested tits and crossbills.

The Flow country of northern Scotland is one of western Europe's last wilderness areas . Its peatlands are now being destroyed by commercial tree planting.

PEATLANDS

Peatlands represent some of the wildest, most desolate habitats in Europe. They take thousands of years to form in places where the soil is waterlogged, which prevents the decomposition of dead plants. The botanical diversity of peatlands is rarely high, but the characteristic plants, such as sphagnum moss, are highly adapted to survival in these specialized conditions.

The most important peatland types in the British Isles are blanket bogs and raised mires. Only about 100,000 sq km (some 39,000 sq mi) of blanket bog are found worldwide, 10 percent of them in Britain, where they include, in the Flow country of northern Scotland, one of the largest expanses of blanket bog in the northern hemisphere – an ecosystem of global significance. They are also found on uplands farther south and on the western seaboard of Ireland. These bogs are increasingly threatened by schemes to drain and plant them with conifers, a development rendered economic only by substantial government subsidies and tax concessions.

Raised, or domed, mires most probably developed some 5,000 to 7,500 years ago, when the climate was wetter than it is now and the bogs grew very rapidly. There are now very few raised mires left in Britain; the central plain of Ireland still supports the finest development of raised mire in Europe. Only about 1 percent of it remains in an untouched state. It is estimated that these mires will all be lost by the end of the century if exploitation for power, industry and horticulture is not halted.

ing philosophy sees conservation as part of a wider landscape planning strategy, and a scheme to declare "heritage zones" has received much support. These zones will be similar to Britain's national parks. The Dublin and Wicklow Mountains will be the first area to be so designated.

Restricted access to land used by the army means that unintentional protection has been given to some areas, particularly in southwest Britain. Clifftop habitats, the heaths and moorlands of Dartmoor and other wild and deserted places have been thus preserved. Conservation in urban areas is also gaining momentum, with the realization that cities have their own special brand of wild places that add greatly to the quality of urban life.

Dartmoor and Exmoor

In the southwest peninsula of Britain, the granite plateau of Dartmoor and the younger sedimentary cliffs, ridges and valleys of Exmoor form the region's two southernmost national parks. Dartmoor comprises the largest and highest area of upland in southern Britain, with an average height of 365 m (1,200 ft), the highest point being High Willhays, at 621 m (2,049 ft). Outstanding features of the landscape are the irregular weathered granite tors (distinctively shaped rocky peaks) that litter the moors, and the extensive valley mires. In late summer the moor is a blaze of color, with purple heather and yellow gorse.

Blanket bog covers the highest parts of the moor, where the annual rainfall exceeds 1,750 mm (70 in). Here heather, cotton grass and sphagnum moss grow in the wet peat. In some areas the blanket bog has been eroded to patches of exposed peat separated by bare soil, thought to be the result of grazing and of longterm climatic change. The drier areas support coarse grassland, which provides grazing for cattle and sheep. In places the land is scarred by granite quarries and china clay pits, whose mounds of white clay create an almost lunar landscape.

Exmoor's moorland is less extensive; it has suffered recently from enclosure and conversion to pasture, and from planting for forestry. But the park is renowned for its wooded gorges, carved out by fast-flowing rivers that are inhabited by otters and salmon, and for its dramatic, rugged coastline. The lower parts of Exmoor form a much gentler, patchwork landscape of scattered farmsteads and small fields rimmed by hedges.

Shaped by humans over several thousand years, Dartmoor and Exmoor do not contain the primeval wilderness normally associated with the world's national parks. Tourism flourishes in both, as visitors are attracted by the wild scenery and by leisure pursuits such as walking, riding, birdwatching and camping.

Its southerly location and upland terrain have given Dartmoor an interesting variety of plant and animal life. Bastard balm and pale butterwort represent southern plant species; northerly species such as crowberry and whortleberry reach their southernmost British limits on the northern tors. It is the most southerly breeding locality in the world for golden

Rugged Dartmoor scenery Its geographical situation makes it a meeting point for southern and northern species in Britain.

Ancient woodland Wistman's wood on Dartmoor is probably the survivor of more widespread woods, which were cleared and burned by generations of tin miners in the past. Although small and contorted, some trees are up to 500 years old. This wood survived because the trees are rooted among blocks of granite. Less than 2 ha (5 acres) in extent, it has doubled its size by natural regeneration in the last century.

Heather and gorse on Exmoor's plateau gives way to woods and fertile farmland in a river valley. The moor, grazed by hardy ponies, sheep and red deer, was being plowed up by local farmers until legislation was passed to protect it.

plover and dunlin, and a number of other continental birds such as the woodlark are found here too.

Exmoor is southern Britain's stronghold for red deer, supporting between 700 and 900 individuals. It also represents the southern breeding limit of the merlin, and of plants such as parsley fern and twayblade. Altogether some 243 bird species have been recorded here, more than a hundred of which breed. Native ponies of ancient stock roam the moors of both national parks.

Managing the place and the people

Responsibility for management is held by the respective national park authorities, whose statutory objectives are to conserve and enhance the quality of the landscape and to promote public enjoyment of it. Both park authorities have large moorland and farmland holdings, which are run on an experimental basis to combine economic profit with wildlife and landscape conservation.

The 1968 Countryside Act recommended that the social and economic well-

being of people living within the park should be taken into account. On Dartmoor these responsibilities have been delegated to the Dartmoor National Park Committee, some of whose members are appointed by central government and the remainder by the regional authority. Since 1976 a comprehensive ranger and information service has been established for the benefit of visitors to the park.

Multiple land use has caused great problems for those concerned with park preservation. Many of these may be attributed to the high proportion of private land in British national parks. Only 19 percent of Exmoor is in public or semi-public ownership. Dartmoor has been a military training ground since 1870 – some 13,355 ha (33,000 acres) are still used for this purpose – while many consider the hunting of red deer by Exmoor's three packs of staghounds to be incompatible with the spirit of conservation. Other difficulties include the recent routing of a local town bypass road through Dartmoor, and the gradual enclosing of areas of moorland for grazing and agriculture. The latter problem has been counteracted to some extent by legislation in 1985, which secured the future of, and legalized public access to, all common land within the park.

ON THE EDGE OF EUROPE

REMNANTS OF THE GREAT FORESTS · CHERISHING THE WILD PLACES · WILDLIFE UNDER THREAT

France contains a remarkable variety of wild places. The dark forests of the Ardennes and the rich oak and beech forests of the lowlands and the Massif Central contrast with the dry thorny scrub of the Mediterranean coast and the gentle rolling cultivated countryside of the plains. The French Alps and Pyrenees have some of the most magnificent mountain scenery in Europe, with alpine meadows full of wildflowers and dense forests that are refuges for some of Europe's rarest mammals. The estuaries and muddy bays of the Atlantic shores and the marshy delta of the Rhône on the Mediterranean coast are important as feeding and wintering grounds for waterfowl and waders, while the rugged cliffs and islands of the Brittany coast in the northwest are thickly populated with nesting seabirds.

COUNTRIES IN THE REGION

Andorra, France, Monaco

Major protected area	Hectares
Aiguilles Rouge NatR	3,200
Armorique RP	65,000
Banc d'Arguin NR	500
Camargue RNP BR RS	85,000
Cerbères Banyuls NR	600
Cévennes NP BR	323,000
Delta de la Dranse	45
Ecrins NP	108,000
Fontainebleau Forest	550
Forêt de la Massane	336
Iroise RP BR	21,400
Lac de Grand Lieu NR	2,695
Marais Poitevin RP	57,800
Mercantour NP	70,000
Pilat RP	60,000
Port Cros NP	2,495
Pyrenees NP	45,700
Sagnes de la Godivelle	24
Scandola WH	920
Vanoise NP	52,800
Vercors RP	135,000
Volcans d'Auvergne RP	281,500
Vosges du Nord RP BR	120,000

BR=Biosphere Reserve; NatR=National Reserve; NP=National Park; NR=Nature Reserve; RNP=Regional Nature Park; RP=Regional Park; RS=Ramsar site; WH=World Heritage site

REMNANTS OF THE GREAT FORESTS

Forests once covered most of France; today about a quarter of the country is still forested. Probably none of the remaining forests has escaped the influence of humans, but many of the original forest species still survive in these seminatural woodlands.

France's wide variety of climate and geology is reflected in the many different types of forests. Over much of the country oakwoods predominate, with beech and poplar wherever the soil is suitable. Near the moist coast of Brittany in the northwest, the clearance and firing of forests for grazing or grouse moors has created heaths dominated by heatherlike shrubs. In the southern heaths, yellow-flowered gorse blends with the purple heather to provide a splendid show of color.

Natural conifer forests of pine and silver fir occur mainly in the mountains, with a ground layer of berry-bearing shrubs such as heathers, cowberry, crowberry, bearberry and alpenrose. In the south, woods of maritime pine grow in sandy coastal areas. Conifers, mainly Scots pine, have also been planted in many places for commercial purposes.

The dry south

Almost no forest survives in the south close to the Mediterranean, where the country has been settled and cultivated for thousands of years. The chief tree in the remaining woodlands is the evergreen holm oak, and sometimes the cork oak. The trees are often widely spaced as a result of cutting, firing and grazing, and the scrubby understorey contains low-growing trees such as juniper, broom, strawberry tree, lentisc, carob and olive, and shrubs like cistus and heather.

These evergreen species are adapted to drought; their small, tough, wax-coated, leathery leaves protect against water loss, and many have spines to fend off grazing animals. The ground layer, containing many herbs, flowers during the rainy seasons of spring and fall. There are many annuals, and other plants often die down below ground or shed their leaves in summer to save water.

In many places these woodlands have been reduced to spiny thickets and scrub

France and its neighbors

called maquis, or to the scattered dwarf shrubs known as garigue. Under pressure of very heavy grazing, this land may degenerate further into a sparse steppe grassland interspersed with patches of stony ground. Even so, it is rich in wildflowers, especially aromatic herbs such as marjoram, rosemary and thyme, and plants that store food in swollen underground roots and stems, such as anemones, cyclamens, asphodels, irises and orchids.

Mountain refuges

The spectacular scenery of France's high mountains attracts many visitors. The dark forests of the Pyrenees in the south-west shelter some of Europe's last surviving bears, and their berries and seeds are food for a wide range of birds and small mammals. The mountain grasslands are home to ibex and chamois, mouflon (a wild sheep) and mountain hares. Red and roe deer and wild boar live in the forests, and in some areas birds such as black woodpeckers, nutcrackers, black grouse and capercaillie can be seen.

The alpine meadows, grasslands and forests support many small mammals, which in turn attract birds of prey, foxes, martens, genets, wildcats and other predators. Above the meadows, alpine rose and dwarf shrubs eventually give way to cushion plants, then to the bare rock and ice of the mountain summits. These southern mountains provided refuges for cold-loving plants as the land warmed at the end of the most recent ice age. In isolated areas they have evolved into many species found only here.

Map of biomes Most of France falls within the temperate broadleaf forest zone of northwestern Europe. On its southern coastal fringe the original evergreen Mediterranean forest has been degraded to open scrubland by centuries of grazing; in the Alps and Pyrenees mixed conifer forest gives way above the treeline to alpine meadows and then to dwarf shrubs.

Biomes
- temperate broadleaf forest
- evergreen sclerophyll forest and scrubland
- mountain and highland system
- lake system

- ◆ major protected area
- ○ Biosphere Reserve
- ≈ Ramsar site
- ✕ World Heritage site

A beechwood in eastern France

Deciduous woodland still covers much of the country – a remnant of the primeval European forest. In summer the branches of adjacent tree crowns meet, creating dense shade. As a result, ground plants are scarce and the woods support few butterflies. They are carpeted with leaf litter that rots slowly.

Shore and clifftop habitats

France's long Atlantic coast has sandy beaches and mudflats, as well as sheer cliffs on the northwest peninsula. The invertebrates of the intertidal shoreline – worms, marine snails, shrimps and molluscs – provide food for scavenging crabs and for many thousands of seabirds and waders. Clifftops that have been left undisturbed by the plow offer narrow refuges for flowering shrubs such as gorse, broom and heathers, all characteristic of Europe's Atlantic seaboard.

CHERISHING THE WILD PLACES

The large size and relatively small population of France masked the need for conservation for many decades, and budding conservationists were deterred by the national passion for hunting. Though the first protected area was created in 1848, when 640 ha (1,580 acres) of Fontainebleau Forest were set aside on the initiative of a group of painters, organized conservation did not begin until 1960, when a government act set up the framework for creating the country's first national parks. The management of these parks was to be flexible, with a specific program defined in a separate decree for each park. Such measures included controls on hunting, forestry, fishing, farming and grazing.

The first national parks were established in 1963 at Vanoise in the Alps, and on Port-Cros, an island off the Mediterranean coast. The first national park in Europe to combine both land and sea areas, Port-Cros protects one of the few remnants of the original Mediterranean forest, as well as a rich and varied underwater environment, and is also a major stopover place for migrating birds. Since 1963 four more mainland national parks have been named, all of them in mountain regions: Ecrins and Mercantour in

the Alps, the Cévennes in the Massif Central, and one in the Pyrenees.

French national parks use a zoning system to control human activity. Each park has a core zone where wildlife is strictly protected. Here access is heavily restricted, and activities such as hunting, building, road construction and camping are forbidden. Surrounding this is a peripheral zone in which recreational and other activities are permitted and tourist facilities may be provided.

National parks are usually open to the

The high limestone plateaus of the Causses in the Cévennes are usually bare of vegetation. In spring they support a magnificent show of wildflowers. Trees grow in the valleys, but much of the countryside has been reduced to grasslands and heath. The area remains one of France's last wild places, and its rugged mountains and steep gorges provide a refuge for some of Europe's rarest birds of prey.

Mediterranean vegetation Much of France's southern coastal areas and the island of Corsica are covered in open scrubland, known as maquis. Olive and fig trees are stunted, and the lowgrowing shrubs such as myrtle and laurel have tough, leathery leaves able to withstand water loss. Many of the ground plants are rich in aromatic oils, possibly to deter insect feeders.

Components of the ecosystem
1 Oak trees, fruits, leaf litter
2 Herbivorous insects
3 Insects
4 Long-tailed field mice, bank voles
5 Earthworms
6 Wild boar
7 Gray squirrel
8 Great tit
9 Sparrowhawk
10 Spider
11 Green woodpecker
12 Tawny owl
13 Badger
14 Weasel
15 Mole

Energy flow
→ primary producer/primary consumer
→ primary/secondary consumer
→ secondary/tertiary consumer
→ dead material/consumer
→ death

A deciduous woodland ecosystem
has many layers. Decomposers live
in the soil and forest floor. Above is
shrubby undergrowth and the leafy
tree canopy.

Primary producers   Herbivores/Omnivores   Carnivores

public, but there are often no roads. The
degree of protection given to both core
and peripheral zones varies. In the
Cévennes, for example, where the landscape itself is the product of centuries of
human occupation, regulations are less
strict than in other national parks.

A variety of reserves
The national parks and their peripheral
zones cover only about 2 percent of
France's land area, but other kinds of
protected areas increase the amount of
protected land to some 15 percent of the
country. Most important among these are
the state nature reserves (NRs), established in 1961, and divided into various
categories, including voluntary reserves
on private land, by further legislation
passed in 1976. By 1987 nature reserves
protected some 1 million ha (2.5 million
acres) of France's wild places. Most of
these protected areas are to be found in
the mountainous parts of the country.

Other kinds of reserves include hunting reserves, in which hunting is controlled, game reserves, in which it is
considerably reduced and where research
into certain game animals may be pur-
sued, and biogenetic reserves. In addition, buffer zones have been established
around national parks, reserves and certain large towns; these are sometimes
financed by local taxes. In certain scenic
zones there are restrictions on building
and other forms of land exploitation.

Protection of the coast is the responsibility of the French Coastal Conservancy,
established in 1976. This is a public
organization that protects coastal areas
and lake shores from urban development
and exploitation by buying up threatened
or valuable tracts of land.

THE CEVENNES NATIONAL PARK

A series of forested granite ridges and
gorges, interrupted by the flat limestone plateau of Causse Méjean, make
up the Cévennes National Park in the
southern part of the Massif Central,
France's central uplands. Centuries of
wood cutting for charcoal and widespread sheep grazing have resulted in
extensive grasslands and heaths. Today
only 25 percent of the area remains
forested with oak, chestnut and beech
and widespread conifer plantations.

This great variety of habitats supports a rich plant life. There are over 40
species of orchids and a number of
interesting alpine plants. Birds of prey
such as golden eagles, peregrines, eagle
owls and Montagu's harriers hunt the
hillsides, and the forests are still inhabited by deer and wild boar, capercaillie and black grouse.

As in many of France's national
parks, there have been several successful schemes to reintroduce species to
the wild. Griffon vultures now breed in
the Tarn gorge, and European beavers
have been reestablished in the Tarn and
Tamon valleys. Following a decline in
sheep farming on the uplands, Corsican
mouflon were introduced to prevent the
landscape reverting to scrub.

The park is more densely settled than
most French national parks, and there
are relatively few restrictions on access.
Some hunting is permitted, and sheep
and cattle farming, as well as forestry,
are encouraged in a landscape that has
been shaped by centuries of traditional
farming. This sympathy between the
park and its human inhabitants ensures
the continued survival of a rich and
varied wild place.

There are various long-established
voluntary organizations that operate
under the umbrella of the France Nature
Environment, a federation of nature conservation societies. The best-known of
these is the National Society for the
Protection of Nature (SNPN). In addition,
these societies establish and manage
nature reserves, sometimes with the aid
of government funding.

Shaped by human hands
Many of France's landscapes have been
shaped by human activity, especially by
traditional farming methods. The rural
parts of France, particularly the upland
areas, are seriously threatened by the
migration of people to the towns. This
presents difficulties for conservation, and
regional nature parks (RNPs) are helping

to counter this problem. By 1988, 24 RNPs
had been established. They are intended
to stimulate local enterprise, including
tourism, and preserve traditional ways of
life, but in the process they afford a
certain degree of protection to local wildlife and landscapes. Each RNP has its own
development plan, designed in consulta-
tion with the people of the area to create a
distinctive image that will aim to attract
business and tourists to the park.

Careful planning is needed to ensure
that these developments do not result in
the degradation of the natural resources
upon which they depend. Nature conservation in France, as in many other countries, often demands a delicate balance
between keeping people on the land yet
protecting wild places against the incursions of humans.

WILDLIFE UNDER THREAT

France has some of the richest wildlife in Europe, with over 500 species of wild vertebrates and more than 4,750 species of plants, including a considerable number of species that are not found outside France (endemic species). But this wildlife is under threat. Since 1900 over 40 species of plants have become extinct in France, 9 of them endemic. Most of these losses can be attributed to habitat loss, arising from a variety of causes.

Agricultural land makes up about 58 percent of France's total land area. Recent "improvements" have led to drastic changes in habitats and loss of wildlife; over 5 percent of this land has been drained, and another 5 percent irrigated. Soil erosion has been caused by uprooting hedgerows and straightening stream channels to enlarge fields. Heavy application of nitrogen and phosphorus fertilizers and the discharge of untreated farm and industrial effluents have polluted groundwater, leading to the build-up of algae on ponds, and stagnation. Forestry schemes, the growth of built-up areas and the development of tourism affect over 300,000 ha (750,000 acres) of farmland every year, as well as destroying or polluting valuable areas of wetland on the coast.

At the end of the 1980s the Loire, one of Europe's relatively unspoiled large rivers, was under threat from a large irrigation scheme. There had been widespread criticism of and public demonstrations against the planned series of dams and dikes, as they would affect the drainage (hydrology) of the whole Loire valley and accelerate the reclamation of land by farmers, with a consequent loss of wetlands and flower-rich meadows. The dams themselves would interfere with the migration route of salmon. This is one of the longest in Europe: the fish travel up to 800 km (500 mi) from the Atlantic coast to their spawning grounds in the upper Allier river, a tributary of the Loire, which rises in the Massif Central.

While human interference has been responsible for the loss of many natural habitats, in some areas the disappearance of the rural population and traditional farming patterns poses an even greater danger to wildlife. In particular, a decline in grazing has led to the degradation of many seminatural habitats.

Exploiting natural resources

The geological formations that attract the tourists attract other forms of exploitation as well, for the hills are rich in minerals beneath the ground, and offer ideal sites for hydroelectric projects and ski developments above it. Foundries located next to bauxite deposits in the Alps and Pyrenees release sulfur dioxide into the air. This gives rise to acid rain, which causes the forests to die back. The noise produced by mining and smelting operations scares away wild animals, while the service roads to the mines and around the foundries create scars on the landscape and encourage the spread of urbanization.

Dams for hydroelectric power generation disrupt water flow in stream systems,

flood habitats in valley bottoms and upset the daily and seasonal migration patterns of animals. Each site acquires its own access roads and strings of unsightly pylons. Spectacular mountain scenery is sought not only by walkers, but by the drivers of all-terrain vehicles. These tear up the fragile upland soils, destabilize stream beds and destroy vulnerable, and often rare, vegetation.

The growth of the skiing industry in the French Alps and the Pyrenees endangers mountain habitats. Ski-lifts and ski-runs intrude upon the wild landscape, together with access roads, avalanche barriers and ski-stations. Large areas of land are covered in concrete or ice that has been packed down by skiers and

The large Mediterranean island of Corsica has at least 20 granite peaks above 2,000 m (6,500 ft), and a rugged coastline of bays, headlands, caves and deltas. The marshes around Scandola are of international importance for their bird life and rich marine environment. Corsica drifted away from the European mainland during the Tertiary period, between 65 and 2 million years ago, and its plants and animals have evolved in isolation ever since, producing many endemic species such as the Corsican nuthatch and Corsican pine. Some 200 to 500 Corsican mouflon, a kind of wild sheep, are protected in special reserves.

The regional park of Corsica covers much of the island's center, an area dominated by maquis on the lower slopes, and hard-leaved (sclerophyll) woodland higher up. Sheep and goat farming has been practiced in Corsica for centuries, but after World War I the largescale migration of the population to mainland France allowed the maquis to encroach on former grazing land. Attempts to control it by burning were often disastrous in this dry climate.

To prevent uncontrolled fires, the regional park authority selectively cleared the maquis by mechanical means. Stock was enclosed and their numbers carefully controlled. Fertilization to improve soil reduced demand for new areas of the maquis to be taken over by grazing. The stock-carrying capacity of the land has risen tenfold, while the amount of maquis burned each year annually has decreased by the same amount. This successful revival of the rural economy has helped to stabilize a valuable habitat.

An island refuge Granite peaks tower over the Désert des Agriates in northern Corsica, preserving a mountain wilderness. The island has many local plant and animal species, including the nimble Corsican mouflon.

High in the Pyrenees, flowering alpenrose, a species of rhododendron, forms a dense shrub layer beneath a stand of pines. The seeds and berries of alpine forest support many small mammals and birds, which are hunted by larger predators such as martens and wildcats, and by birds of prey.

piste-making machines. The soil becomes impermeable, and this increases the erosion of stream beds during spring melt and summer storms. Waste water gives rise to considerable pollution, especially in the winter. The supply of water to the ski resorts often depletes local sources by the middle of winter.

Skiers often show little understanding of the vulnerability of the mountain plants and wildlife. Many people ski off the ski-runs, injuring bushes and young trees. The practice of arriving at high-altitude, deep-snow sites by helicopter or in small airplanes creates enormous disturbance for the wild animals. The proposal to cross the national parks with ski-lifts in order to improve the access to ski-runs has raised fears that a precedent may be set, which could well lead to the national parks themselves being opened up to wholesale development by the ski industry. The national parks are seen as last wilderness strongholds, to be defended against those keen to exploit them for profit or pleasure.

The Camargue

At the delta of the Rhône on the Mediterranean coast is the renowned Camargue, an enormous wetland that extends over some 145,000 ha (362,500 acres). It is the first Ramsar site in France designated under the 1971 international convention on wetlands, and has been an internationally recognized Biosphere Reserve since 1977. It is an intricate landscape of associated beaches and dunes, fresh, brackish and saltwater ponds, seasonally flooded or diked pasture lands, reed beds, grasslands and woods as well as cultivated fields, commercial salt beds and even industrial sites. Its location gives it additional stature as a resting point for migrating birds winging their way across Europe to and from Africa, while the mild Mediterranean climate makes it an ideal winter refuge and breeding territory in the early spring.

The particular mixing of the Mediterranean Sea and the fresh water of the Rhône gives the Camargue its unique character, and contributes to the rich variety of its plant and animal life. It is one of only two places in Europe where the greater flamingo regularly breeds, and this rare and beautiful bird has become the symbol of the Camargue.

Millions of ducks, geese and waders pass through the Camargue each year. They come from as far afield as northern Europe and Siberia, and use the Camargue's many different habitats for different activities – feeding, roosting, breeding and nesting – according to the time of day or the season. Ducks, for example, rest in the ponds and *sansouires* (extensive areas of succulent, salt-tolerant vegetation) during the day, moving to the nearby

Horses of the Camargue, an ancient, semi-wild breed possibly dating back to the Stone Age, gallop through the shallow water of a lagoon. The wetlands of the Camargue are internationally recognized as an important habitat for breeding and overwintering birds.

marshes to feed in the early morning and evening. Although their roosting places are protected, many of the marshes are privately owned. Hunters lie in wait here as the birds move to their daily feeding places; the number of wintering waterfowl in the Camargue has decreased by 40 percent over the last ten years.

Among the waders that nest in the Camargue, such as Kentish plover, stone curlew and avocet, is the rare collared pratincole, which nests nowhere else in France. The permanent marshlands that are not pastured are covered with a thick reed growth, an ideal nesting habitat for species such as the great bittern, purple heron, marsh harrier, bearded reedling, reed bunting and various warblers.

When the regional nature park was set up in 1970 it took in three existing reserves: the Camargue, which includes the Etang de Vaccares, one of France's first nature reserves, covering 13,117 ha (32,412 acres), state-owned but managed by the private National Society for Nature Protection; the Etang des Imperiaux, 2,777 ha (6,860 acres), a reserve owned by the commune of Les Saintes-Maries-de-la-Mer, the main settlement in the area; and the Tour du Valat, a private reserve managed by the public Sansouire Foundation. The Coastal Conservancy has acquired control over a further 2,000 ha (5,000 acres). The protection of the park extends some 22 km (14 mi) out to sea to cover the marine environment.

Preserving the wetland Much of the Camargue lies within protected areas. The park authorities are able to minimize the impact of human activity within their boundaries, but elsewhere the Camargue is threatened by the encroachment of rice farming. As well as reducing the habitat available for wildlife, the paddies have altered the natural variations of water levels in the remaining wetlands.

Salt-tolerant trees Tamarisks are usually thought of as desert plants, but they are tolerant of salt as well as drought. Salt and fresh water mix in the Rhône delta, and tamarisk trees play an important part in the ecology of the salt and brackish pools that fringe the marshes.

Pressure on the Camargue

Rare local breeds of black cattle and of white horses, thought to be the last descendants of primitive horses, have been reared on the pastures of the Camargue for centuries. However, a rise in cultivation and grazing in recent years has increased pressure on this unique and irreplaceable wetland. By 1984 fields, salt extraction beds and industrial sites covered 56 percent of the land – an increase of some 40,000 ha (100,000 acres), including 33,000 ha (82,000 acres) of wetlands, over the previous 40 years. The most dramatic changes came with the establishment of rice paddies after World War II, which seriously affected drainage.

Reconciling all these activities and preserving the remaining wetland is a great challenge for the park authorities. They have to minimize the impact of human activities while revitalizing traditional agricultural practices, fishing and stock-raising, and helping to coordinate the provision of facilities for tourists. An information center at Saintes-Maries provides exhibitions, audiovisual displays and literature, and organizes hiking, riding and even cycle trails. Research is carried out by the Biological Station of Tour du Valat and the Ecology Center for the Camargue. They have introduced artificial nest mounds of clay and mud to encourage breeding flamingoes.

WETLAND RICHNESS

In their battle to reclaim their land from the sea by constructing dykes and polders, the people of the Low Countries have created vast areas of coastal wetlands. These are valuable havens for waterfowl and waders, and nurseries for fish and crustaceans. The reed-fringed canals and ditches that criss-cross the waterlogged lowlands provide nesting sites for waterside birds and fishing grounds for herons and otters. In the upland areas farther inland, land cleared of forest for grazing has turned into peat moors covered with sphagnum mosses and diverse marsh plants. Centuries of grazing and peat-cutting have produced a mosaic of heathlands, lakes, meadows and marshes. These are rich in wild birds, but in this densely populated and extensively farmed region there is little room left for mammals.

COUNTRIES IN THE REGION

Belgium, Luxembourg, Netherlands

Major protected area	Hectares
Amsterdamse Waterleidingduinen NR	3,400
Biesbosch NP	5,800
Blankaart NR RS	160
De Weerribben NRS	7,200
German-Luxembourg NP	72,500
Groote Peel/Mariapeel NR	2,000
Hautes Fagnes NR	894
Hoge Veluwe NP	5,700
Kalmthoutse Heide NR RS	4,045
Kampina Heath NR	1,100
Kennermerduinen NP	1,200
Lesse et Lomme NR	970
Mechelse Heide NR	545
Miejendel NR	1,300
Naardermeer NR	750
Noordhollands Duinreservaat NR	4,800
Oostvaardersplassen NR	5,600
Savelsbos NR	170
Schiermonnikoog NP	2,500
Terschelling NR	5,300
Texel NR	2,400
Vlaamse Banken NR RS	1,900
Waddenzee BR RS	260,000
Zwanenwater NR	590
Zwin NR RS	550

BR=Biosphere Reserve; NP=National Park; NPR=Nature Park; NR=Nature Reserve; RS=Ramsar site

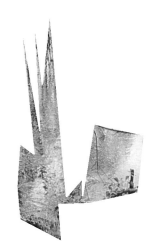

LOWLAND HABITATS

The Low Countries are aptly named: no less than one-third of the Netherlands is below sea level, and the most extensive wild habitats of the huge deltas formed by the rivers Rhine, Meuse and Schelde are mudflats, estuaries, dunes, saltmarshes, enclosed seas and fens. Off the Dutch coast, extending north into Germany and Denmark, the Frisian Islands form a protective barrier along the coast, enclosing the Waddenzee, one of the most important wetlands in Europe. The islands' saltmarshes and flats, scrub and woodland provide nesting and feeding areas for huge flocks of wildfowl, and their tidal shallows are home to the common seal. The coastal reserves of the Netherlands and Belgium have a rich dune vegetation, and are inhabited by hares, rabbits, stoats and polecats as well as by large populations of birds.

Vanishing woodland

When, some 10,000 years ago, the ice-sheets of the ice age retreated, they left behind a waterlogged environment of impenetrable peat bogs and marshes. As the climate grew milder forests of oak, lime, alder, hazel and beech gradually covered the land. Then people started to explore the land and settled there to cultivate it. And so the great felling began. Today only 8 percent of the Netherlands is covered with forest. In the upland area of southern Belgium and northern Luxembourg (the Ardennes) the situation is better; some 34 percent of Luxembourg remains forested.

When the forests disappeared, so did the large mammals such as brown bears, wolves, moose and beavers, along with the lesser spotted eagle, black stork and other bird species. Where stretches of ancient forest survive in the south, game animals such as red and roe deer and wild boar still exist and are protected.

The human factor

The activities of the early agricultural communities enriched the environment, creating new habitats. As oak and birch forests were cleared on sandy upland soils, heathlands formed, maintained by

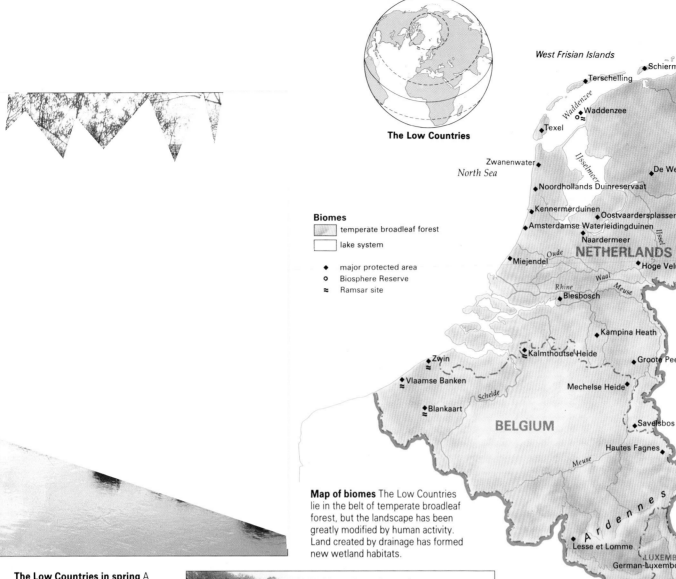

The Low Countries

Biomes

- temperate broadleaf forest
- lake system

- ◆ major protected area
- ○ Biosphere Reserve
- ≈ Ramsar site

West Frisian Islands
Schiermonnikoog
Terschelling
Waddenzee
Waddenzee
Texel
North Sea
Zwanenwater
De Weerribben
Noordhollands Duinreservaat
Kennermerduinen
Oostvaardersplassen
Amsterdamse Waterleidingduinen
Naardermeer
Miejendel
NETHERLANDS
Oude
Hoge Veluwe
Waal
Rhine
Meuse
Biesbosch
Kampina Heath
Zwin
Kalmthoutse Heide
Groote Peel/Mariapeel
Vlaamse Banken
Mechelse Heide
Schelde
Blankaart
Savelsbos
BELGIUM
Hautes Fagnes
Meuse
Ardennes
Lesse et Lomme
LUXEMBOURG
German-Luxembourg
Moselle

Map of biomes The Low Countries lie in the belt of temperate broadleaf forest, but the landscape has been greatly modified by human activity. Land created by drainage has formed new wetland habitats.

The Low Countries in spring A stream winds through a pleasant landscape that owes a great deal of its character to human management. The waterside willows are pollarded and the more distant trees, part of a plantation, stand in uniform lines. Yet this countryside still has a wide variety of wildlife, particularly birds.

Heathlands in the sandy country of the eastern Netherlands have replaced the original forest cover, removed by clearance for agriculture. They are threatened by development – some 90 percent of heathland in Belgium and the Netherlands has been destroyed, with the loss of a rich, specialized habitat.

extensive sheep grazing and the cutting of heather sods. These sods, mixed with cattle manure, were plowed back to enrich the farmland. As a consequence of this, a splendid variety of plants established themselves and flourished.

The heathlands have been invaded by steppe birds such as black grouse and curlew, and the small meadows support black-tailed godwits, ruffs and common snipe – birds that originated in eastern Europe. In Belgium, limestone grasslands

created by clearing and grazing have been colonized by numerous different plant species, including some that are more typical of the southern mountains and the eastern European steppes.

Peat extraction from fens and moorlands left behind water-filled hollows, lakes and marshes. These were soon settled by birds such as the spoonbill, purple heron and little bittern. One such ancient wetland, De Weerribben in the Netherlands, has become a stronghold for

otters and is today a national park. Peat- and reed-cutting continue to maintain this valuable habitat.

Over the centuries new land has been created along the coast: dyke systems enclose areas of water that are then drained and the lowlying land, or polders, reclaimed for cultivation. In some of the new polders reedbeds have been sown; these provide nesting and feeding areas for ground-feeding birds such as the marsh harrier.

Despite the accelerating encroachment of agriculture on natural habitats, the Low Countries still possess an impressive number of plants and bird species. In the Netherlands more than 1,450 plant and 400 bird species have been recorded, and of the latter 165 species breed regularly. The Netherlands and Belgium are internationally important staging posts for migrating birds. Up to 500,000 geese overwinter in the Dutch wildlands, and they are the only refuge in northwestern Europe for some species of birds.

SAFEGUARDING THE WILD PLACES

The Low Countries are the most densely populated region of Europe. This density has been accompanied by extensive farming and industry, which has meant that natural habitats have been reduced to tiny patches or squeezed out altogether.

State-led and private conservation

By the end of the 19th century, the enormous pressure on the land in the Netherlands alerted people to the need to acquire wild areas for effective protection. In 1906 the Naardermeer, an important site for spoonbill and purple heron colonies in the center of the region, was bought privately to preserve it for the future. Two years later the first state nature reserves were created by the National Forest Service. Established in 1899, the Service had been set up to manage some of the remaining patches of natural forest, by then a mere 5 percent of the country's total land area.

Following a rapid expansion of farmland during the 1970s, the relationship between farmers and conservationists became very strained. The government introduced a scheme to provide farmers using ecologically sound farming practices with government grants to compensate them for reduced yields. Today, about 9 percent of cultivated land is managed under such agreements.

In 1985 the government initiated a new approach to prevent further fragmentation of the rural environment, the aim of which was to create five experimental national landscape parks, each covering at least 10,000 ha (25,000 acres). When set up, these will encompass rural settlements and farmland as well as natural habitats, where small-scale farming using traditional methods will be practiced. In addition, landscape zones of at least 5,000 ha (12,500 acres) will be set aside, and there are plans to add a further 20 national parks to the existing six. By the end of the century a threefold increase in the number of protected areas is expected: they will eventually number about 3,000. Already some 4.5 percent of the Netherlands receives some form of protection.

In Belgium the establishment of protected natural areas, mainly of botanical interest, has mostly been in the hands of private organizations. In 1952 many of them came together to form the National Union for the Protection of Nature. It was not until 1957 that the first state-owned reserves were set up, and 1973 before legislation was provided for the establishment and subsidizing of protected areas. To date, barely 3 percent of the country receives some form of protection.

In Luxembourg, too, reserves have been established only recently. In 1965 the German–Luxembourg Nature Park was set up, the very first park in Western Europe to cross international frontiers. It covers an area of 7,250 ha (18,000 acres), of which about a third is forested with deciduous trees and conifer plantations. Its meadows, woodlands and gorges have a rich plant and animal life, including at least 36 species of orchids. However, the park is managed primarily for recreational use.

This is the only natural park or state-owned reserve in Luxembourg. However, about 20 voluntary organizations are concerned with protecting habitats. They all belong to the Luxembourg League for the Conservation of Nature and the Environment, which manages several small reserves. As in the Netherlands and Belgium, private conservation movements are given state assistance. Protected-area legislation was formalized in 1981, and it is intended eventually to extend protected status to more than 40 percent of the country's land area.

Managing the environment

The habitats of the Low Countries, created by human activity, are not stable environments, and their management has to take account of this. Without continued grazing or firing, for example, the heathlands and peatlands would gradually be invaded by shrubs and trees, and eventually revert to woodland.

The coastal wetlands are in an even more dynamic state of change. Lagoons can silt up, and reedbeds can quite rapidly become dry land as their dead remains crumble and fall to the ground in winter. Reserve managers usually cut the reeds regularly to prevent this happening, and to keep the water open. Even salt-marshes and salt grasslands may lose some of their plant species in the total absence of grazing.

A manmade habitat that actually benefits from less human intervention is the roadside verge. In the Netherlands, in particular, where spraying with herbicides is less prevalent than in Belgium and Luxembourg, such verges are full of wildflowers, and provide a rich habitat for insects. By cutting verges instead of spraying them, and by delaying mowing until after the plants have flowered and set seed, many species of wild flowers survive here for all to see.

HAUTE FAGNES NATURE RESERVE

The Haute Fagnes Nature Reserve, established in 1957 in eastern Belgium, is one of the Low Countries' most remarkable areas. Covering 3,894 ha (9,600 acres), it forms the northern part of the Haute Fagnes–Eifel Natural Park, which extends into West Germany.

The reserve's most striking features are its extensive formations of high peatland, some of which reach a depth of 7 m (23 ft) and are truly unique in Europe. Most of the area is situated more than 550 m (1,804 ft) above sea level, and it receives 1,400 mm (55 in) of rainfall a year, twice the Belgian average. Low average temperatures allow snow to lie for up to two months each year, making the ground extremely boggy. Marshes and peat bogs, found mainly on the plateau and in the valleys, alternate with oak, beech and spruce forests on the higher ground.

The peat bogs harbor several plant species typical of more northerly latitudes, such as marsh gentian and

The high peatland bogs and marshes of Haute Fagnes are a rich habitat for marsh plants.

marsh andromeda. The mammals, on the other hand, are more typical of central European mountain areas, and include wildcat, wild boar and red deer. The reserve also supports breeding populations of black and middle spotted woodpeckers and Tengmalm's owl that are of European importance.

A haven for waterbirds A lighthouse marks an island in the IJsselmeer, a freshwater lake created by damming the Zuiderzee. Large areas have been drained in recent years to create polders of rich farmland, but the lake remains a vitally important staging post for migratory populations of duck and waterfowl.

A heathland ecosystem Small mammals feeding on low plants and shrubs provide good hunting for birds of prey.

Components of the ecosystem

1 Dwarf gorse
2 Heather, heath
3 Woodrush
4 Grasses
5 Broom
6 Insects feeding on heather
7 Short-tailed vole
8 Rabbit
9 Wheatear
10 Kestrel
11 Stonechat
12 Violet ground beetle
13 Red ant
14 Adder

Energy flow
⇨ primary producer/primary consumer
➡ primary/secondary consumer

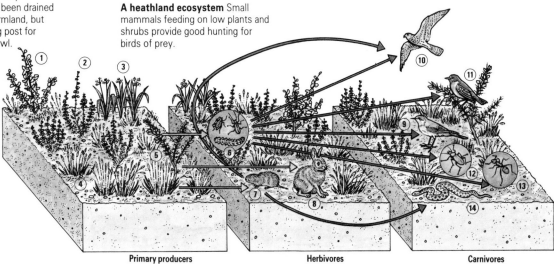

Primary producers   Herbivores   Carnivores

NATURE UNDER PRESSURE

The most outstanding threat to nature in the Low Countries is the ever-increasing pressure on land to cater for the population's growing need for housing and food. The rich clay soils of this delta region have readily accommodated intensive agricultural practices and now the people are having to pay the price – the region affords one of the smallest areas of natural habitat per person in the world.

Threatened by pollution
Today, the land is still used mainly for agriculture: the Netherlands ranks as the world's second largest exporter of agricultural produce. Cattle- and pig-breeding are intensive: the 15 million pigs outnumber the human inhabitants of the country. This causes gigantic problems with the disposal of manure, and has led to the poisoning of the subsoil and underground water supplies.

Modern methods of farming have undoubtedly impoverished the environment of the Low Countries. Mechanized farming results in large-scale monoculture (a single crop instead of mixed crops), with a consequent loss in the number and diversity of habitats as the last remaining patches of woodland, hedgerow and wetland that once enriched these farmlands are destroyed. Excessive use of pesticides and other chemicals have further reduced the variety of animals and plants. Since the beginning of the century, at least 48 and possibly as many as 78 species of plants have dis-

Flowering hedgerow trees and meadows make a bright patchwork against a forested hillside in the Ardennes, in the south of the region, where poor soil and high rainfall have restricted settlement and limited agriculture to grazing.

appeared; some 700 are considered to be in danger of extinction.

Farming practices have had a detrimental effect on the region's birds, too. A key example is the sad plight of the white stork. Changes in land use have reduced the wet meadows that were the storks' feeding habitat. Some 400 breeding pairs were recorded in 1913; in 1984, the last year when wild storks are known to have bred, only two pairs remained.

Forestry has also been responsible for the loss of wild places. In the economic depression of the 1930s, the National Forest Service of the Netherlands purchased large areas of so-called wasteland such as peat bogs and heather moorlands for largescale forestation schemes. Today these would be regarded as highly valuable wilderness areas.

The rivers Rhine, Meuse and Schelde pass through some of Europe's most heavily industrialized areas, including West Germany's Ruhr valley, before they drain into deltas in the Netherlands and Belgium. In the 1960s, uncontrolled industrial waste discharges, particularly from chemical plants along the Rhine and Schelde, had a devastating effect on some bird and animal populations breeding in the Waddenzee. High mortality rates particularly affected royal terns and eider ducks, and between 1968 and 1973 the area's population of common or harbor seals plummeted from 1,500 to 500.

Beechwoods in spring provide plenty of insect life for nesting birds. Their plentiful seeds (or beechmast) in the fall are food for mammals and migratory birds.

Birch trees are beginning to invade a heathland of heather and juniper. Without careful management species-rich wildernesses like this will revert to woodland; many have already been lost to forestation.

A brighter future
Today a large number of nongovernmental conservation organizations help to foster public concern for natural history and conservation. Some are nonpolitical in aims and methods, others highly active in changing political decisions. A public campaign that achieved great success was the movement, led by the Waddenzee Association, to dissuade the government from converting the entire Netherlands' Waddenzee into polders in 1975.

On 1 February 1953 a storm surge caused widespread damage in the province of Zeeland, in the southwesterly part of the Netherlands, drowning 1,835 people. As a result, the government decided that the tides should be brought under control. All the waterways in the province were dammed up and the open estuary, the Oosterschelde, was finally closed in 1987. This scheme had a profound impact on Zeeland's tidal environment, turning highly productive tidal ecosystems into freshwater lakes, 2,200 ha (5,436 acres) of which became the Markiezaat.

But the government soon labeled the area as wasteland, and the city council of Bergen op Zoom announced a plan to reclaim the area for development. Conservation organizations attacked the scheme vigorously, pointing out that not only was there no need for further residential development, but also that the Markiezaat had developed a remarkably high diversity of plant species and represented a wetland of international importance for migrating and overwintering waterfowl. It also attracted such rarities as the black stork, spoonbill and osprey. They called for the lake and marshland to be designated a nature reserve.

The development plans were deferred, but then a new threat arose – a plan to construct a series of windmills to generate electricity, with possible disruptive effects on waterfowl. The Dutch Society for the Protection of Birds was prepared to take legal action to prevent the project going ahead.

Another powerful private organization is the Dutch Society for the Protection of Birds (Vogelbescherming). Established in 1899, it is one of Europe's oldest conservation bodies. Today it has over 55,000 members and campaigns vigorously for effective legislation to protect birds. It owns and manages eight reserves with a total area of 800 ha (1,976 acres), and has drawn up management strategies for threatened species such as the spoonbill and the barn owl. It is also involved in international conservation projects.

As people have more leisure time, outdoor recreation and activity holidays have gained in popularity. The managers of protected areas are becoming increasingly aware of this, providing information centers and nature trails for visitors.

In 1987, visitors to parks in the Netherlands exceeded 2 million.

Recognizing the value of environmental education, the Netherlands government finances the Institute for Nature Conservation Education. This private association enlists the help of some 3,000 well-trained nature guides to involve the public in activities such as restoring habitats, pollarding (a special form of pruning) willows, or cleaning up nature areas; it also organizes nature trails.

For a long time conservation of the last wilderness refuges was seen as incompatible with development, but this attitude has changed dramatically. Now, a policy of selective growth is being followed, even when this entails landscape conservation at the taxpayer's expense.

Schiermonnikoog

Schiermonnikoog (popularly known as Schier) is one of the five inhabited islands of the Waddenzee, a shallow coastal lagoon of about 8,200 sq km (3,166 sq mi) that extends along the coast of the Netherlands and West Germany. The sea is undoubtedly western Europe's most important wetland and tidal area. In May 1984 the island was designated as a national park of international standing. Covering 2,500 ha (6,170 acres), it contains a wide variety of habitats that includes dunes, saltmarshes and managed meadows and farmland.

The diversity of Schier's landscapes and habitats, and the history of its cultivation, are typical of the islands of the Waddenzee. Between 8,000 and 4,000 years ago, a large part of the mainland consisted of peatlands, with sandy embankments fringing the coast. When the sea level dropped about 3,000 years ago, these sand deposits became dunes and, eventually, islands. Today, the prevailing sea currents cause extensive dune erosion on the western sides of the islands, while sand banks are deposited on the eastern side; the average easterly land shift of the islands is currently one kilometer (0.62 mi) per century.

The island's marshy dune valleys support a fascinating wealth of plant communities of great botanical importance, including many species of orchids and the uncommon grass of Parnassus. The mild climate allows several plant species of sub-Mediterranean origin to grow here, such as sea spurge. Other species, such as alpine bearberry, are survivors from the ice age, when the climate was colder and arctic species could flourish here. Two reserves have been established – Westpunt, a marshy dune valley with a freshwater lake, and Kobbeduinen, which has brackish and freshwater lakes, dune and marsh valleys and stretches of sand plains – to preserve these habitats.

The island is of crucial importance to migrating waterfowl. Birds that pass through and overwinter in the Waddenzee come from as far afield as Canada, Greenland, Iceland and eastern Siberia. The abundant invertebrates in the tidal flats and shallows provide them with a rich food source. Brent and barnacle geese can be found on the marshes and in the meadows. More than fifty bird species breed on the island, including such rarities as the hen harrier, eider duck and ruff. Harbor seals haul themselves out of the water along the sandy beaches and on the tidal mudflats; they are protected on the reserve at Obbedvinen.

Pressures on the habitats

As long ago as the 17th century, fishing and whaling industries flourished on the island. When they declined at the turn of the last century, the growing fashion for beach holidays helped to rescue Schier from oblivion. Schiermonnikoog Natural Park was established to counter the threat of tourism to the island's wildlife. The use of yachts, motor boats and overnight moorings was curbed, and only residents' cars allowed.

Since the park was established, a network of roads has been built, allowing cyclists and pedestrians to explore the saltmarshes. Most of the island is open to the public all year round, though some parts of the Kobbeduinen in the east of the island are closed during the breeding season to protect nesting birds.

Despite the international importance of the Waddenzee as a principal staging area for migratory birds, in 1983 the government allowed surveys and large-scale test-boring for natural gas on Schier. The Dutch nature conservation organizations are forced to remain ever-vigilant against such threats.

Shifting sands The dunes on Schier are continually moving, but may be stabilized by marram grass, sea buckthorn and other hardy plants, and colonized by a surprising number of species.

A line of dunes against the sea Part of the long barrier formed by Schiermonnikoog and the chain of neighboring islands that close off the Waddenzee from the North Sea. The shallow waters of the coastal lagoon drain at low tide to expose rich feeding grounds for birds.

Centuries of grazing have created the distinctive character of the Frisian landscape. Grazing prevents the dune vegetation being swamped by scrub and woodland.

A LAND OF CONTRASTS

MOUNTAIN AND PLAIN · PROTECTING THE WILD PLACES · THREATENED BY "PROGRESS"

Isolated from the rest of Europe by the mountain barrier of the Pyrenees, and from the African landmass by the Strait of Gibraltar, the Iberian Peninsula has blended the alpine and tropical elements of each to create a unique and unmistakable character of its own, almost unequaled in the temperate regions of the world for its variety of habitats. In the Sierra Nevada of the south, Mulhacén, the highest peak of the Iberian Peninsula, shelters the southernmost glacier in Europe, while a few kilometers to the east the Almerian badlands constitute Europe's only true desert. Not far to the north, arid salt steppes represent another habitat found nowhere else in Europe. Inland lakes and marshes, such as the Laguna de Gallocanta, are of international significance for their bird and plant communities.

COUNTRIES IN THE REGION	
Portugal, Spain	
Major protected area	**Hectares**
Aigües Tortes NP	22,400
Arrábida NaP	10,800
Covadonga NP	16,900
Caldera de Taburiente NP (Canary Is)	4,690
Cañadas de Teide NP (Canary Is)	13,570
Delta de Ebro NatR	64,000
Doñana NP RS BR	75,700
Estuário do Tejo NR	22,850
Fuente de Piedra NR RS	1,364
Garajonay NP WH (Canary Is)	3,980
Ilhas Desertas (Madeira)	proposed
Ilhas Selvagens (Madeira)	3,400
Las Tablas de Daimiel NP RS BR	1,800
Monfragüe NaP	17,800
Ordesa NP BR	15,700
Paul do Boquilobo NR BR	395
Peneda-Gerês NP	60,000
Picos de Europa NR	7,600
Ponta de Sagres NR	6,000
Ria Formosa NR	10,500
Serra da Estrêla NaP	55,200
Serra de Malcata NR	21,760
Serranía de Ronda NR	22,000
Sierra de la Demanda NR	73,800
Sierra Nevada BR	120,000
Sierras de Cazorla NaP BR	214,340
Timanfaya NP (Canary Is)	5,100
Tortosa y Beceite NR	29,300

BR=Biosphere Reserve; NaP=Natural Park; NatR=National Reserve; NP=National Park; NR=Nature Reserve; RS=Ramsar site; WH=World Heritage site

MOUNTAIN AND PLAIN

Except for a few industrial areas in the northeast and around Madrid in the center, the Iberian Peninsula is sparsely populated. Rugged mountains, vast barren plains and primeval forests are interspersed with traditional agricultural landscapes. The bulk of the peninsula consists of the Meseta, an ancient plateau mostly more than 600 m (2,000 ft) above sea level, which endures scorching summers and bitter winters. Centuries of grazing and cultivation have reduced the vegetation to scrub or steppe grassland.

Mountain refuges

The sunbaked, almost treeless, plains of the Meseta are bordered by the Cantabrian Mountains, culminating in the dramatic limestone turrets of the Picos de Europa in the north and the Pyrenees in the northeast. Deciduous woodlands, mostly of oak and beech, broken by meadows and pastures, flourish in the moist maritime climate of the northern ranges, while in the Pyrenees coniferous forests extend up to the treeline. The higher peaks have a typical alpine vegetation consisting of lowgrowing, wind-resistant cushion plants. These mountain ranges and their forested valleys are the

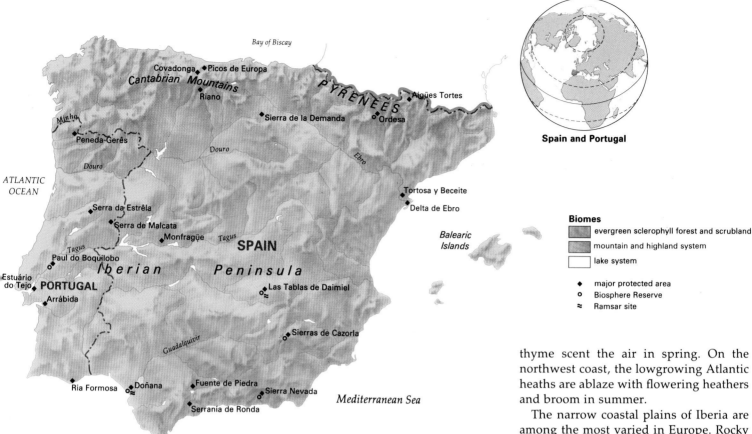

Bay of Biscay

Covadonga ◆ Picos de Europa

Cantabrian Mountains

Riaño

◆ Aigües Tortes

PYRENEES

◆ Sierra de la Demanda   ○ Ordesa

Minho

Peneda-Gerês

Douro

Douro

Ebro

ATLANTIC
OCEAN

Serra da Estrêla

◆ Tortosa y Beceite

Serra de Malcata

◆ Delta de Ebro

Monfragüe   *Tagus*

Tagus   SPAIN

*Balearic
Islands*

Tagus

Paul do Boquilobo

Iberian   Peninsula

Estuário
do Tejo ◆ PORTUGAL

◆ Arrábida

◆ Las Tablas de Daimiel

Guadalquivir

○ Sierras de Cazorla

Ria Formosa   ○ Doñana

◆ Fuente de Piedra

◆ Sierra Nevada

Mediterranean Sea

Serranía de Ronda

Strait of Gibraltar

Spain and Portugal

Biomes
- evergreen sclerophyll forest and scrubland
- mountain and highland system
- lake system

◆ major protected area
○ Biosphere Reserve
≈ Ramsar site

Map of biomes Spain falls within the zone of hard-leaved evergreen sclerophyll vegetation. Its arid southeast contains Europe's only warm desert, while its high mountain ranges support an alpine vegetation.

The snow-capped volcanic cone of Pico de Teide in Cañadas del Teide National Park, Tenerife. Brightly colored flowering shrubs, many endemic, grow wherever soil has been able to accumulate on the lava and volcanic ash.

Evergreen cork oaks stripped of their bark for use as cork. This takes place every nine years, and the trees can survive for up to 500 years. Much of Spain's once extensive sclerophyll forest has been cleared to make way for arable farming.

last refuges of Iberia's large mammals – the brown bear, wolf and ibex.

On the southern border of the Meseta are the Sierra Nevada and the Andalucian mountain ranges, which extend into the Mediterranean as the Balearic Islands. Although their summits are often snow-clad, the foothills support a subtropical vegetation more typical of northern Africa than of the European mainland. The zone between is home to the Spanish lynx.

Away from the heat of the plains and the rigors of the alpine zone a typical hard-leaved sclerophyll vegetation has developed in response to the Mediterranean climate. It consists of herbs and evergreen shrubs and trees with tough, leathery leaves, adapted to drought. These plants often lie dormant during the summer months.

Lengthy human settlement of the land has resulted in the formation of secondary plant communities such as maquis scrub, heathlands and steppe grasslands. These now cover large parts of the peninsula and the Balearic Islands. The thorny trees and spiny bushes of the maquis have evolved to cope with the grazing animals, and the reduction of their leaves to thorns serves to reduce water loss during the frequent droughts. Fragrant aromatic herbs such as marjoram, rosemary and thyme scent the air in spring. On the northwest coast, the lowgrowing Atlantic heaths are ablaze with flowering heathers and broom in summer.

The narrow coastal plains of Iberia are among the most varied in Europe. Rocky promontories are interspersed with tiny coves and bays backed by extensive dune systems. Their moist valleys and marshy lagoons are oases of flowering plants and wildlife. The precipitous cliffs and off-shore islands of the north play host to large seabird colonies. By contrast, the southern Atlantic coast of Portugal – the Algarve – and the shore of the Mediterranean comprise sweeping sandy beaches and an occasional river delta, often extending into lowlying salt and freshwater marshes, such as those at Aiguamolls de L'Empordà in Catalonia in the northeast, Ría d'Aveiro on the Portuguese coast, and the internationally important wetlands of Doñana at the mouth of the Guadalquivir river on the southwest coast of Spain.

The Atlantic islands
The fertile subtropical island groups of the Atlantic Ocean – the Azores, Madeira and the Canary Islands – have evolved their own distinctive habitats, plants and animals. Volcanic cones, some snow-capped, rise above barren plains of lava and sculptured pumice. The islands once supported laurel forests, fragments of the extensive woodlands that covered much of southern Europe and northern Africa some 65 million years ago. Today the inhabited islands support a thriving agriculture and little forest remains. A significant remnant is protected by the Garajonay National Park on Gomera in the Canary Islands, which has been designated a World Heritage site.

PROTECTING THE WILD PLACES

The protection of wild places began early in Spain. In 1916 the General Law of National Parks was passed, which designated Covadonga on the western flank of the Picos de Europa and Ordesa in the Pyrenees as protected areas. Although it was one of the first laws in Europe that gave legal protection to national parks, it was allowed to lapse until 1954, when the national parks of Cañadas de Teide and Caldera de Taburiente on the Canary Islands came into being.

The most important step forward in Spanish landscape protection came in 1975, when a law was passed to give protection to natural sites. This set out four categories of protected areas. Afforded the greatest degree of protection are the reserves of scientific interest, where access is forbidden so as not to disturb wildlife, especially breeding birds and mammals. Less strictly protected are the national parks, natural monuments and natural parks.

Nevertheless, the growth of protected areas has been slow. By 1990 there were still only five national parks on mainland Spain, covering a total area of 95,000 ha (235,000 acres). In the ten years from 1978 to 1988 30 new natural parks were established, and several more were planned. There were 16 reserves of scientific interest and 10 natural monuments. In total some 600,000 ha (480,000 acres) were protected to some degree – no more than 1 percent of Iberia's land area.

Hunting has a long history in Spain, and remains a popular sport. This has led to the creation of some 46 national game reserves, where hunting is strictly controlled and restricted to certain game species, such as wild boar. Covering over 15,000 sq km (5,800 sq mi), the game reserves are mostly in mountainous areas.

A growing awareness

By 1983 17 self-governing regions had been established in Spain under the new constitution of 1978. These have the power to introduce supplementary legislation on environmental affairs and to set up protected areas such as natural parks; they also manage the game reserves, which had previously been controlled by the central government.

By 1985 the regions had given protected status to a total area over three times greater than that covered by Spain's national parks. Most regional designations are natural parks.

Unfortunately, there is little financial support for the management of these sites, and all too often a natural park is a paper designation only. But the decentralization of conservation has had other advantages: local interest in wildlife has grown enormously, and voluntary groups increasingly apply pressure at local government level.

Two state bodies in Spain are involved in nature conservation – the Department of the Environment (DGMA), and the National Institution for Nature Conservation (ICONA). DGMA is broadly responsible for environmental planning and legislation, while ICONA is responsible for national parks.

Portugal has been slower to participate in such conservation measures. The sole national park is in the mountains of Peneda-Gerês in the northwest. There are also five natural parks and seven natural reserves on the mainland, and a natural park on Madeira. Two additional reserves are proposed in the Madeira island groups – one to protect the seabird colonies on the Selvages, and the other on the Desertas, home of a small population of the endangered Mediterranean monk seal. There are 24 permanent hunting reserves in Portugal, covering an area of

MONFRAGÜE NATURAL PARK

A naturalist's paradise of Mediterranean forest, scrublands and *dehesa* landscape, the Monfragüe Natural Park is centered on the dammed rivers Tagus and Tiétar in west-central Spain. *Dehesa* – a parkland of cork and holm oaks – is restricted to the Iberian Peninsula and northwest Africa. It supports a characteristic plant and animal life that is unknown elsewhere in the world. Despite a long history of human occupation and cultivation, remnants of the original Mediterranean woodland still survive on the steeper slopes.

The park is famous for its exceptional bird life, and is home to over 20 species of birds of prey. A substantial proportion of the world's population of endangered Spanish imperial eagles and black vultures nest here, as well as several pairs of the rapidly declining black stork. Other interesting Monfragüe residents include the black-shouldered kite, the azure-winged

A favored breeding site A surviving fragment of Spain's ancient *dehesa* landscape of holm and cork oak within Monfragüe Natural Park. Many birds of prey, including griffon and black vultures, breed here, and efforts are being made to extend the areas of *dehesa* parkland.

magpie and the eagle owl; European cranes visit in winter. Among the many species of mammals breeding at Monfragüe are the genet, mongoose, wild cat, wild boar and Spanish lynx.

In many parts of Spain and Portugal the ancient *dehesa* and Mediterranean forest are being replaced by rapidly growing eucalyptus plantations, which do not support many of the native plants and animals. Efforts are being made to reverse this trend: within the park, eucalyptus plantations are being returned to their former state, and a number of experimental areas have been set up in an attempt to demonstrate that traditional *dehesa* can still be farmed economically.

Subtropical vegetation, including dwarf fan palms more typical of northern Africa, grows on a rocky promontory in southern Spain. The mingling of northern African and Mediterranean species is not found elsewhere in Europe.

Components of the ecosystem

1 Alpine grasses and heathers
2 Chamois
3 Spanish ibex
4 Snowy vole
5 Brown hare
6 Ptarmigan
7 Golden eagle
8 Lynx
9 Carrion and regurgitated matter
10 Griffon vulture

Energy flow

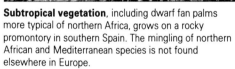

primary producer/primary consumer
primary/secondary consumer
dead material/consumer

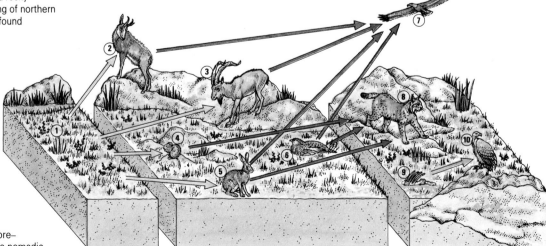

A mountain ecosystem showing herbivore–carnivore relationships. The herbivores are nomadic, according to seasonal rainfall and local water holes.

Primary producers　　　　**Herbivores**　　　　**Carnivores**

34,650 ha (85,600 acres), and 14 "conditional" hunting reserves (41,500 ha/ 102,500 acres). There is a state agency for nature conservation, and the government's forestry department also has a small section dealing with the protection of game species.

Success breeds conflict

So much of the Iberian Peninsula is still in a wilderness state that few people see the need for reserves. When livestock is destroyed by protected species such as wolves, bears, lynx and birds of prey, rural communities resent the fact that their livelihood is being threatened. City people visit the reserves for recreation purposes, but the growth of tourism brings hotels, restaurants, car parks, cable cars and the inevitable lines of pylons that spoil even the most scenic skyline.

The successful protection of wild species sometimes results in new areas of conflict. The 30,000 common cranes that winter at the Laguna de Gallocanta forage mainly in the surrounding cultivated land, causing considerable damage to autumn- and winter-sown crops, and attracting poachers. Attempts are being made to solve the problem by using different methods of planting and by firing rockets to scare the birds. Similar problems are encountered in the rice fields of the Ebro Delta and in the Albufera lagoon near Valencia on the east coast, where breeding waterfowl and shore birds congregate in large numbers.

THREATENED BY "PROGRESS"

The Iberian wilderness is under threat. Some of the greatest pressure comes from the intensification of farming methods, which often conflicts with conservation interests. In the past, Spain and Portugal's traditional agriculture achieved a balance with wildlife. This began to change after both countries joined the European Community (EC) in 1986, and came under pressure to increase agricultural yields.

Farmers in the south were encouraged to uproot their old, unproductive olive trees, since there is a surplus of olive oil in the EC. Unfortunately, these ancient groves provide an ideal habitat for the rich bird communities typical of the area. Similarly, the *dehesa* is being denuded of its oaks so that the machines required for modern cereal cultivation can be brought on to the land. The longterm effect of these tree clearances will be to exacerbate soil erosion.

Further north, livestock-rearing in the Cantabrian Mountains is becoming less profitable because cheap meat and dairy products are being imported from elsewhere in the EC. One proposal for economic recovery is to forest the area with eucalyptus and non-native pines for the paper and timber trade. There is no doubt that the superb mountain haymeadows, currently some of the richest Atlantic grasslands in Europe, will suffer if this measure goes ahead.

TOURISM AND COASTAL DESTRUCTION

From Lloret de Mar in the northeast to Gibraltar in the south, few stretches of the Spanish coastline are free of the "concrete wall" of hotels and villas. The environmental consequences of coastal tourism are ruinous. Woodland and scrub habitats behind the beaches are devastated, leading to soil erosion, and heavy trampling destroys dune ecosystems, as at Maspalomas (Fuerteventura) and Matalascanas, near Doñana National Park. Often coastal wetlands are drained, resulting in severe habitat loss. This has happened to an extensive area just west of the Guadalquivir estuary, where as much as three-fifths of the coastal marshes have been lost over the last 30 years.

Beach and marine pollution are rife, especially on the Mediterranean shores, where tidal motion is barely sufficient to stir the waters. Some of the most heavily polluted shores on the entire Iberian coast lie between Mataró and the Llobregat delta in the northeast; in one year – 1984 – bathing was prohibited at 30 beaches in this area.

The fact that tourism eventually destroys the very features that attracted visitors to the area in the first place has not gone unnoticed. An indication of this is the attempt by the Spanish government to control coastal tourism with a "coast law" that halts development within 100 m (330 ft) of the high-water mark.

Unfortunately no such restraints operate yet in the Portuguese Algarve. Portugal's pressing need for foreign capital means that tourism is still being heavily promoted here, and every year sees the disappearance of another headland with its rich variety of maquis plants and animals.

The Doñana National Park in the Guadalquivir delta is one of Europe's most important wetland reserves and is visited by almost every known species of European waterfowl during the winter and in the migration seasons. Many rare birds breed in the marshes. The park also contains dunes, heaths and woods, which remain relatively unspoilt because it was once a royal hunting reserve. These habitats are rich in reptiles, as well as many small mammals.

Ancient woodlands A remnant of the evergreen laurel forest that covered much of southern Europe and northern Africa 65 million years ago survives on the islands of the Azores in the Atlantic. The forest has been extensively cleared for agriculture, but has also dwindled naturally through climatic change. Strenuous efforts are being made to conserve the remaining forest, which supports many endemic plants, insects and birds that have evolved in isolation from the mainland. The mosses and ferns that cover the trunks give these forests the appearance of a jungle.

Other habitats under threat

The region's coastal marshes and wetlands are among its most threatened habitats. Portugal's Ría d'Aveiro is suffering from industrial and domestic pollution and the Santoña salt marshes, possibly the most important site for bird observation on the north Atlantic coast, are menaced by development. Santoña has been proposed as a natural park, and intense lobbying by local conservationists may yet save its marshes.

Water extraction for crop irrigation has seriously affected the water levels of various inland lagoons and marshes, especially in the south. The Daimiel National Park in south-central Spain was once of international ornithological importance. However, it was almost destroyed by an irrigation scheme in the 1970s, and as a result, waterfowl ceased to visit its dried up marshes. The public outcry that followed led to stringent water conservation measures and it is possible that the wetland will recover one day.

The inland lagoon of Fuente de Piedra in Andalucia, which is home to the peninsula's only sizable breeding population of greater flamingoes, is now suffering from the same symptoms. Even the Doñana National Park in the southwest, conserving one of Europe's largest wetlands, is threatened by the diversion of its water supply into nearby agricultural projects. For the moment this danger has been overshadowed by the threat to breeding birds posed by the legitimate hunting of the introduced Californian crayfish. In 1986 an estimated 30,000 chicks and eggs were accidentally destroyed by crayfish hunters.

More conspicuous is the threat posed by the controversial reservoir that extends across two river valleys near Riano on the southern side of the Cantabrian Mountains. The waters have inundated some of the richest limestone and riverside grasslands of the region, depriving the white stork of its northernmost breeding site in the Iberian Peninsula and also curtailing the territory of some of the last Iberian brown bears in existence, preventing them from moving from one seasonal feeding area to another.

Skiing, already a popular pastime in Spain, is increasing. Picturesque valleys in the Pyrenees (Valle de Arán) and the Sierra Nevada (the resort of Solynieve), Spain's two most spectacular mountain ranges, are already disfigured by cable

A mountain reserve Ordesa National Park protects some of the most magnificent scenery in the Pyrenees. This display of ramonda is typical of the many wildflower species found here, which include the endemic Pyrenean saxifrage with flower spikes nearly 1 m (3 ft) tall.

cars, chair lifts and chalets, while deforestation of slopes to provide ski runs causes soil erosion, giving rise to landslides in extreme circumstances.

Pressure groups

The ecological movement was initially rather slow to awaken to the dangers threatening Spain and Portugal's vast tracts of sparsely populated wilderness. Today, however, Spain has more than 200 voluntary conservation groups. These range from the Spanish World Wide Fund for Nature (WWF) to a whole host of regional organizations. Portugal has a national ornithological society, and there are voluntary conservation groups in Oporto and Braga.

With so much wilderness remaining, the creation of nature reserves is not necessarily the best way to protect the environment. Instead, these various societies and groups are concentrating on public campaigns and political pressure focused on specific threats to wildlife and habitats. This strategy is proving highly effective, as it is reinforced by a strong sense of regional identity. Today, nature is increasingly being protected for its own intrinsic value, rather than simply conserved as a resource for hunting.

Covadonga National Park

A herd of chamois graze on a hillside. They are extremely agile and spend the spring and summer among the alpine meadows and rocky mountain tops, but the females descend to the wooded slopes in winter. Several thousand chamois in Covadonga National Park live out of range of human disturbance.

In the mountains of northern Spain close to the Atlantic coast lie the Picos de Europa, a small limestone range of mountains divided into three blocks or massifs by precipitous gorges. The Covadonga National Park covers most of the Cornión, the western massif.

The park is outstandingly beautiful. The high mountain peaks are clad on their upper slopes with beech forest, blending into mixed oak woodland lower down. Above the treeline the bare peaks support lowgrowing alpine plants, which include several species found nowhere else in the world. Interspersed among the woods are traditional haymeadows, an important and increasingly threatened seminatural habitat. Two scenic lakes, Enol and Ercina, formed during the ice age, are extremely popular with tourists.

While Covadonga's plant communities are unique, it is most famous for its animals. The endangered Iberian subspecies of the wolf and brown bear, both shy and secretive mammals, breed in the nearby mountains and forage within the park. They are liable to be disturbed by increasing tourism. The higher slopes are home to several thousand chamois. Almost all the European representatives of the weasel family live here, and some of the last *asturcón* (wild horses) still run free around the lakes.

The birds, too, are exceptional. Among the birds of prey are Egyptian and griffon vultures, goshawks, short-toed, golden and booted eagles, kites and harriers. The woodlands are one of the last strongholds of the handsome black woodpecker and the Iberian race of the capercaillie. On the mountain peaks and cliffs wall creepers, snowfinches and alpine choughs may be spotted. Rare reptiles and amphibians include Alpine and Bosca's newts, the Iberian rock lizard, the golden-striped salamander and the Iberian viper. Several endemic butterflies are known to inhabit the mountains.

Managing the land

Much of the national park cannot strictly be called a wilderness as it is a farmed landscape, though few people live within its boundaries. In summer the upland pastures are grazed by livestock from the surrounding villages, while lower down the meadows are cut for hay.

The national park suffered greatly during the civil war in Spain (1936–1939), when much wildlife was destroyed. During the postwar recovery period, widespread exploitation of the park's resources took place in the form of mining, hydroelectric schemes on the Cares and Dobra rivers, timber extraction and fishing. However, since the appointment of a new director in 1984, Covadonga is once again being managed primarily for its wildlife and landscape value.

The state-run National Institution for Nature Conservation (ICONA), based in the nearby provincial capital Oviedo, is responsible for park policy and management. It has imposed a prohibition on burning the scrublands. This practice, carried out for centuries by local farmers, has been shown to reduce the acidity of the soil and remove dead and degenerate vegetation, so promoting the regeneration of a nutrient-rich "graze" for the livestock. Prohibition has led to more frequent "accidental" fires: huge areas have suffered from uncontrolled burning.

The centuries-old laws controlling grazing, previously self-administered by the local farmers, have been usurped by ICONA. Absentee landlords now have a say in allocating grazing, and overgrazing has thus become a severe problem, preventing woodland regeneration. In

The clear waters of Lake Ercina lie in a landscape created by the glaciers of past ages. This part of the Covadonga National Park is easily accessible by road, and is very popular with visitors. There are two hotels and a campsite nearby: alpine choughs have been seen scavenging for scraps around the carpark.

Beech forests replace oak on the higher slopes of the mountains of Covadonga. Where forest has been cleared, a scrubland of gorse, broom and heather develops. Recent policies to prohibit controlled burning of the scrubland by farmers has added to problems of overgrazing. It prevents woodland regeneration and reduces the variety of wildlife.

other matters relations with local residents have improved, and current proposals to extend the national park to the south and east, to take in the valleys and villages of Valdeón and Liebana respectively, are welcomed by all concerned.

Some 2 million people visit the shrine of the Virgin at Covadonga and the glacial lakes every year. So far, there has been little attempt at an education program, though there is a new visitor information center at Valdéon. Abandonment of traditional agriculture in favor of the greater profits to be made from tourism will certainly affect the landscape and wildlife of the park in the future.

Winter and summer on the mountains

Meadows above the treeline on many high mountains are valuable refuges for rare plants of various origins, many of which retreated there during the ice age. The inaccessibility of the higher mountain peaks enabled some species to evolve in isolation into new endemic forms. For most of the year these alpine meadows are covered with snow: in the spring they burst into life and the growth of lush grasses supports many grazing animals, including nimble-footed chamois and ibex as well as smaller marmots and hares.

At the end of summer, hibernating mammals such as marmots start collecting hay for the winter: during the summer feeding season they have built up layers of body fat to see them through this period, which may last eight months. Not all the small mountain mammals hibernate. Snow provides good insulation against the cold: voles and pygmy shrews forage beneath its blanket for roots, worms and insect larvae right through the winter. Other animals, too big to live under the snow, grow thick, white coats for warmth and camouflage. The larger mammals and birds migrate down the mountain to the shelter of the valley forests for the long winter months.

After the winter snows have melted, a high alpine meadow in the Pyrenees is carpeted with flowers.

MEDITERRANEAN WILDLIFE REFUGES

FROM THE MEDITERRANEAN TO THE MOUNTAINS · SAVING NATURE'S DIVERSITY · THE ENVIRONMENT IN JEOPARDY

Italy and Greece have a distinctive blend of habitats and wildlife. Behind their long Mediterranean coastlines, dry lowland scrub gives way to mountains forested with broadleaf trees such as oak, beech and hornbeam and, at higher altitudes, with conifers typical of the Alps. The islands of Crete and Sicily have hot, arid environments that are reminiscent of Africa, while the wildlife of the numerous small Aegean islands has similarities with the plants and animals of the Turkish mainland. The mountains and islands of Greece and Italy provide remote refuges for plants and animals from earlier times, sheltering them from changes in climate. The variety of plants found here is one of the richest in Europe, and the region continues to protect rare mammals and birds of prey that have disappeared elsewhere.

COUNTRIES IN THE REGION

Cyprus, Greece, Italy, Malta, San Marino, Vatican City

Major protected area	Hectares
Abruzzo NP	40,000
Ademello–Brenta NaP	44,000
Aenos NP	2,900
Akamas Peninsula NP	15,000
Calabria NP	17,000
Circeo NP BR RS	8,400
Etna NP	60,000
Evros Delta RS	10,000
Gennargentu NP	100,000
Gran Paradiso NP	73,000
Gulf of Arta RS	25,000
Lake Kerkini RS	9,000
Lake Mikri Prespa NP	19,500
Maremma RP	7,000
Monte di Portofino NaP	1,200
Mount Oeti NP	7,200
Mount Olympus NP BR	4,000
Mount Parnassos NP	3,500
Northern Sporades MP	100,000
Pindus NP	12,900
Pollino NP	50,000
Punte Alberete NR	480
Rodopi Forest NM	550
Salina di Margherita NR	3,800
Samaria Gorge NP BR	4,800
Stagno di Cagliari RS	3,500
Stelvio NP	137,000
Umbra Forest NR	10,000
Zafer Burnu	5,635

BR=Biosphere Reserve; MP=Marine Park; NaP=Natural Park; NM=National Monument; NP=National Park; NR=Nature Reserve; RP=Regional Park; RS=Ramsar site

FROM THE MEDITERRANEAN TO THE MOUNTAINS

The landscape and vegetation of Italy and Greece have been greatly altered by human activities. Italy is the more densely populated, with major industrialization and a high concentration of people in the north and on the coasts. However, areas of the mainland, and Sardinia and Sicily, still possess large tracts of unspoilt and remote mountainous country. Greece, on the other hand, being a land of mainly rural people (except around the capital city of Athens), has many natural and seminatural habitats. These are, in fact, increasing because many of the Greek islands have been depopulated this century. Greece boasts more than 6,000 species of plants, about 680 of which are unique to the country (endemic).

The coastal lowlands

The wetlands on the coasts, where they are not urbanized, are a major wildlife resource. In particular, the deltas of the larger rivers attract huge numbers of birds – waders, wildfowl and herons. Most of Europe's white and Dalmatian pelicans live in the Gulf of Arta and other inland waters in northwestern Greece. In the islands of the Aegean hundreds of endangered monk seals are thriving.

Near the coasts, the lowlands have been extensively cleared for cultivation. The agricultural landscape that has replaced the natural vegetation is dominated by olive groves, together with vines and cereals. There are said to be 15 million olive trees between Delphi and Itea on the Gulf of Corinth in central Greece, the so-called "sea of olives". The groves may be part of the human landscape, but they are a rich habitat for wildlife.

Modern farming methods, which are developed to a high degree in northern Italy, are unsympathetic to wildlife. However, the relatively unmechanized techniques of southern Italy and many parts of Greece have allowed native plants and some animals, such as the jackal and the brown bear, to survive. Birds and mammals are at greatest danger from enthusiastic Italian hunters, but many plants still survive even in quite well-cultivated areas.

The maquis scrubland of lowland and coastal areas (in Italy known as macchie) is typical of the Mediterranean. It is dominated by dwarf oaks such as Kermes oak and by spiny shrubs with small, leathery, drought-resistant leaves, which may be hairy or waxy. This scrub grows to several meters in height and is thought to have replaced native holm oaks and Aleppo pines.

The scrubland began to develop after the original forests had been cleared; repeated cycles of cutting, burning and grazing prevented their subsequent growth. Where grazing and burning have been excessive, the shrubs grow in hummocks not more than a meter (3 ft) tall. This stick-like vegetation is known as garigue in Italy, and phrygana in Greece. Despite their aridity, the macchie and phrygana are rich habitats.

The insect life is abundant: grasshoppers chirp by day, crickets and cicadas by night. Snakes, lizards and scorpions, and lesser predators such as tarantulas and praying mantids, thrive here. Many of these creatures, like the lizards, are strikingly camouflaged, a necessity in such an open habitat.

The insect-eating birds include the colorful bee-eaters, rollers and hoopoes. Tortoises graze on the sparse vegetation, and quails and partridges hunt for seeds. The lack of cover makes the macchie and garigue a good hunting ground for birds of prey – buzzards, hawks, short-toed eagles (snake eaters) and scops owls.

Mediterranean mountains

The valleys and slopes of the mountainous interiors of Italy and Greece and the Greek islands are covered in forests of deciduous oak, beech, hop hornbeam and elm. In Sicily moist winds penetrate

Mediterranean maquis This scrub vegetation has colonized lowland and coastal areas. The forests were cleared and the ground burnt periodically to improve the grazing. Maquis vegetation includes dwarf oaks and shrubs such as the strawberry tree and tree heath.

from the western Mediterranean, and there are stands of holly and pedunculate oak, more typical of western and northern Europe. Above about 1,500 m (4,920 ft), the mountain slopes are snowcovered in winter, and here coniferous trees, notably black pines and several species of fir, predominate.

On the eastern fringes of Greece, and on Rhodes and Crete, there are mountain forests of cypresses with wide, spreading branches. Cyprus still has a small number of cedars. On the higher slopes, above 2,000 m (6,560 ft), where there is rocky grassland, typical mountain and alpine plants and animals are found.

The forests and rocky gorges of the mountains provide the last refuge in this part of Europe for some of the larger mammals that are now so rare elsewhere. The brown bear, for example, survives in the Abruzzo National Park of central Italy and in the Pindus Mountains of northern Greece; the wolf in the Sila Mountains of Calabria in southern Italy and more widely in Greece, especially on the northern borders and in the Pindus Mountains; the wild pig in Italy; the lynx in the remote gorges of the Greek–Albanian border; and the ibex and chamois more generally in several parts of the region.

Classic alpine scenery in the Gran Paradiso National Park, Italy. Both Italy and Greece have high mountain areas with stands of coniferous forest and, above the treeline, alpine meadows. These relatively undisturbed areas are the habitats for larger mammals such as bears, lynx and wolves which have become extinct elsewhere in Europe.

Map of biomes Both Italy and Greece have long stretches of coastline dotted with islands. Much of the coastal lowland has been urbanized or cultivated, but there are some valuable wetland habitats. Behind the coastal zones lowland scrub predominates; farther inland are mountains forested with broadleaf and coniferous trees.

Biomes
- evergreen sclerophyll forest and scrubland
- mountain and highland system
- lake system

- ◆ major protected area
- ○ Biosphere Reserve
- ≈ Ramsar site

135

SAVING NATURE'S DIVERSITY

Centuries of human settlement in Italy and Greece, and the clearing, grazing and burning that change the landscape, mean that few habitats remain that can strictly be called "wildernesses". Even where protected areas have been established, few reserves are actively managed, and protection is in name only.

From royal reserves to national parks

The first national parks in Italy, Gran Paradiso in the Alps and Abruzzo National Park in the Apennines east of Rome, were both established in the 1920s and had both previously been royal hunting preserves. Two more national parks, Stelvio in the central Alps and Circeo on the west coast south of Rome, were established soon afterward. A fifth park, in the Calabrian mountains of the south, was set up in 1968. Stelvio was expanded in 1976 to its current area of 137,000 ha (338,400 acres); it is the largest of Italy's national parks and contains the country's biggest glacier. Several endemic plants are found here.

It is proposed that a number of regional parks and forest reserves, at present without strong legal protection, should be designated national parks. One such area is the great volcanic cone of Etna in eastern Sicily, which supports several endemic species of plants. If these plans are carried out, the resulting network of national parks will cover a wide range of climates and habitats.

Most of these areas are in mountainous regions, but there is also a strong need to protect Italy's dwindling wetlands. It has therefore been proposed that part of the Po delta be given national park status. At present, only one site in the delta is

protected – Punte Alberete, created by the World Wide Fund for Nature (WWF) in 1969. It is now managed as a nature reserve. The area's marshy woodland serves as a feeding and nesting ground for herons and egrets.

The various forest and nature reserves in the lowlands and smaller mountain

A mountain pool in the Abruzzo National Park in the central Apennines. The park is the last home for many species of plants and animals that have disappeared elsewhere in the Apennines. It is particularly famous for its ancient beech forests which lie below the rocky summits of the mountains.

The maquis ecosystem has large populations of insects, which support many birds, small mammals, snakes and lizards.

ranges include the great beechwoods of the Umbra Forest Nature Reserve on the Gargano peninsula of southeastern Italy, and the coastal sands and marshes of Maremma, on the coast of Tuscany. These reserves are not always respected by local people. Italy's woods and mountains are widely exploited both by hunters and by people who gather edible fungi and culinary herbs. The huge tourist industry causes additional damage to fragile mountain habitats.

Greek strongholds

Many of Greece's conservation problems arise from its having a predominantly rural population that is trying to raise its standard of living through tourism – and with some success. Despite this, the country has a relatively long history of nature reserves, and several new ones have been set up recently.

The first national parks were established in 1938 – on Mount Olympus, in northeastern Greece, whose scrubland and forest contain some of the finest mountain plants in the Balkans, and on Mount Parnassos in central Greece.

Since World War II, eight other national parks have been established in Greece, covering an area of 7,000 ha (17,500 acres). As in Italy, most of them are in the mountains. They are a precious haven for endemic plants and provide remote haunts for the threatened larger mammals and birds of prey.

More recently established reserves include Mount Oeti in central Greece, whose rich plant life includes several Turkish species as well as endemic plants; the forested and mountainous limestone

Components of the ecosystem
1 Olive
2 Thyme
3 Spiny burnet
4 Spiny broom
4a Seeds
5 Dormouse
6 Goat
7 Ants
8 Herbivorous insects
9 Red kite
10 Common gecko
11 Bonelli's eagle
12 Crested lark
13 Praying mantis

Energy flow
⇨ primary producer/primary consumer
⇨ primary/secondary consumer
⇨ secondary/tertiary consumer

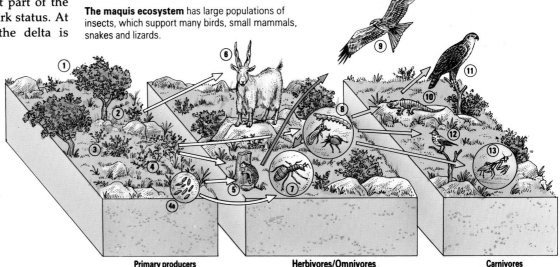

Primary producers   Herbivores/Omnivores   Carnivores

CLIFFHANGERS IN THE SOUTHERN AEGEAN

The many islands of the southern Aegean consist of rugged, dissected limestone with soaring coastal cliffs and, on Crete, gorges that plummet from the mountains down to the sea. These great walls of rock are accessible only to rare plants, animals such as the native Cretan ibex, and birds, notably birds of prey such as the lammergeier (bearded vulture), Europe's largest bird. Over the millennia the cliffs have protected the wildlife from harsh climatic changes and have prevented the development of a dense tree cover. Agriculture and other human activities have not been able to encroach on these valuable refuges.

Many of the plants that grow on these cliffs have beome specialized, adapting themselves to life in the precipitous limestone crevices. They include small shrubs or plants that are woody at the base, able to live in narrow cracks in the rock. Many are related to more widespread herbaceous species of the region. Some of these plants are very beautiful, such as the shrubby pink, a species of flax, and a relative of the bellflowers, with great spikes of blue flowers, which is found only on Crete. Another is a pale-flowered cabbage, found in southeastern Greece and Crete, which is a forerunner of the garden cauliflower.

The Abruzzo National Park The thickly wooded mountains of the park provide a secluded habitat for several rare and protected species. There is a remnant population of the Apennine wolf, the largest brown bear population in Europe and an Apennine subspecies of chamois. Originally a royal hunting reserve, the national park was established in the 1920s and has since become a model for park management.

The Vikos-Aóös gorge, west of the Pindus Mountains in Greece. The unusual ridged cliffs at the top of the gorge support many specially adapted plants that gain a hold in the rock crevices. The rocky ledges also provide ideal safe nesting sites for birds of prey.

island of Cephalonia in the Ionian Islands; the shallow Lake Mikri Prespa, over 19,500 ha (48,160 acres), which has one of the most important colonies of waterbirds in Europe, including a large population of pelicans; and the great Samaria Gorge on Crete, famous for both its endemic plants and its population of the rare Cretan ibex. The gorge is one of the few reserves in which tourist activities have been restricted.

Of lesser status as protected areas are the marshes and lagoons of the Evros Delta on the Turkish border, the wetlands around the Gulf of Arta on the west coast, and the forests of the Rhodope Mountains on the Bulgarian border, with the only stands of Norway spruce in Greece.

Perhaps the most significant nature reserve in Greece in terms of planning for the future is Lake Kerkini in northeastern Greece, formed only 40 years ago by the damming of the river Struma. This new habitat shelters threatened birds, such as spoonbills, and other wildlife.

THE ENVIRONMENT IN JEOPARDY

Threats to the richest and least disturbed habitats of Italy and Greece continue to increase, despite the establishment of national parks. Many smaller sites have little or no protection, and the national parks themselves are subjected to pressure from tourism and from industry and agriculture. Even quite remote places in the region are affected by air and water pollution, and the Mediterranean Sea is now seriously poisoned by effluents.

Old and new dangers

Fire poses a particular danger to Mediterranean habitats – it consumes vast tracts of forest and scrubland throughout the region every summer. In recent years extensive scrub communities on the island of Rhodes and forests of Calabrian pine on Thasos in the Aegean Islands have been devastated. Natural fires caused by electrical storms and volcanic activity are part of the history of both Italy and Greece, but they are small in scale compared to the conflagrations that people cause. Often fires are started deliberately to clear the land for development. Scrub vegetation is able to regenerate to some extent after being burned, but forests are less likely to do so and take longer; and the thin soils of the region can take years to recover from fire damage.

Modern farming methods create many environmental problems, often leading to the wholesale destruction of habitats. The removal of trees and vegetation causes erosion and the runoff of nutrients from the land. In the flat river plains and deltas of Italy and Greece, most of the wetlands have been drained for agriculture, or are being encroached upon. The Evros Delta on the Greek–Turkish border has been proposed as a national nature reserve and yet large areas have been drained for farming, cattle are allowed to graze where birds nest, and new roads and settlements have created further damage. Fortunately sandy offshore islands survive undisturbed and so can provide a breeding ground for more than 600 pairs of terns, including the rare sandwich tern.

Ancient Greece Mountain scenery near the ancient archaeological site of Bassae in Arcadia, southern Greece. Greece has the richest plant life in Europe, with over 6,000 species – of which nearly 700 are endemic. The massive increase in tourism in recent years has put pressure on natural habitats as millions of feet have damaged vegetation and disturbed the wildlife.

Pseudo-steppe landscape in southern Macedonia in Greece. This unusual habitat is found in areas of low hills where overgrazing and tree clearance has exposed the ground. The thin soil is then easily washed away by rainwater, leaving a stony surface. Pseudo-steppe is rich in herbaceous plant species.

Major industrial processes inflict increasing environmental damage on the landscape. On Mount Oeti in central Greece bauxite is being extracted by unsightly strip mining within the boundary of the reserve. This directly threatens a group of snow meltwater spring pools that are the only site for a tiny, unique plant, the Mount Oeti speedwell.

Road construction for mining and forestry in the remote parts of Italy and Greece is bringing about further change, to both good and bad effect. Locally it has created diversity in the habitat, allowing endemic plants to colonize new rock faces, for instance. But the effects on animals are more detrimental. Nesting sites are disturbed and easier road access allows more animals to be hunted.

The spectacular gorge of the river Voidomatis lies at the heart of the Vikos-Aóös Gorge National Park in Greece. Where the gorge is less steep, the dense vegetation includes forests of beech, fir and pine, which support populations of bears, wolves and jackals.

The most recent, and perhaps the most potentially damaging, threat to habitats comes from tourism, the major industry of Greece and much of Italy. Millions of town-dwellers invade the countryside at the weekends, and visitors from abroad flood the national parks in summer and the mountain areas in winter, damaging vegetation and disrupting wildlife. The facilities that go with tourism – ski-slopes and ski-lifts, hostels, hotels and access roads – put enormous pressure on the environment. On Zakinthos in the Ionian Islands, for example, café and hotel proprietors caused major damage to turtle nesting sites in order to develop the beaches for tourism.

The way ahead

The most effective way of scaling down the rate of injury to the environment is to make local inhabitants aware of the value of wildlife as a tourist attraction, and to encourage them to take steps to preserve it. Education programs are also needed for tourists if they are to understand the fragile nature of the habitats, and it may be necessary to implement some degree of control over where they go. Already in Italy, on busy holidays, the roads into sensitive areas of national parks are closed to vehicles.

A number of sporting and outdoor organizations take some interest in conservation and often educate their members about the threats their activities pose to wildlife. The Hellenic Alpine Club, for example, maintains mountain refuges and paths in several of the Greek mountain national parks, including Olympus and Parnassos.

The exploitation of certain rare species still needs to be controlled. Thousands of Greek tortoises were exported as pets to European countries until the trade was banned by law. A Greek presidential decree of 1980 set out to protect more than 700 plants from collection, sale or export, including many threatened endemic species. This was an important step forward, but ultimately education is needed to discourage local people – particularly hunters and mushroom enthusiasts – and collectors from farther afield from destroying wildlife habitats.

ALGAL BLOOM IN THE BAY OF VENICE

In the summer of 1989 the Adriatic coast of Italy and Yugoslavia was blighted by a stinking, yellow-brown slime which, at its worst, was 9 m (30 ft) thick and stretched for 650 km (406 mi). It deprived other marine life of light and oxygen, killing quantities of fish.

The cause of the slime, produced mainly by the single-celled diatom *Chaetoceros* (an alga), is thought to be twofold. The Po river, the largest river in Italy, discharges into the Adriatic the sewage of 15 million people and as many farm animals, industrial effluent, and the runoff of phosphates and nitrates used as fertilizers.

In the past, a freezing winter wind – the bora – cooled the waters and stirred them up, preventing large colonies of algae, which feed on the nutrients, from forming. But the wind's frequency has decreased, and now, aided by warmer winters and summers, the almost static Adriatic provides ideal conditions for massive algal growth.

Despite protective measures such as the use of floating booms to contain the algae, and poisonous chemicals and even flamethrowers to destroy them, there is every sign that the slime will return, damaging the region's ecology, and its economy, further. The Italian government is prepared to respond with alacrity in order to safeguard the tourist industry, from which it receives valuable revenue.

Gran Paradiso National Park

Gran Paradiso is the longest-established and perhaps the most famous of the national parks of Italy and Greece. Situated in the northwestern corner of Italy on the border with France, it covers an area of 73,000 ha (180,300 acres) and continues across the border as the Vanoise National Park. It was set up in 1922 by King Victor Emmanuel III as a hunting reserve in order to protect the population of ibex, which had been hunted almost to extinction in Italy by the early 19th century.

The park is dominated by a complex of mountain peaks, the greatest of which, at around 4,060 m (12,180 ft), is Gran Paradiso itself. These mountains form the park's rugged central core, an inaccessible wilderness of glacier, cliff, scree and alpine grassland habitats. The fringes of the park, which have permanent inhabitants and also cater for tourism, are affected by poaching.

Only some 6 percent of the park is covered by woodland, but it is nevertheless a widespread habitat that provides a home for birds such as the nutcracker and black woodpecker, and rare and handsome flowers such as the martagon lily. The valleys contain broadleaf deciduous trees interspersed with conifers. Spuce, larch and pine dominate the slopes higher up, and beneath them grow various dwarf shrubs, species of the heath family such as cowberry, which has red berries in late summer, alpenrose, and the delicate twinflower, a creeping shrub with paired pink bell-like flowers.

Protecting the ibex

It is above the treeline, from some 2,000 m (6,562 ft) upward, that the real glories of the park are found. Despite the inroads into animal numbers made by hunting in the past, Gran Paradiso still contains a substantial variety of species. A large population of chamois lives among the high rocks and cliffs, as well as marmots, red foxes and rare birds such as the golden eagle and the eagle owl. But the pride of the park is its 3,000 or so ibex. These animals are remarkably fearless, and it is possible to approach within a few meters of them without causing alarm. Their numbers are sufficiently stabilized to allow animals from the park to be reintroduced into other areas of the Alps. However even in Gran Paradiso the ibex are vulnerable. In winter they move down to the lower slopes, which lie outside the park's boundaries, where they are often hunted or poached.

The pressure from the many summer visitors is damaging the fragile habitats of the park's five main valleys. As in Europe's other parks, it may be necessary to control the flow of tourists if the Gran Paradiso is to survive unspoilt.

Outside the summer season, the local population numbers only a few hundred. The traditional way of life of these mountain communities is as vulnerable as the wildlife. For centuries the local people have cut the meadows and grazed animals in the park, activities that diversify rather than damage the environment. However they are under increasing pressure to adopt new methods of farming, which may have the opposite effect.

An Alpine ibex surveys the Valnontey valley. The park provides the last refuge for the ibex, which was hunted almost to extinction. The ibex live on the lush mountain pastures in summer and descend to lower regions in winter.

The oldest Italian national park, Gran Paradiso is one of the most spectacular in Europe. Here a fallen tree is allowed to lie, providing a rich habitat for the decomposers that break down the dead tissues, releasing nutrients into the food chain. The park's popularity with tourists brings with it problems of litter and noise pollution, and there is controversy over some property developments in the valleys.

The Great Wall is another name for Gran Paradiso mountain, the huge rock dome that forms the center of the national park. Around the central massif are a variety of habitats including grassy valleys, dense woodland, alpine meadows and mountain tundra giving way to rocky crags and ledges. Apart from some African reserves, few other parks in the world support as much wildlife.

REMNANTS OF A FORESTED PAST

A VARIETY OF FORESTS · CONSERVATION AND CONTROVERSY · THREATENED BY INDUSTRY

The center of Europe was once almost completely covered in forest. It was broken only by rivers and lakes, a few large expanses of moorland and by mountains where they rose above the treeline. Today, genuine wilderness areas are rare. Vast stretches of the forests have been cleared and the moorlands plowed up to make way for agriculture and industry. Still more trees are being lost today to the ravages of acid rain, caused by air pollutants from industry and automobile exhaust. Expanding human settlement has decimated the rich wildlife of the forests, and few large mammals remain. But extensive tracts of forest still survive in the mountains of the south. The wilderness areas of the North Sea coast include some important European wetlands, supporting thousands of waterfowl and waders.

COUNTRIES IN THE REGION

Austria, Liechtenstein, Switzerland, West Germany

Major protected area	Hectares
Ammergebirge NR	27,600
Augsburg Forest NR	1,600
Bavarian Forest NP BR	13,100
Berchtesgaden NP	20,800
Bodensee NR	776,900
Chiemsee RS NR	8,660
Derborence LPA	1,000
Diepholzer and Moorniederung NR	15,060
Donau-Auen NPA	38,500
Hohe Tauern NP	250,000
Holloch Karst NR	9,200
Karwendel NPA	72,000
Lüneburger Heide NR	19,700
Monte San Giorgio NR	2,500
Neusiedlersee/Seewinkel NR RS BR	60,000
Niedere Tauern NP	75,000
Nordfriesisches Wattenmeer NP	285,000
Oberharz NR	7,000
Ostfriesisches Wattenmeer NR	240,000
Pfälzerwald NR	1,000
Piora LPA	3,700
Rothwald NR	600
Ruggeller Riet NR	90
Starnberger LPA RS	5,720
Swiss NP BR	16,870
Vallée de Joux/Haut Jura NR	22,000

BR=Biosphere Reserve; LPA=Landscape Protected Area; NP=National Park; NPA=Nature Protection Area; NR=Nature Reserve; RS=Ramsar site

A VARIETY OF FORESTS

Europe's ancient forests reached their greatest extent in West Germany; in Austria and Switzerland they were limited by the high altitude of the Alps. Today, most of this surviving forest has been altered by human activity. Only in a few relatively inaccessible areas, mainly in the mountains, have remnants of virgin forest been conserved.

As the central European landscape gradually rises southward to the Alps different types of forests are encountered. Lowland forests of oak and beech give way to beech forest on higher ground, dominated by red beech mixed with silver fir and spruce. These forests support large herds of deer, which are maintained in great numbers for the enormously popular sport of hunting. Even where hunting has low priority, the large populations of deer, no longer contained by predators such as lynx and wolf, graze young tree seedlings and cause problems for the managers of nature reserves. There are also pine forests in the northeast and spruce forests in the Black Forest, which are inhabited by wildcats, pine martens and forest birds such as the capercaillie.

These forests merge into the higher subalpine silver fir and spruce forests, with Swiss stone pine and larch in the drier inner Alps. They are home to a wide variety of mammals and birds – foxes, badgers, polecats, squirrels, Arctic hares, and white partridges, woodpeckers and various species of owls. At these altitudes there are also heaths of broom, juniper, heather, bilberry and bracken.

Above the treeline the vegetation consists mainly of low growing, shrubby mountain pines. Higher still, where soil lies exposed between the rocks, there are rhododendrons, grasses and alpine wildflowers. The habitat supports nimble footed ibex and chamois, rodents such as marmots, mountain hares, ptarmigan and many other mountain birds. The abundant small mammals provide a plentiful food supply for a variety of birds of prey, including the rare golden eagle.

Forests also once grew along the banks of the big rivers. One of the last tracts of riverine forest in Europe survives on the banks of the Danube in the far east of the region. It covers an area of 8,000 ha (20,000 acres) and contains all the stages of succession from swamp forest to true terrestrial forest. The wetter parts are dominated by species of willow, the remainder by oak, ash, elm and other deciduous species such as alder. This forest shelters 109 species of breeding birds, 41 species of mammals, 20 amphibian species and 40 kinds of fish. A similar riverine forest is found along the banks of the March a little to the north.

From the coast to the plain

Much of the North Sea coast of West Germany consists of reclaimed marshes, bogs and heaths turned to grassland by human use. But a large natural ecosystem – the Waddenzee – is also found here. This unique habitat extends into Denmark and the Netherlands. It is separated from the North Sea by a string of offshore islands. Its mudflats, saltmarshes, flanking dunes and brackish marshes are an important breeding, nesting and feeding ground for hundreds of thousands of birds, among them the rare Sandwich tern and the cormorant. Their food source is an abundant population of crabs, worms and mollusks and other small invertebrates. In addition, the waters of the Waddenzee serve as a nursery for the young of many North Sea fish.

Inland lies the North German Plain, a highly cultivated landscape that retains pockets of heath, moor and bog. However, these habitats are being reclaimed for pasture and cultivation. The surviving heathland is colonized by heather, grasses, mosses and lichens, and in places by juniper; boggy areas support sphagnum mosses, club mosses, sundews and cotton grass.

The purple heather of the Lüneburger Heide (Lüneburg Heath) in northeast Germany. In this lowlying area are found heathers, grasses and mosses with sundews, sphagnum mosses and cotton grass growing in the wetter parts. The low vegetation is broken only by scattered juniper trees and occasional birch woods.

Central Europe

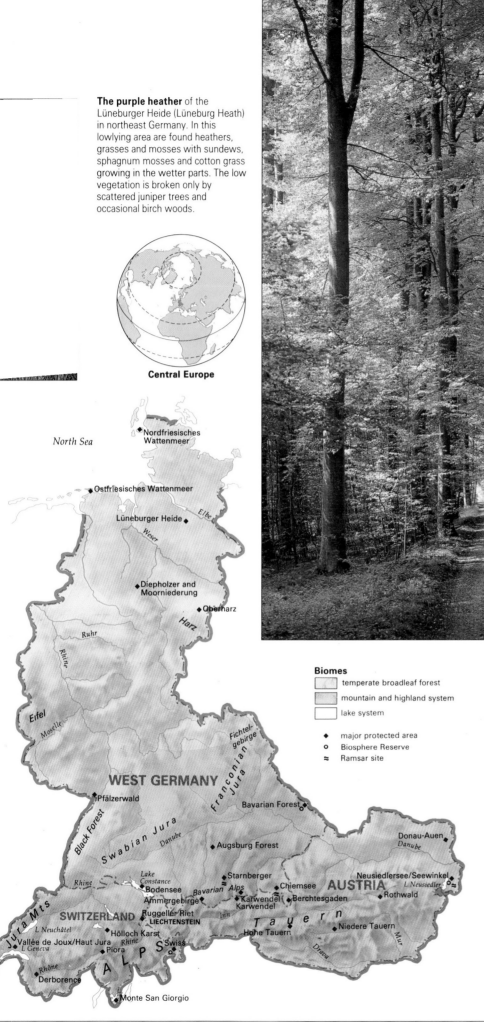

North Sea

Nordfriesisches Wattenmeer
Ostfriesisches Wattenmeer
Lüneburger Heide
Elbe
Weser
Diepholzer and Moorniederung
Oberharz
Harz
Ruhr
Rhine
Eifel
Moselle
Fichtelgebirge
Franconian Jura
WEST GERMANY
Pfälzerwald
Bavarian Forest
Black Forest
Swabian Jura
Danube
Augsburg Forest
Donau-Auen
Danube
Lake Constance
Bodensee
Starnberger
Chiemsee
Neusiedlersee/Seewinkel
L Neusiedler
AUSTRIA
Bavarian Alps
Ammergebirge
Karwendel
Karwendel
Berchtesgaden
Rothwald
Ruggeller Riet
LIECHTENSTEIN
Inn
Tauern
Rhine
SWITZERLAND
L Neuchâtel
Hölloch Karst
Hohe Tauern
Niedere Tauern
Vallée de Joux/Haut Jura
L Geneva
Rhine
Swiss
Jura Mts
Piora
Mur
A L P S
Drava
Rhône
Derborence
Monte San Giorgio

Biomes

- temperate broadleaf forest
- mountain and highland system
- lake system

- ◆ major protected area
- ○ Biosphere Reserve
- ≈ Ramsar site

Summer in the forest These typical European mixed woodlands include beech, hornbeam, oak, cherry, maple and ash. The dense tree canopy provides little light for undergrowth, but some herbaceous perennials will grow in the sunnier spots.

Lakeland habitats Bodensee (Lake Constance) on the Swiss–German border has several nature reserves, important for their populations of ducks and geese. More than 200,000 overwinter on the lake.

Map of biomes Central Europe's habitats range from coastal marshes, mudflats and dunes in the north to heathland and deciduous forest in the south. Conifers and alpine meadows appear above the treeline.

CONSERVATION AND CONTROVERSY

On the face of it, there is no shortage of protected areas in central Europe. The Swiss National Park, created in 1914, was one of the first strictly protected areas in Europe. In West Germany and Austria a great number of nature parks were established after World War II; today there are more than 60 in West Germany alone.

Despite this apparently good record, there is a lack of understanding among the general public of the importance of preserving original, naturally developing ecosystems, which hinders the work of conservationists. Even official conservation regulations hardly make the distinction between a strictly protected reserve and a protected area with traditional land use. At best, conservation in Austria and West Germany has focused on particular species. The conservation of ecological processes has been largely ignored until very recently.

Consequently, most nature parks are not primarily concerned with the conservation of nature, natural landscapes and wildlife. They are areas used for agriculture and forestry, which have a high recreational value because of their managed landscapes. The main objective is to conserve the variety of landscapes within these parks, together with their village communities, and to make them available for recreational purposes.

The Bavarian Forest National Park, established in 1969, was one of the first to adopt a wider approach to conservation. There the policy of the park managers is to leave the forests to develop naturally, to conserve natural succession, not to remove trees blown down by the wind, and not to fight insect predation. The aim is to protect the whole environment, as it is in the Berchtesgaden National Park in the Bavarian Alps.

In Switzerland the situation is better. Here, official nature conservation is more successful and is enhanced by influential private nature conservation organizations such as the Swiss Nature Conservation Association and the World Wide Fund for Nature (WWF) of Switzerland. Except for a short hunting season, alpine animal life is strictly protected. In Swiss nature reserves, nature conservation is afforded the top priority and most park managers have the means and the will to enforce it.

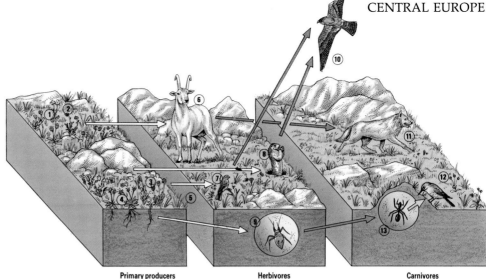

Primary producers   Herbivores   Carnivores

Components of the ecosystem
1 Sedge
2 Orchid
3 Groundsel
4 Plantain
5 Grass
6 Chamois
7 Yellowhammer
8 Marmot
9 Root-living aphid
10 Peregrine falcon
11 Gray wolf
12 Gray-headed woodpecker
13 Yellow hill ant in ant-hill

Energy flow
⇨ primary producer/primary consumer
⇨ primary/secondary consumer
⇨ secondary/tertiary consumer

An alpine meadow ecosystem
Winter snow forces many animals to
hibernate or migrate down the
mountain for food.

A mountain refuge The
Berchtesgaden National Park in
southeast Germany provides a safe
home for chamois, red deer and the
once heavily hunted golden eagle.

The hills are alive In early summer
the alpine meadows above the
treeline in the high Alps are a
profusion of color as flowering plants
such as orchids, saxifrages and
gentians come into bloom.

THE SWISS NATIONAL PARK

Founded on the initiative of the Swiss Nature Conservation Association, the Swiss National Park spreads over 16,870 ha (41,670 acres) of the mountainous far eastern corner of Switzerland. One third of the park is covered by coniferous forests, another third by alpine meadows and pastures, and the remainder by scree slopes and massive outcrops of dolomitic limestone.

The park belongs to four local communities, which have leased their land to the state for an unlimited period and receive an annual payment for it. Its objective is that it should be "a reserve where nature is fully protected against all sorts of human intrusions and where the entire fauna and flora is totally left to its own free natural development".

Compared with reserves in other European countries, its regulations are extremely strict. Walking is permitted only on marked trails. Visitors are not allowed to take anything away from the park; nor may they leave anything behind. Dogs, fires and camping are also prohibited. A team of rangers enforces the regulations, and informs and helps the 500,000 people who visit the park each year.

However, "natural" development has its problems: the natural regeneration of the mountain forests is disturbed in many parts of the park by the large populations of red deer. In former times they would have been kept in check by natural predators such as the wolf, or by human hunters.

Pressure from hunting and forestry

The present West German law for nature conservation states that "orderly agricultural and forestry uses do not disturb the balance of nature". From an ecologist's point of view this is an absurd statement; it shows the extent to which the powerful agricultural and forestry lobbies have influenced conservation legislation.

Since agriculture and forestry are practiced in all the large and most of the smaller nature reserves in West Germany without any limitations, it is not surprising that unrestricted hunting is allowed in them too. The damage inflicted by hunting, particularly in mountainous areas, is very considerable. In the interests of the shoot, huge populations of roe and red deer as well as chamois are protected. But their presence in large numbers alters the character of the forests and prevents their rejuvenation. Here and there, traditional livestock grazing causes additional pressure. The result is species-poor forests that are susceptible to all kinds of threats, instead of naturally diverse and stable forests.

The situation in Austria is little different; in places, the loss of forest to pasture is even more widespread. This threat extends even to national parks. Indeed, it is the main reason why the Hohe Tauern National Park in the southern Alps is not officially recognized as a protected area by the International Union for the Conservation of Nature and Natural Resources (IUCN).

Acid rain damage Waste chemicals from industrial sites and motor vehicles are carried by the wind and then deposited as acid rain on soil, plants and streams. The extreme stress on coniferous trees produces needle loss and eventually the death of the tree.

THREATENED BY INDUSTRY

Central Europe is very heavily industrialized, and its natural habitats are threatened by the pollution that is a by-product of its factories. The most extensive form of pollution in central Europe is airborne, in the form of acid rain. Acid rain results when poisonous chemicals in the air, particularly oxides of sulfur and nitrogen, are brought down in solution by the rain and are then deposited on plants, the soil, and in river and stream systems. Concentrations of nitrogen oxides in the atmosphere, caused by automobile exhaust, and of sulfur emissions from industrial processes, are still increasing. The destruction of the forests is not only harmful to the trees; it also threatens the landscape leaving it vulnerable to soil erosion and avalanches that occur more often.

The first warnings of acid rain damage were sounded in 1960, when the acidity of the soil in the German mountains was found to be rising. Visible damage to vegetation began to appear around 1975. The trees most affected were those that grew on the crests of mountain ranges, where precipitation is high and fog and frost are frequent. Their crowns were found to be thinning out and open glades were developing in the forest. Silver firs were the first to be affected, then came spruce, and finally deciduous species, especially oak, were discovered to be suffering damage.

By the early 1980s, the increase in soil acidity had risen a hundred-fold since 1960, and vast areas of trees were starting to die in the Bavarian Forest, the Black Forest, the Harz mountains and the Fichtelgebirge. In the Alps the damage began some years later, but today the forests here are also badly affected. In 1987 more than 50 percent of West Germany's forests were registered as "affected"; in Austria and Switzerland the percentage of damage is only a little lower.

River pollution

Other forms of pollution include industrial waste, which is frequently carried by rivers. The Rhine has become increasingly salty since industry and mining have increased; this affects agriculture and natural habitats farther downstream. A third of the Rhine's salt content comes from potash mines in Alsace.

Wetlands under threat Infilling, agricultural and industrial pollution and recreational activities are destroying wetlands in Europe and around the world. However, the need to conserve and manage these important habitats has now been recognized.

Another consequence of river pollution is the devastation of the riverine forests. Neither the Austrian capital, Vienna, nor other towns situated along the upper course of the Danube, have adequate waste treatment plants. As a result, large amounts of mud and waste are able to enter the river, destroying the aquatic habitats in the river as well as those along the river banks.

OBERHARZ – THE DYING FOREST

The Oberharz Nature Reserve is situated on the border of West and East Germany and forms part of the Harz Nature Park, a popular holiday area visited by several million tourists a year. Nearly all of its 7,000 ha (17,000 acres) are covered with forest, although centuries of forestry and mining have made great modifications to its nature. Today spruces dominate, growing at a height of 700–800 m (2,300–2,600 ft) above sea level.

In the last ten years, however, this area has been endangered by air pollution. A yellow and brown discoloration was first seen in the oldest conifers in 1979–1980. Then the needles died off early on the tree tops. Older trees and trees on the forest periphery, and on the highest mountains, were especially threatened, as they were more susceptible to damage by storms and high winds.

By 1985, 75 percent of the forest in the reserve was affected. It is a frightening possibility that the reserve will suffer the same catastrophe that has already struck the forests of the Ore Mountains, a few hundred kilometers away in East Germany, and the Riesengebirge situated on the border between Poland and Czechoslovakia east of the region. In 1970 the trees on the crests of these mountains were still green; by 1985 they were all dead.

An environmental crisis This area in the Harz mountains was once a healthy forest. Air pollution reduces the trees' resistance to disease; they become weak and blow down in high winds.

The valleys and floodplains of the March and the Danube contain wetlands of international importance, and conservation organizations such as the WWF, IUCN and UNESCO have pressed strongly for the establishment of the Donau-Auen National Park. But they are competing with influential industrial and political interests, which want to construct a power station with a huge dam in the area. This would undoubtedly destroy large stretches of this extensive forest system. At the beginning of the 1990s it remained unclear whether the conservationists would win the battle.

The conflict with tourism

The tourists, mountaineers, skiers and weekenders who pour into the countryside, including the nature reserves, cause widespread disturbance that has led to the virtual disappearance of some sensitive bird species, such as the wood grouse and the capercaillie. The expansion of cross-country skiing in the 1980s increased the damage to wild places even in winter. Downhill skiing continues to use up more and more mountain slopes in the Alps. Already 50,000 ha (150,000 acres), nearly 0.6 percent of Austria's total land area, are used as ski runs. As a result, nature lovers seeking peace and quiet put pressure on still more remote areas, especially nature reserves.

Existing conservation regulations are insufficient to control and channel such vast numbers of people. In West Germany there is a trend toward increased protection, but current attempts at regulation are still met with strong opposition from local farmers, who are under pressure to modernize alpine farming techniques yet further. Their continued use of pesticides and fertilizers, and the construction of more trails, hampers the creation of new nature reserves.

The Bavarian Forest

The Bavarian Forest or Bayerischer Wald, on the border between West Germany and Czechoslovakia, is the largest and least-disturbed mountain forest in central Europe. The woodland on both sides of the border covers about 200,000 ha (500,000 acres) and is crossed by few public roads. Exploitation of the forests started only 150 years ago, with the result that some areas of original forest have survived to the present day.

The Bavarian Forest National Park was established in 1969. More than 98 percent is forested. The higher, mountainous parts are covered with spruce, which can survive the severe frosts and heavy snow of the winters. A ground vegetation of couch and wavy hair grass, bilberries, ferns and mosses grows thickly beneath the trees in summer.

Most of the rest of the forest is of mixed red beech, mountain elm, maple, lime and ash. The plants of this forest floor include thorny fern, lady fern, woodruff and Turk's cap lilies. On the moist valley bottoms a meadow spruce forest flourishes, together with bog birches. The park also encloses areas of ancient sphagnum bog in which dwarf spruces and pines and other bog plants grow.

The Bavarian Forest National Park is especially rich in forest birds such as the capercaillie and hazel grouse, four species of woodpeckers and Tengmalm's and pygmy owls. The eagle owl, Ural owl and raven are being reintroduced. There are 55 breeding bird species, 14 of which are threatened. Other rare forest animals include the lynx and the fish otter.

Managing the forest

After the park was formed, hunting was stopped and forestry practices were gradually reduced. By 1988 8,000 ha (20,000 acres) were protected as strict reserves, where human intrusion is forbidden. The forests in these are thus left alone to develop naturally.

The speed with which managed forests revert to their natural state is astonishing. Trees blown down by storms or felled by insect invasions are left to lie. This accumulation of dead wood is an essential condition for the natural development of the forest, and it is constantly increasing.

Public access throughout the park is strictly controlled. Over about half of its area visitors are forbidden to leave the marked trails. However, the policy of the park is to educate. A large national park center offers an extensive program of exhibitions, films, slide shows and guided tours. Of the roughly 1.2 million visitors each year, some 30,000 take part in excursions led by experts.

As an important tourist attraction, the national park contributes vitally to the economy of the region. These financial benefits make it easier to win the cooperation of the people who live locally for its aims. The park's success is recognized internationally: in 1972 it was officially registered as a national park of international significance; in 1984 it was designated the first German Biosphere Reserve; and in 1986 it was awarded the European Diploma.

Forest wilderness About one quarter of the Bavarian Forest National Park has been left as wilderness area. Human interference is forbidden to allow the forest to develop naturally.

Red deer graze in a forest clearing. When the national park was established as a protected area in 1969 hunting was banned, and the red and roe deer populations have increased dramatically as a result. They are responsible for much damage to young trees.

A raised bog in the Bavarian Forest, formed from moss and plant remains. It supports plants such as bilberry, cranberry, carnivorous sundews and cotton grass.

MARSHES, FORESTS AND MOUNTAINS

REMNANTS OF AN ANCIENT WILDERNESS · PRESERVING THE LAST WILD PLACES · A POISONED LAND

Eastern Europe is a very complex region. The great North European Plain is bounded to the north by the Baltic Sea, with its drowned coastline of silted lagoons, spits of sand and somber conifer forests. To the south are the forested Carpathian Mountains and the Balkans, and beyond them the island-fringed Adriatic Sea. To the east lie the Ukrainian steppe, the Black Sea and the wetlands of the Danube delta with their rich wildlife. Great rivers such as the Danube and Tisza meander across the plain, bordered by dense forests and swampy floodplains. Despite centuries of human settlement, significant areas of almost primeval wilderness remain. Although unrestrained industrialization has caused enormous environmental damage, efforts are being made to protect these diverse habitats.

COUNTRIES IN THE REGION

Albania, Bulgaria, Czechoslovakia, East Germany, Hungary, Poland, Romania, Yugoslavia

Major protected area	Hectares
Aggtelek NP BR	19,595
Bialowieza NP BR WH	5,317
Biebrzanski NP	20,000
Dajti NP	3,000
Danube Delta NR	38,450
Djerdap NP	64,000
Durmitor NP WHS	33,000
Hortobagy NP	52,000
Kampinowski NP	35,486
Karaburun NR	12,000
Kiskunsag NP BR	30,630
Krkonose NP	38,500
Lake Ferto LPA BR	12,550
Middle Elbe BR	17,500
Ohrid WH	38,000
Pelister NP	12,500
Pilis LPA BR	23,000
Pirin NP WH	27,500
Plitvice NP	20,000
Retezat NP BR	20,000
Rosca–Letea R BR	18,150
Rügen/Hiddensee NR RS	25,800
Srebarna Lake R BR RS WH	600
Sumava NP	167,117
Tatra NP	21,164
Tatransky NP NR LPA	76,883
Trebonsko LPA BR	70,000
Vitocha/Bistrichko Branichte NP BR	26,500
Wolinski NR BR	4,844

BR=Biosphere Reserve; LPA=Landscape Protected Area; NP=National Park; NR=Nature Reserve; R=Reserve; RS=Ramsar site; WH=World Heritage site

REMNANTS OF AN ANCIENT WILDERNESS

The great crescent of the Carpathian Mountains contains some of Eastern Europe's best conserved forests and alpine habitats. Beech and spruce forests give way to upland peat bogs, broad subalpine meadows and precipitous peaks. Clearance for cultivation has reduced these forests to a mere fraction of their former size, and tourism and aerial pollution have had a harmful effect as well. However, the magnificent and ancient beech and spruce forests of Bulgaria, Romania and Yugoslavia still provide a sanctuary for bears, wolves, wildcats and lynx, though their numbers have been greatly reduced.

Scots pine and Norway spruce grow on the sands and light loam soils of the north. Beech forests are found in the lowlands, as well as in the upland areas of Czechoslovakia in the center of the region; and oriental beeches grow in Bulgaria and Romania. Hungarian and Turkey oaks are common in the south and species of lime are characteristic of the Balkans, while coastal maquis (degraded evergreen oak woodland) is widespread along the Adriatic coast of Yugoslavia and Albania. Forests of pines, silver fir and spruce survive in the southern mountains, with unique forests of Macedonian pine in Bulgaria.

Habitats shaped by human activity

A long history of cultivation and grazing has had a profound influence on the region's landscape and vegetation: there was little deciduous forest left in the north even by 3000 BC. Today, the northern plains have an agricultural and industrial landscape. Even the semi-natural habitats of the farmed plains of Hungary and Romania are maintained by cutting and grazing. The steppe grasslands of the Hungarian plain, called the puszta, were created by centuries of grazing. Both they and the traditionally farmed lowland meadows of the river valleys are rich in wildlife, including the great bustard, an increasingly rare bird.

Rivers, lakes and caves

The rivers of the region create dramatic landscapes. In Hungary, the great bend of the river Danube curves around a limestone region topped by the forests of the

A complex lakeland system stretches across the Wigry National Park in the Mazurian lakeland area of northeast Poland. There are over 9,000 lakes, ranging from large bodies of water to small woodland pools and marshy bogs. The lakes are home to beaver colonies that affect the wetland ecology with their dams and burrows. Huge populations of waterfowl are also common to the lakes.

Woodland habitats Eastern Europe was once extensively forested, but human agricultural and industrial activity over the centuries has reduced the forest areas dramatically. Today, the more remote areas of woodland are still relatively undisturbed and sustain populations of mammals such as bears, moose, lynx and beavers. These species have been unable to survive in Western Europe.

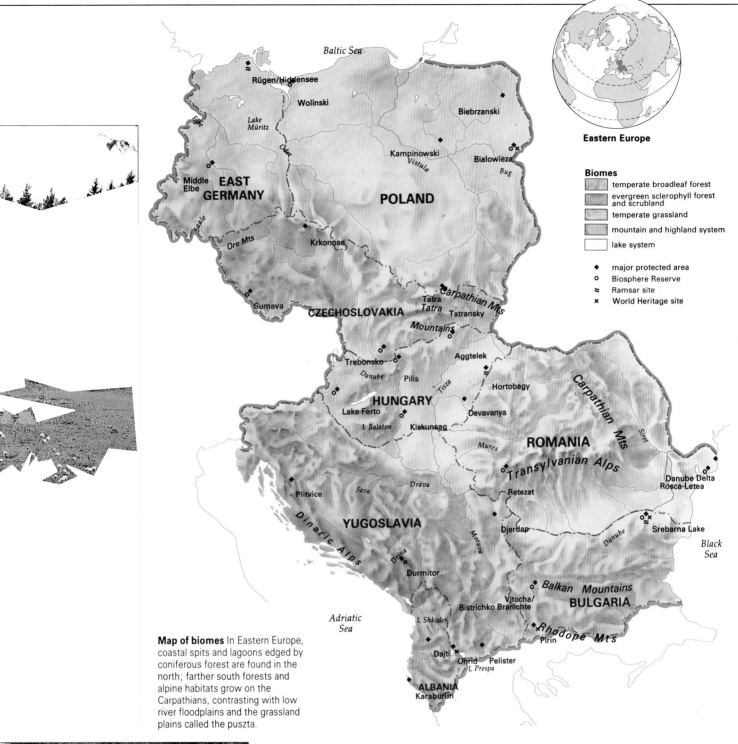

Map of biomes In Eastern Europe, coastal spits and lagoons edged by coniferous forest are found in the north; farther south forests and alpine habitats grow on the Carpathians, contrasting with low river floodplains and the grassland plains called the puszta.

Eastern Europe

Biomes

temperate broadleaf forest
evergreen sclerophyll forest and scrubland
temperate grassland
mountain and highland system
lake system

◆ major protected area
○ Biosphere Reserve
≈ Ramsar site
✕ World Heritage site

Pilis Landscape Reserve. The Tara river at Durmitor National Park in Yugoslavia forms a canyon 60 km (37 mi) long and 1,000 m (360 ft) deep.

The limestone uplands of Czechoslovakia and Yugoslavia conceal the world's third largest underground cave system, 23 km (14 mi) long and extending 8 km (5 mi) below the international frontier between Czechoslovakia and Hungary. The caves have evolved unique specially adapted animals, such as the blind olm (a kind of salamander).

The northern part of the European plain is a land of peat bogs, marshes and extensive lakes, whose pine-fringed reed beds provide food for fishing ospreys and large gatherings of geese and swans. High among the mountains of the south, the deep ancient lake Ohrid, located in the southern tip of Yugoslavia (a World Heritage site with a unique community of threadworms and snails that date back about 20 million years), contrasts with the shallow lakes of Prespa and Skadarsko. Their rich and varied fish species support birds such as pelicans, egrets and pygmy cormorants.

The northern, southern, western and eastern limits of many continental plant and animal species occur within the region. Despite its long history of human settlement, Eastern Europe has managed to retain many wild animals that have almost vanished from the west. Wolves, brown bears, moose, bison, lynx and beavers are still found in its deeper forests and remoter mountains.

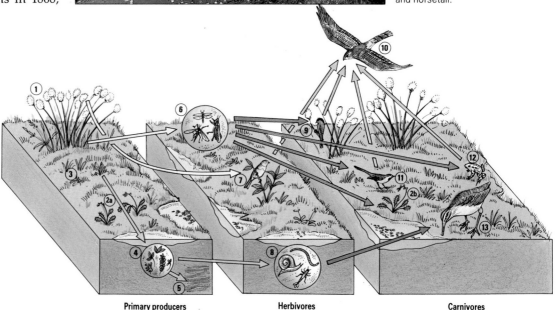

PRESERVING THE LAST WILD PLACES

Eastern Europe's great diversity of natural features are paralleled by differences in population densities, social and economic conditions, and cultural and historical backgrounds. This means that there are at least as many different approaches to the establishment and management of protected areas as there are countries within the region.

The idea of protected landscapes is popular in Eastern Europe, yet protected area coverage, strictly in terms of national parks, is insignificant. Because very few natural ecosystems have survived in the region's extensive lowland areas, as a result of farming, most national parks are in forested mountain areas. In Poland they cover only 0.4 percent of the land. But the situation is much improved when all protected areas are taken into consideration, including lowland semi-natural habitats which are less disturbed and certainly more extensive than similar habitats in the rest of Europe. By this computation, some 9 percent of the land area of Poland was protected in 1986, and there are plans to increase this figure to 28 percent by the year 2000.

The rise of conservation

For centuries conservation in Eastern Europe was sporadic, a byproduct of unrelated religious, economic or strategic considerations. Two species of alpines were given protection in the Tatra Mountains of the central Carpathians in 1868,

The river Biebrza meanders across northeast Poland. This area is of international importance for its meadow and peat bog habitats. The landscape is, in effect, managed by the traditional farming methods that continue to be practiced, and is famous for its winter populations of waterfowl and birds of prey.

Freshwater wetlands such as swamps, marshes, ponds and pools are mainly bird habitats as they form a barrier to terrestrial predators. The plant life includes submerged and floating vegetation such as pondweeds and duckweeds as well as sedge, rushes, reeds, water lilies and horsetail.

Components of the ecosystem
1 Cotton grass
2a and 2b Sundew
3 Sphagnum moss
4 Plant detritus
5 Peat
6 Herbivorous insects
7 Linnet
8 Worms, larvae, mollusks
9 Willow tit
10 Hen harrier
11 Gray wagtail
12 Common frog
13 Snipe

Energy flow
➩ primary producer/primary consumer
➩ primary/secondary consumer
➩ secondary/tertiary consumer
➩ dead material/consumer or decomposed
➩ death

A bog ecosystem is rich in invertebrates, the food supply for a great variety of specially adapted wetland birds.

Primary producers   Herbivores   Carnivores

and proposals to place the High Tatra, then within the Austro-Hungarian empire, under protection were put forward in 1876, a mere four years after the creation of the world's first national park, at Yellowstone in the United States.

Systematic policies of nature conservation developed only after World War I, when the region's present-day states first began to take shape. The emergence of the global conservation movement in the last 40 years, and sponsorship from organizations such as UNESCO, the United Nations Environment Program and, since 1980, the United Nations Economic Commission for Europe, have encouraged the process.

Eastern Europe was one of the first regions to develop national conservation programs, no doubt aided by its centralized planning systems. It has taken the lead in creating transborder protected areas such as the Sumava/Bavarian Forest National Park (Czechoslovakia/ West Germany) and the Tatra/Tatransky National Park (Poland/Czechoslovakia). Several countries – for example, Bulgaria, Czechoslovakia, East Germany and Poland – have reserves in the thousands; others have several hundred.

In all but two countries (Yugoslavia and Poland) protected areas are all state-owned. The more extensive reserves tend to be used for ecosystem and landscape protection and for limited recreation, while the smaller areas are used for scientific research. Sites of international importance have been designated under the World Heritage convention, the Ramsar Convention on wetland sites and UNESCO's Man and Biosphere Program.

The growth rate of protected areas in Eastern Europe has been impressive, but in this heavily farmed and industrialized landscape there is a limited number of even semi-natural sites to be preserved. However, progress can still be made in drawing up plans to protect areas that are influenced by human activity, but are nonetheless rich in wildlife.

A unique northern habitat
The meadows and peat bogs of northeast Poland, comprising some 80,000 ha (200,000 acres) of the valley of the river Biebrza, are important on a continental scale. Their value is maintained through a unique blend of traditional low-intensity farming (hay production and grazing of livestock in small, privately owned hold-

THE HUNGARIAN PUSZTA

In the center of Eastern Europe large tracts of forest give way to the flat sweep of the great plains of the Hungarian puszta, an area of 50,000 sq km (20,000 sq mi) of seemingly endless steppe grassland, an ancient salt desert that is unique in Europe. Within this expanse lie the famous Hortobagy National Park, the Kiskunsay National Park, the great bustard breeding areas of Devavanya and the major bird migration routes along the Danube and Tisza valleys. Steppes have existed for at least 10,000 years, and once stretched 8,000 km (5,000 mi) from the Danube basin to Mongolia and China. As a result of intensive agriculture and overgrazing they are now deteriorating and disappearing altogether.

The puszta has a characteristic landscape of treeless plains, saline steppes and salt lakes, and includes scattered sand dunes, low, wet forests and freshwater marshes along the floodplains of the ancient rivers. Mirages are a com-

mon sight, and mosquitos abound. Some 300 species of birds are found here, 140 of them breeding, and up to 5,000 cranes pass through on migration. A very rich diversity of invertebrates includes the steppe tarantula and many different species of dragonflies.

In ancient times the area served as an immense grazing land for the animals of nomadic herdsmen; indeed it was grazing that transformed the once wooded steppe into the present-day dry expanses of tussocky grassland. A number of ancient domestic breeds, such as the gray cattle, the Hortobagy sheep and the Hungarian sheep (the *racka*) still exist. Tourism in the area is increasing but the pastoral population, already small, is decreasing. This may prove to be a major problem in preserving the puszta's unique habitat.

The puszta is a unique grassland habitat now threatened by overgrazing and intensive agricultural practices.

ings) and seasonal flooding. The area shelters impressive populations of both breeding and overwintering waders, ducks, terns and birds of prey. Certain species such as the great snipe and the moustached warbler have their strongholds here, and a 2.5 ha (6 acre) plot of marsh has been privately purchased to protect the great snipe's display grounds. It forms Eastern Europe's first truly private nature reserve.

The most serious gaps in protection at present are in Albania, which is not party to any international conventions; and Czechoslovakia and Romania have yet to agree to the Ramsar wetland and World Heritage conventions. Important wetlands such as those of the Danube Delta, the Sava river wetlands in Yugoslavia and the forests along the Danube and Tisza rivers remain unprotected. Birds such as

cormorants, night herons, egrets, spoonbills and storks breed in these sites, as do several increasingly rare species of large birds of prey.

Tourism in Eastern Europe's national parks is especially popular since travel abroad has, until recently, been difficult. As many as 3.5 million people may visit Czechoslovakia's Tatransky National Park in a single year. Around 10 percent of each national park in Yugoslavia is designated for recreation development. At Plitvice National Park a large proportion of the park's running costs are financed from local activities. Although such use causes disturbance, some animal populations have benefited greatly from their protected status: bison herds have increased considerably, and beavers and white-tailed eagles have been successfully reintroduced to the region.

A POISONED LAND

Pollution is a major environmental issue in Eastern Europe, the result of unrestrained industrialization and the uncontrolled exploitation of water, minerals, soil, fossil fuels and timber. The scale of the disaster is enormous: life expectancy is becoming shorter, children's diseases and urban illnesses are increasing, industry is unable to function and nature is gradually dying.

Czechoslovakia and East Germany are the world's greatest per capita producers of sulfur dioxide, one of the prime contributors to acid rain, which can cause the death of trees. In 1983 it was estimated that two-thirds of Poland's forests have been seriously damaged. Added to this, over 1 million ha (2.5 million acres), one-third of Czechoslovakia's forests, have been seriously damaged; 75 percent of trees in the country's four national parks are dying; a quarter are already dead.

The threat to the rivers

Water pollution has turned the Vistula, Poland's largest river, into an open sewer; and the Oder, Elbe and Danube are not much better. Heavy metal pollution of rivers is a threat to human health, and also poisons aquatic and riverside life.

The fate of the Gabcikovo/Nagymaros stretch of the Danube near the Czech-Hungarian border is typical of what is happening to most of Eastern Europe's large rivers. Termed "Europe's Amazon", it was once a lush forest area of low-lying dense willow and poplar woods crisscrossed with small channels and ditches, islands and backwaters, giving way to marshes and meadows downriver. Now it is flushed not by the river's waters, but by waste of every type imaginable. Over 50 percent of the industrial and domestic waste from the nearby city of Bratislava is dumped untreated into the Danube. Pollutants include sulfuric acid and oil, which forms a sediment on the river bed, stifling life. As a result of irrigation schemes the water table has dropped by as much as 10 m (30 ft), resulting in the drying out of rare riverside forests; as much as 700 ha (1,750 acres) of forest have already disappeared.

Plans are already in progress to construct a hydroelectric power station here, together with a 20-km (12-mi) long lake, two dams, and a canal to re-route the

The island of Cres off Yugoslavia's Adriatic coast has an open landscape ideal for griffon vultures. The lack of cover allows them to spot their food source, carrion, over wide areas.

The acidification of lakes by acid rain pollutes the water, depriving wetland birds of their food source. Added to this, in Eastern Europe, industrial and agricultural waste is disposed of directly into the rivers.

Danube. The wildest part of the Danube will dry up, while the reduced river flow will prevent pollutants from being flushed away. The project will destroy habitats along more than 200 km (124 mi) of the Danube. Similar schemes on the Tisza in Hungary and the Iron Gates Gorge on the Yugoslav/Romanian border have caused huge problems.

Fighting back

New ministries for environmental affairs were created in Hungary, Poland and Bulgaria in 1989, and in Czechoslovakia in 1990, partly as a result of the pressure for political change in Eastern Europe, which finally brought an end to the communist monopoly of power. Pollution and environmental disasters – whether it be the fallout from the Chernobyl nuclear disaster of 1986, the poisoning of the Danube and Oder rivers, acid rain or coastal degradation – do not respect international or ideological boundaries. Co-operation between countries from both East and West to reduce pollution has increased, and is likely to grow even further. For example, West Germany has

THE DANUBE DELTA

The mighty Danube, after flowing 2,800 km (1,740 mi) through six countries, enters Romania and empties its waters into a 60 km (37 mi) wide fan of channels. These meander through willow forests and meadows, tracts of dense reed beds and shallow lakes, before flowing out through large coastal lagoons and around sand dunes and offshore islands into the Black Sea. This delta complex forms one of the largest wetland wildernesses in Europe.

There are three reserves located within the delta: Rosca-Letea, St Gheorghe-Palade-Perisor and Periteasca-Leahova, covering some 56,600 ha (139,800 acres) in all. The place is a paradise for water birds, with 160 breeding species. The luxuriant forest of the Rosca-Letea Biosphere Reserve is the home of a number of rare reptiles, including Orsini's viper, the Aesculapian snake and the eremias lizard. The floating islands of reeds contain 10 percent of the world's population of Dalmatian pelicans (150 pairs) and the world's largest colony of pygmy cormorants (12,000 pairs). The delta is a critical resting and feeding area for northern birds, with at times over 1 million migrants occupying the area at once. In some winters the delta holds the entire world population of the red-breasted goose.

The buildup of acidity in soils and water, due to the building of polders and the creation of fish farms covering 63,000 ha (156,000 acres), threaten the habitats of the delta. Plans to convert 20 percent of the wetland to polders have fortunately been delayed by financial and technical difficulties. Pollution of the Danube is also a serious problem. The port town of Sulina, which handles 3,500 ships a year, has been declared an ecological disaster area since 4,000 tons of toxic polychlorinated biphenyls (PCBs) were dumped in the delta.

Conservation across frontiers International cooperation has led to the establishment of the Tatra National Park, one of several East European transborder protected areas. It covers both Czechoslovakian and Polish land and is an important site for alpine plants.

recently supplied power station filters and other forms of technology to East Germany and Czechoslovakia.

Numerous ecological pressure groups have been formed, and have achieved some notable successes. For example, they have forced the closure of the Skawina aluminum plant in Poland and halted the construction of the Nagymoros nuclear power station in Hungary. Despite these initiatives, the environmental crises look as if they will continue for the foreseeable future while the economies of Eastern Europe remain in stagnation, international hard currency debts rise even higher, and social and political unrest persists. The cost of the environmental cleanup in East Germany is estimated at several hundred billion dollars. Ironically the best preserved natural areas are in the east of the region – precisely those areas worst affected by contamination from the Chernobyl nuclear disaster.

Today, the prospect for conservation appears far brighter than at any time since before World War II. The greatest need now is to promote scientific and technological solutions to environmental problems and to apply expert ecological advice to the highest levels of policy making. However, the rapid expansion of nuclear energy programs, put forward by the governments of Eastern European countries as the way to ensure a cleaner environment for the future, may – according to others – herald an even worse environmental nightmare.

Bialowieza National Park

Bialowieza Forest is situated in northeast Poland near the border with the Soviet Union, a region now known as the "green lung" of the country: its unique qualities were first recognized in legislation as early as 1541. The forest, which has been under the patronage of kings and tsars, is today a national park, a United Nations Biosphere Reserve and an internationally important World Heritage site.

The national park, covering only 5,316 ha (13,300 acres), lies in the center of an extensive forest complex of over 1,200 sq km (470 sq mi) and is recognized as a unique fragment of primeval lowland forest. There are 113 different plant communities in Bialowieza Forest, containing over 550 species of flowering plants (over a quarter of Poland's flora). Of the 26 tree species, lime, oak, hornbeam, pine, ash and alder are the most common; beech and yew are absent. Peat bogs, marshes,

meadows and aquatic communities form open spaces in the forest. The forest is a mixing ground for species, since it lies on the divide between the rivers that drain into the Baltic and the Black Sea. Consequently species of northern (Lapland willow), southern (silver fir), western (durmast oak) and eastern (dwarf birch) European trees are all found there.

The woodland is in a virgin state, never having been felled. In summer, the canopy is closed and little light filters to the forest floor. Individual trees may be between 250 and 400 years old and over 40 m (100 ft) in height. The completeness of the tree cover warms the climate within the forest, where there is less wind.

A conservation success story

The forest is probably best known for its European bison, which number some 260 individuals. Their success story is a result

The last European bison, or wisent, shelter in the woods of the Bialowieza National Park. Bison once roamed the continent in vast numbers, but loss of habitat and hunting decimated the populations. An international rescue program was launched in the 1920s and today there are over 250 individuals.

Ancient woodlands The Bialowieza National Park encompasses the last tract of primeval lowland woodland left in Europe. The forest is over 500 years old and some parts have never been felled or subjected to forest management. The forest contains a wealth of plant and tree species including pines, oaks, hornbeam, limes and ash. The animal life is equally rich and includes bison, wolves, otters, beavers and birds such as woodpeckers and eagles.

The Bialowieza Forest is a site of international importance. It was once a royal hunting ground, and has also provided shelter for Napoleon's troops and a hiding place for partisans during World War II. Now the forest is well-protected and has been designated a National Park, a World Heritage site and a United Nations Biosphere Reserve.

of international cooperation. When the last bison in the forest was killed by a poacher in 1919, joint British, German, Polish and Swedish efforts were made to locate captive animals originating from Bialowieza. Two cows (from Sweden) and a bull (from Germany) were brought home to the forest in 1929. Ten years later their numbers had increased to 14. In 1952 the first animals were released into the wild, but kept under constant observation. Every newborn bison is given a name and is entered into a pedigree book, a record maintained since 1932. Together with the 200 in the Soviet section of the forest, Bialowieza contains the world's largest herd of these forest giants.

The success shown with the bison is being repeated with the wild horse, the tarpan, which has been reintroduced into an area of the forest. In addition, the forest is home to wolves, lynx, moose, otters, beavers and 228 species of birds, over 160 of which breed. These include a large number of eagles, owls (9 species) and woodpeckers.

Just to the northeast of Bialowieza are the most extensive peat bogs remaining in Eastern Europe. They are home to the great snipe, aquatic warblers, Montagu's harrier, several eagles and wildfowl in their thousands. To the east and west of these wetlands is the dark and dense Mazurian forest, home of six species of woodpeckers, the Tengmalm's, Ural, pygmy and eagle owls and a range of birds of prey. This is without doubt a truly forgotten corner of Europe.

A VAST WILDERNESS

The Soviet Union contains some of the greatest stretches of untamed wilderness left on Earth. Extending from the Baltic Sea in the west to the fringes of the Pacific Ocean, from the frozen Arctic Ocean to the mountains of the Tien Shan and the Kara Kum desert in the south, this vast country encompasses an enormous range of habitats. They include broad, flat steppes, ice-packed Arctic coasts, desolate and remote tundra, seemingly endless forests, soaring mountains, hot springs and active volcanoes, scorching deserts, and numerous lakes and rivers. The Soviet Union is a land of superlatives. It has the world's largest forest, the taiga, which contains a third of the world's timber resources; the world's largest and deepest freshwater lake, Lake Baikal; and four of the longest rivers in Europe and Asia.

COUNTRIES IN THE REGION

Mongolia, Union of Soviet Socialist Republics

Major protected area	Hectares
Altai SR	863,861
Astrakhan SR BR	63,400
Baikal Region NP SR BR	1,564,019
Central Black Earth SR BR	4,795
Central Forest SR BR	21,348
Central Siberian SR	972,017
Chatkal Mountains BR	71,400
Chernomor SR BR	87,348
Kavkaz SR BR	263,477
Kronot SR BR	1,099,000
Kuril NR	65,365
Lake Sevan NP	150,000
Lapland SR BR	278,400
Matsalu R	48,634
Oka River BR	45,845
Pechoro-Ilych SR BR	721,322
Repetek SR BR	34,600
Sayano-Shushen SR BR	389,570
Sikhote-Alin SR BR	340,200
Sokhondin SR BR	31,053
Taymyr SR	1,348,316
Voronezh SR BR	21,348

BR=Biosphere Reserve; NP=National Park; R=Reserve; SR=State Reserve

THE CHANGING LANDSCAPE

Much of the Soviet Union consists of vast featureless plains stretching, monotonous and unchanging, as far as the eye can see. From the Ural Mountains west to the Baltic Sea, south to the Black Sea, and east to central Siberia they represent some of the largest expanses of flat land in the world. Plant and animal life is restricted by the climate, which over much of the northern plains is harsh and cold. Winter lasts more than half the year and permafrost covers nearly half the country. In the north the barren tundra is hard and frozen in winter, a mudflow-covered swamp in summer.

The tundra herbs are mostly lowgrowing shrubby plants that shelter beneath an insulating layer of snow in winter. In the most exposed areas only lichens, mosses and algae survive, except in sheltered cracks and hollows where snow accumulates and a few flowering plants can gain a foothold.

Polar bears, seals, walruses and reindeer inhabit the ice-fringed shores, treeless cold deserts and tundra of the northern mainland and the Arctic islands. In summer the tundra teems with breeding birds – ducks, geese and waders that have migrated north to take advantage of the long summer days, the plentiful supplies of fish and the insects that swarn around the pools of melted snow.

Further south, the tundra gives way to a great zone of coniferous forest, the taiga. This dark forest harbors a wide variety of animals, including moose, lynx, brown bears, wolverines, sables, capercaillies and hazel grouse. Dominated by pine, spruce, fir and larch for much of its range, the taiga in the far east contains many species more typical of China and Japan: the Korean pine, Khingan fir, Yeddo spruce and Japanese stone pine.

In the drier south, the trees of the taiga thin out into the Eurasian steppe, a vast, usually treeless and flat expanse that is dominated by drought-resistant grasses and sedges. In the moister western parts of the Russian plain the taiga is replaced by broadleaf forests of oak and beech, while in the far southeast, along the border with China, lying in the path of moisture-bearing trade winds, there grow broadleaf forests of Mongolian oak, hornbeam, lime and maple draped with lichens and climbing plants.

Much of the steppe grassland is now under cultivation, for this is the granary of the Soviet Union. Beyond the steppe, the grasses give way to semidesert and desert in the south, or are interrupted by the great ranges of the Caucasus, Pamir,

The taiga in winter Snow highlights the bare branches of larches, which shed their leaves in winter, unlike the other coniferous trees of the taiga zone. Larches are found mainly in eastern Siberia; dark, evergreen pines and spruce predominate elsewhere. The sloping shape of the trees prevents their branches breaking under the weight of snow.

A land of quietly flowing rivers In the lowlying plains of central Asia the northern forests give way to empty stretches of open, undulating steppeland interlaced by long, winding rivers and dotted with reed-fringed pools and marshes. These provide valuable breeding and feeding grounds for cranes, as well as waders and waterfowl.

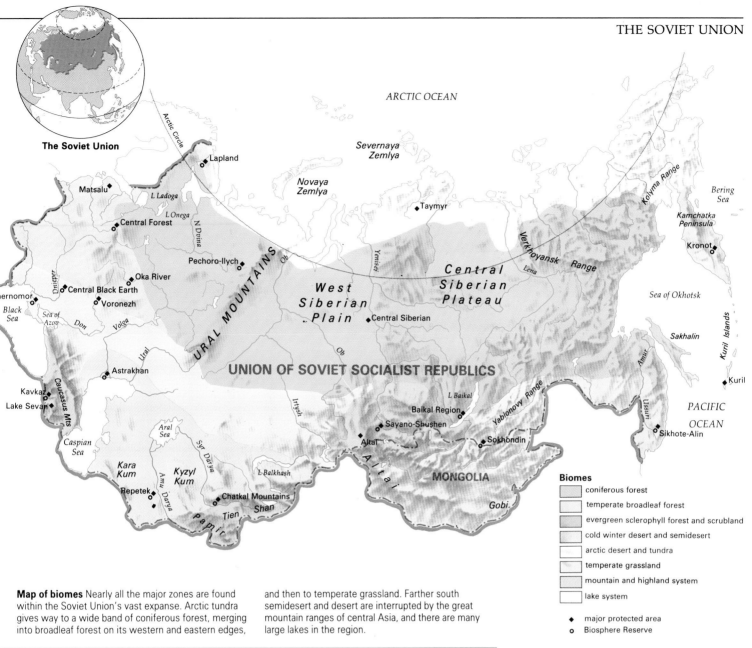

The Soviet Union

ARCTIC OCEAN

Lapland
Matsalu
L Ladoga
L Onega
Central Forest
Pechoro-Ilych
Oka River
Central Black Earth
Voronezh
Chernomor
Black Sea
Sea of Azov
Don
Volga
Astrakhan
Kavkaz
Lake Sevan
Caucasus Mts
Caspian Sea
Kara Kum
Repetek
Aral Sea
Kyzyl Kum
Syr Darya
Amu Darya
Chatkal Mountains
Tien Shan
Pamir
L Balkhash

Severnaya Zemlya
Novaya Zemlya
Arctic Circle
N Dvina
Ural
URAL MOUNTAINS
Ob
West Siberian Plain
Central Siberian
Irtysh
Ob
Altai

Taymyr
Yenisei
Central Siberian Plateau
Lena
Verkhoyansk Range
Kolyma Range
Sea of Okhotsk
Sakhalin

UNION OF SOVIET SOCIALIST REPUBLICS

L Baikal
Baikal Region
Sayano-Shushen
Sokhondin
Yablonovy Range
MONGOLIA
Gobi

Bering Sea
Kamchatka Peninsula
Kronot
Kuril Islands
Amur
Ussuri
PACIFIC OCEAN
Sikhote-Alin
Kuril

Biomes

- coniferous forest
- temperate broadleaf forest
- evergreen sclerophyll forest and scrubland
- cold winter desert and semidesert
- arctic desert and tundra
- temperate grassland
- mountain and highland system
- lake system

- ◆ major protected area
- ○ Biosphere Reserve

Map of biomes Nearly all the major zones are found within the Soviet Union's vast expanse. Arctic tundra gives way to a wide band of coniferous forest, merging into broadleaf forest on its western and eastern edges, and then to temperate grassland. Farther south semidesert and desert are interrupted by the great mountain ranges of central Asia, and there are many large lakes in the region.

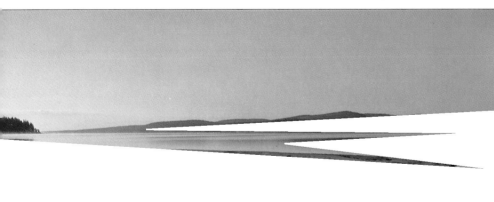

Tien Shan, Altai and Yablonovy mountains, with their mixed woodlands, alpine meadows and their high-altitude deserts. Beyond the peaks lie subtropical Georgia in the southwest and the barren lands of Mongolia and the Gobi desert in the east.

Centers of diversity

The Soviet Union contains over 21,000 species of plants, of which 6 percent are found nowhere else in the world. During the last ice age, which ended about 10,000 years ago, ice covered much of the northern part of the country, and plant and animal species migrated south.

As the land warmed after the ice age, other species migrated north into the region from Europe and Asia. In the mountains of the far southeast, and in the volcanic Kamchatka Peninsula and the Kuril Islands, these invading plants and animals found a wide range of habitats – mountains, river valleys, islands, moist monsoon-soaked coasts and volcanic hillsides. As a result, they diversified to form many new, or endemic, species.

159

PRESERVING THE WILDERNESS

By the late 1980s, the Soviet Union boasted a network of 150 or so nature reserves. Covering over 175,000 sq km (67,550 sq mi), they are carefully distributed throughout the 15 republics in a deliberate effort to preserve the best representative examples of the region's typical ecosystems. Huge though this figure is, it represents less than 1 percent of the Soviet Union's total land area. An increase of 40 percent in the area of reserves has been promised by the year 2000. The Taymyr Nature Reserve in Siberia, created in 1979, is the largest in the country, covering 1.3 million ha (3.2 million acres); second in area is the Central Siberian Reserve, which covers some 972,000 ha (2,400,000 acres).

The 1977 Soviet constitution outlines the duty of both state and citizens to protect nature, and reflects a long tradition of interest in conservation. The country's earliest environmental laws were passed during the 17th century in the form of hunting, land-use and forestry regulations. The first conservation committees emerged in 1909, and the All-Russian Society for the Protection of Nature was created in 1924. Today the state's ecological services cover most of the Soviet Union.

Degrees of protection
There are more than 60 types of protected areas in the Soviet Union. The most extensive are the nature reserves, which are primarily used for scientific research. They have permanent staffs of between 25 and 50 specialists and researchers working in their laboratories, museums and monitoring stations. Within the nature reserves, economic activity is forbidden, and tourists and nonspecialist visitors are refused admission.

Many nature reserves have been established to preserve threatened or endangered species, such as the bison in Oka River Valley Nature Reserve (which also has a breeding station for birds of prey and Siberian cranes). Other nature reserves in the Soviet network extend protection to the Russian desman (a mammal related to the mole, that has been hunted close to extinction for its fur), the onager and goitered gazelle, red-breasted goose and Far East skink, a species of lizard. Yet others specialize in

Steam rises from a hot spring in a waterfilled crater in the Kamchatka Peninsula. Many endemic species flourish in this volcanically active area, adding to the diversity of wildlife in the Soviet far east.

A waterlogged landscape Surface water accumulates after the spring snowmelt and cannot drain away if permafrost lies beneath the surface. Only tussocky grass grows in the permanently waterlogged ground.

research into such diverse topics as agricultural land use, recreational use and desert ecology. International cooperation between the Soviet Union and the United States has led to the twinning of reserves for comparative research purposes.

Nature sanctuaries are areas where protection is extended on a temporary basis at certain times of year, especially if it contains breeding colonies of birds. There are more than 2,700 of them, covering more than 4 million sq km (1.5 million sq

MIXING THE SPECIES

The wildlife of the eastern edge of the Soviet Union is particularly diverse and has been given special protection by a network of state nature reserves. This narrow belt of land, 4,500 km (2,800 mi) long, stretches north from the Siberian tiger's stronghold in the Sikhote-Alin Mountains near China to the Koryak Mountains and the Kamchatka Peninsula bordering the Bering Sea. It encompasses a great variety of habitats, from taiga and tundra to meadows, warm temperate woodland and marshes, and from moist coastal plains to volcanic springs and high mountaintops. The area is the meeting point of Siberian, Chinese, Indo-Malayan, Mongolian and European species.

The taiga meets the subtropics in the area of the Ussuri river where a unique plant life is found. Ginseng, Amur lilac,

Amur maackia and a wealth of lianas all thrive in the forests of monumental black fir, Korean pine, Manchurian ash and walnut and Amur cork oak. The forests are inhabited by leopards, bears, the mysterious musk deer and over 340 species of birds.

In the shadow of volcanoes in the Krono State Nature Reserve on the eastern side of the Kamchatka Peninsula, 20 active geysers add to the diversity of habitats. Here endemic species such as the Kamchatka fir and the 3-m (9-ft) tall Kamchatka nettle are present.

More endemic species, such as the Far East skink and the Kunashir grass snake, are protected in the recently created Kunashir Nature Reserve in the Kuril Islands, which extend south of the Kamchatka Peninsula between the Sea of Okhotsk and the Pacific Ocean.

mi). Economic activity is permitted but controlled within their boundaries.

The Soviet Union's 18 national parks cover 147,400 sq km (56,900 acres), and are designed to combine conservation objectives with recreational activities. There are also national hunting reserves and thousands of natural monuments. In addition to protected areas, green zones are being declared, which will raise the total area of reserved lands to 240,000 sq km (92,660 sq mi).

The administrative bureaucracy is being streamlined and simplified. In 1988 a super-office for nature conservation was created – the State Committee for Nature

Conservation (Goskompriroda) – which took over tasks previously spread among more than 300 different organizations. Nongovernmental conservation organizations are springing up, with the creation of the Ecological Project Center, the Moscow Center of Ecological Youth Movement and the Association for the Support of Ecological Initiatives.

An international heritage

The Soviet Union has many protected areas of international importance. At the end of the 1980s there were 18 Biosphere Reserves, covering 92,000 sq km (35,500 sq mi), designed to preserve the genetic

Components of the ecosystem
1 Plant plankton
2 Animal plankton
3 Herbivorous protist
4 and 5 Freshwater shrimps
6 Osprey
7 Lake Baikal seal
8 Pike
9 Carp
10 Transparent bottom-dwelling fish

Energy flow
⇨ primary producer/primary consumer
⇨ primary/secondary consumer
⇨ secondary/tertiary consumer
⇨ dead material/consumer

A freshwater lake ecosystem, such as that of the famous Lake Baikal, is based on algae and detritus from lakeside plants.

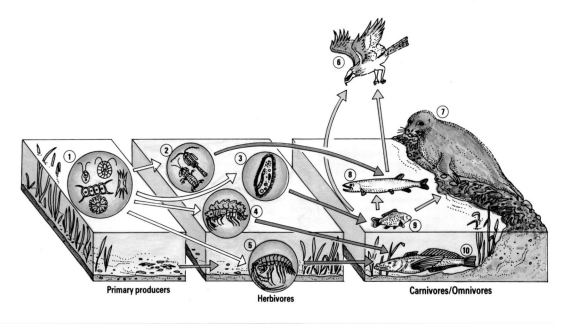

Primary producers

Herbivores

Carnivores/Omnivores

diversity of the plants and animals living there, and strictly controlled to facilitate scientific research programs. These Biosphere Reserves range from the Kavkaz Reserve, which protects the many species endemic to the area, to the Repetek Reserve in the sandy wastes of the Kara Kum, center of much research into desert ecology, and the Chatkal Mountain Reserve high in the lofty Tien Shan range on the border with China, another area rich in endemic species.

Under the international Ramsar Convention, 12 important wetland sites have been proclaimed as protected areas. These include the vast reed beds of the Matsalu Reserve in Estonia in the northwest and the Astrakhan Reserve situated on the Volga delta where it drains into the Caspian Sea. The latter is the summer molting retreat for many ducks and geese and an important spring and autumn stopover point for 5–7 million migrating waterfowl, as well as being the breeding ground for huge colonies of birds and a spawning ground for fish.

In addition, the Soviet Union boasts a number of successful captive-breeding programs, and considerable progress has been made in reintroducing captive-bred animals such as musk oxen into areas where the wild populations have been reduced and are close to extinction.

The Caucasus mountains are the most western of the great ranges that form a climatic barrier between the north and the south. Their highest slopes are a refuge for plants isolated after the last ice age, and lower down they are covered with forests, which include oak, chestnut and Caucasian fir.

PROBLEMS ON A GRAND SCALE

The 1986 nuclear accident at Chernobyl, near Kiev, brought home the danger of the complacency with which people treat nature. In less dramatic ways humans intrude into even the most remote parts of the Soviet Union. The untouched Siberia of popular imagination is today only an illusion, and with the building of the Baikal–Amur Mainline (BAM) railroad even the east of the Soviet Union has been opened up for exploitation. Seas and lakes have been poisoned, forests covering entire mountain ranges burnt, and the air in over a hundred major cities is polluted. Widespread soil erosion has caused desertification and the creation of massive dustbowls that affect an area equal in size to western Europe.

An indication that attitudes are beginning to change comes with the creation of the State Committee for Nature Conservation at the highest level of Soviet bureaucracy. Its task is to ensure that environmental protection extends throughout the Soviet Union. Despite this growing awareness, however, the nature reserve laws are still broken. Oil prospecting, domestic animal grazing, overfishing, trapping, hunting, tourism and illegal building encroach on several reserves.

The threat to the taiga

One problem that is likely to intensify is the economic exploitation of that greatest of wildernesses – Siberia. A highly visible sign of this is the aftermath of logging in the taiga forest. As much as one-fifth of the newly exposed soil may be washed away by rain after felling, and uncontrolled fires abound in the recently cleared areas. In the mid-1980s more forest was burnt in Irkutsk province alone than was planted throughout the whole of the Soviet Union. In 1987 more than 600 fires took hold and blazed for a month and a half before being checked.

In recent years oil spillages have greatly increased pollution in Eurasia's largest freshwater basin, Lake Baikal, and the general opening up of the Siberian wilderness has led to increased air, soil and water pollution. One result has been the southward retreat of the forest zone: loss of lichen and moss, plants that are highly vulnerable to airborne pollution, results in cooling of the soil, and a consequent fall in air temperature and

increase in wind, conditions that are less favorable for the growth of trees.

These developments are connected with the great push to complete the building of that vital artery of expansion into eastern Siberia, the BAM railroad. The BAM has a development zone of 300–500 km (185–310 mi) on either side of its 4,670 km (2,900 mi) of track. The railroad's construction will affect an area equivalent to the Amazon Basin. Exploitation zones for oil, gas, gold and minerals cover approximately 3.5 million sq km (1.35 million sq mi) – an area the size of Europe. The scale of Siberia's oil and gas exploitation is now being questioned by the Soviet parliament.

In Siberia waste disposal also poses difficulties. The low temperatures of the area slow down bacterial action so that the natural breakdown of waste takes longer. Siberian rivers need to flow some 1,500 km (930 mi) before they are cleansed, compared with 200–300 km (125–185 mi) for rivers elsewhere.

"Desert forest" in the Repetek Nature Reserve. The roots of shrubby trees, predominantly black saxaul, retain the shifting sands of the Kara Kum desert. Many mammals and reptiles find shelter and food beneath their branches; in spring the desert becomes a sea of color as the shrubs burst into flower.

The silver trunks of birches are reflected in spring floodwater. They grow extensively on the margin between the tundra and taiga where conditions are too wet or altitude too great for conifers to flourish. They are often the first trees to colonize cleared woodland spaces.

REPETEK NATURE RESERVE

Quiet reigns in the empty corner of the Kara Kum desert where the Repetek Nature Reserve is situated. The main visitors are scientists, for the reserve – which includes more than 30,000 ha (75,000 acres) of unspoilt sand desert with 10-m (33-ft) high sand dunes – is the Soviet Union's foremost center for research on desert ecology.

Repetek Nature Reserve was created in 1928 as a place to study quicksands. Investigations now cover the entire desert ecosystem, so that the work carried out here will benefit desert protection more generally: the strict habitat protection afforded by Soviet nature reserves creates the ideal conditions for such research. Studies initially conducted as part of the international Biological Program, and later under the UNESCO Man and Biosphere Program, brought the center worldwide attention. In 1978, on the 50th anniversary of its foundation, Repetek was designated a Biosphere Reserve.

The reserve's expert staff of about 20 people run sessions for overseas scientists as part of the United Nations Environment Program (UNEP). Courses deal with the control of shifting sands and the ecology of desert pastures.

Repetek's desert habitats include creeping sand ridges hundreds of meters long that are entirely free of vegetation. Where there are valley depressions in the dunes, leafless, drought-resistant black saxauls grow, creating a unique "desert forest".

More than a third of the 211 plant species recorded in Repetek are found nowhere else in the world. These "forests" provide shade during the day for large, rare mammals such as the goitered gazelle, the caracal and the marbled polecat. Smaller jerboas, ground squirrels and shrews stay in the cool of their underground burrows during the day and emerge at night. The reserve boasts 23 reptile species including the desert monitor.

Re-routing the rivers

No less epic than the BAM undertaking was the plan to divert north-flowing Siberian rivers to flow south to irrigate the dry southern deserts of Central Asia – the unprecedented long-distance transfer of an estimated 120 cu km (29 cu mi) of water a year. Although the scheme was apparently abandoned in 1987, political and nationalist pressures have forced some compromises, and limited diversions of the rivers may yet take place. The consequences of such a gigantic project are unknown, but similar, smaller projects affecting the Aral, Caspian and Azov seas have had devastating environmental effects, causing drastic shrinkage of these lakes and widespread soil erosion and desertification.

A growing awareness

In the Soviet Union all land and resources are under the control of the state, but the new thinking under *perestroika* has encouraged greater environmental awareness at all levels. Pressure groups have proliferated and the issues are now fully discussed and examined in the media. Scientists and academics are also free to voice their concerns. There is a Soviet Greenpeace movement, and environmental groups, lobbying against local centers of industrial pollution, were a powerful force in the development of nationalism in the Baltic states.

As in the West, natural history is popular with the general public: each showing of a natural history program on television is watched by a quarter of the vast Soviet audience.

Lake Baikal Region Biosphere Reserve

Lake Baikal lies in the heart of Siberia. Russians call it "a part of their souls"; its water is of exceptional purity and clarity. Formed over 25 million years ago, Lake Baikal is the world's deepest (1,620 m/ 5,314 ft) and largest inland body of fresh water with an area of 31,500 sq km (12,200 sq mi). Its 23,000 cu km (5,500 cu mi) of water are equivalent to a fifth of the world's fresh water and four-fifths of the Soviet Union's surface fresh water. More than 330 rivers enter the lake's catchment area of 600,000 sq km (231,600 sq mi), but only one, the Angara, leaves it.

There are three state nature reserves on its shores, the Baikal and the Barguzin Reserves, both created in 1916, and the Baikalo-Len Reserve, designated in 1986. In the latter year the Pribaikal and Zabaikal National Parks were added.

Between them the reserves cover a large area of the shores around the lake and the surrounding mountains.

Numerous hot springs are refuges for plant species that remained after other species were forced to retreat south with the onset of the last glacial period some 60–70,000 years ago. Much of the area is covered in taiga, dominated by larch, pine and Siberian stone pine. These dense forests are inhabited by sables, Siberian red deer, musk deer, bears, lynx and wolves. Woodlands of poplar, willow and cherry grow in the more sheltered valleys. Above the forests, elfin cedar woodland gives way to alpine meadows and wind-resistant plants.

The climate is severe. In the high parts temperatures remain below freezing for more than 200 days a year. For five

months of the year the lake is covered in 2.5 m (8 ft) or more of ice, though many of the streams that supply it, warmed by the geothermal springs, remain free of ice.

Lake Baikal Biosphere Reserve contains around 1,000 species of animals and 400 species of plants that are not found anywhere else in the world. There are also some 50 species of fish, many of them endemic too. Its most famous inhabitant is the small Baikal seal, but the key to the ecology of Baikal is the epishura, a tiny crustacean unique to the lake. The epishura comprises up to 98 percent of all the minute animal life (zooplankton) in the water. It feeds on algae and is a major link in the food chain that supports Baikal's unique forms of life. In addition, the epishura acts as a filter capable of extracting 250,000 tonnes of calcium each

The "blue eye of Siberia" The sun sets over Lake Baikal, which is covered in thick ice during the winter. The largest body of fresh water in the world, it supports thousands of unique species of animals and plants.

Dense taiga forest comes right down to the unspoiled shore. The reserves surrounding the lake protect many rare mammals, especially the sable, a large relative of the weasel hunted for its fur.

The Baikal seal, an endemic species, is related to the Arctic ringed seal. It is thought to have arrived in Lake Baikal before it was cut off from the sea.

year: water as pure as that of Lake Baikal is almost unknown elsewhere on Earth. Forests of green freshwater sponges and no fewer than 258 species of shrimps also contribute to the filtration process.

The Barguzin Nature Reserve was created specifically to protect an endemic species of sable, much hunted for its highly valued fur, which earned it the name "soft gold". Today the population has risen to a healthy 2,500 animals. However, the rapid growth in tourism has brought new dangers: the shores of the lake are littered with waste, while sheep have taken over in areas once grazed by moose and wolves. Fire and soil erosion, the inevitable accompaniment of human intrusion in the taiga, are a constant hazard. Pollution of the lake is blamed for a decline in the numbers of

birds that use it as a stopover point or breed along its reed-fringed shores, and for an even sharper fall in birds of prey.

Apart from the research staff, there are no people living in the reserves, and no visitors. So remote is this wilderness that it took more than four years to lay the first tracks for the BAM railroad. But isolation is no protection against the pollution of industrial development.

The building of paper and pulp factories in Ulan-Ude and along the Selenga and Barguzin rivers southeast and east of the lake, together with thermal power plants, chemical and petrochemical installations and the BAM railroad, have destroyed life on some 2,000 ha (5,500 acres) of the lake bottom. The consequent reduction of the annual Baikal fish catch has undermined the traditional way of life

carried on by the local people, which was based on fishing.

In the 1980s a comprehensive protection plan for the lake was put through by the Central Committee of the Communist Party. As a result, production facilities began to be closed, lead and zinc waste was diverted, and the inflow of polluted water was completely stopped. A number of government officials were removed from their posts for grave environmental negligence. A special zone 1.6 km (1 mi) wide was created on either side of the BAM to protect air, water, soil and vegetation from pollution. The threat is by no means ended, but the greater freedom of expression now allowed to environmental scientists, pressure groups and the media means that the future of Lake Baikal looks brighter than it has done in the past.

LIVING DESERTS

HARSH ENVIRONMENTS · CONSERVATION AND TRADITION · A VARIETY OF PRESSURES

The habitats of the Middle East range from the rich coral reefs and mangrove forests of the Red Sea and The Gulf – among the finest in the world – to the snow-covered mountains of eastern Turkey and Iran and the Hindu Kush in Afghanistan. In the south and east the landscape is dominated by a vast desert, part of the extensive semiarid to extremely arid area of the Indo-Saharan desert belt that stretches from the Canary Islands off the west coast of Africa all the way to the Indian subcontinent. In the north and western parts of the region, where a more temperate climate prevails, a lusher vegetation is found; evergreen thorny scrub fringes the Mediterranean coast. Oases and lakes in this arid region provide important refuges for migratory birds making their way from Europe to eastern Africa.

COUNTRIES IN THE REGION

Afghanistan, Bahrain, Iran, Iraq, Israel, Jordan, Kuwait, Lebanon, Oman, Qatar, Saudi Arabia, Syria, Turkey, United Arab Emirates, Yemen

Major protected area	Hectares
Ajar Valley WiR	40,000
Al-Areen WiR	800
Asir NP	415,000
Azraq/Shaumari R	32,000
Bande Amir NP	41,000
Bentael NP	810
Beydaglari NP	69,800
Dilek Peninsula NP	10,985
Ein Gedi NR	2,780
Farasan Archipelago PA	60,000
Golestan NP RS	91,895
Harrat al Harrah R	137,750
Jabal al-Ara'is R	None given
Jebel Hafit R	11,700
Jiddat al Harasis NNR	2,750,000
Kavir NP PA BR	420,000
Kuscenneti NP	52
Mount Carmel NR	30,900
Mount Meron NR	10,117
Munzur NP	42,800
Qurm (Sultan Qaboos) NNR	1,000
Ramon (Negev) NR	100,000
Touran PA BR	1,295,400
Uludag NP	11,338
Uromiyeh Lake NP RS BR	1,295,400
Wadi al-Azib (proposed) NR	24,200
Wadi Serin-Jabal Aswad R	53,000

BR=Biosphere Reserve; NNR=National Nature Reserve; NP=National Park; NR=Nature Reserve; PA=Protected Area; R=Reserve; RS=Ramsar site; WiR=Wildlife Reserve

HARSH ENVIRONMENTS

Deserts are not so uniform and unvarying as their popular image suggests. Those of the Middle East include the plateaus of central Iran and Afghanistan, which are subjected to extremely cold winters, the more temperate deserts of Syria and Iraq, the extreme heat and aridity of the so-called Empty Quarter of central Saudi Arabia and Oman, and the flat, humid Red Sea coastal belt known as the Tihamah. These areas are clothed with a sparse shrub cover of saltbush and artemisia, and with doum palms and acacia groves near oases or watercourses (wadis) where the water table is close to the surface. Whenever rain falls in the desert, spectacular if short-lived shows of annuals such as poppies, daisies and bromegrass flower, as do brightly-colored perennials like irises, tulips and gladioli. These will have survived the drought by storing food in swollen roots or stems beneath the ground.

Water in an arid region

There are only three substantial rivers in the Middle East, which are fed by rainfall and melting snow from highland areas. All are associated with ancient periods of human occupation and cultivation: the Euphrates and Tigris in Iraq, and the Amu Darya along the northern border of Afghanistan. A lesser river, the Jordan, flows for only 205 km (125 mi) from the Sea of Galilee south to the highly saline and lowlying, inland Dead Sea.

There are also huge inland lakes and impressive oases, justly renowned for their spectacular scenic beauty. They range from Lake Van in Turkey, whose brackish waters support only one species of fish, to the torrid marshes and water-filled basins of the Euphrates delta and Yabrin oasis in Saudi Arabia. These are fringed by thick vegetation, surmounted by the dark fans of palm trees. The greenery contrasts stongly with the rugged ground surrounding the oases.

The presence of open water in the desert attracts huge numbers of migratory birds such as ducks, swallows, warblers and thrushes, seeking rest, food and shelter on their long journey between Europe and eastern Africa. In recent years birds such as quails and buntings have benefited from the increasing number of artificial oases created by irrigation schemes, notably around Al-Qassim in Saudi Arabia. These pump up water from permeable rocks, or aquifers, a kilometer or more below the surface, and use it for growing crops. Much farther to the north, the Bosporus in Turkey is a crossing point

Spring in the mountains An upland valley in Syria is carpeted with flowering herbs and fresh grass growth. Not all the Middle East is desert. These uplands on the Mediterranean fringe of the region are high enough to catch rainfall, and support lush pastureland.

A typical desert scene Bushes are strung out like necklaces along the courses of dried up streams. These remain moist after the surrounding land has become parched, enabling the bushes – an important source of food and shelter for animals – to survive.

The Middle East

Map of biomes The Middle East is predominantly desert and semidesert. There are mountains on its northern edge and evergreen scrub along the Mediterranean coast. The natural forest cover was once more extensive over much of the region than it is now, particularly in the northwest.

Biomes

- temperate broadleaf forest
- evergreen sclerophyll forest and scrubland
- warm desert and semidesert
- cold winter desert and semidesert
- mountain and highland system
- lake system

- ◆ major protected area
- ○ Biosphere Reserve
- ≈ Ramsar site

for large migratory birds such as eagles and storks; they can thereby avoid the open Mediterranean Sea on their journeys between Europe and Africa.

The green fringes

The Middle East habitats that are richest in plant and animal species are the thorny, drought-tolerant Mediterranean scrub that fringes the coasts of Turkey, Syria, Lebanon, Jordan and Israel, and the cool, shady, deciduous woodlands of beech, oak and hornbeam found on the plains and hills of west and central Turkey. Higher ground in this part of the region also supports significant woodlands of evergreen juniper, cedar and pine, though these have been badly ravaged by clearance for agriculture and grazing over the centuries.

In the southwestern part of the Arabian Peninsula, the Asir mountains and the Yemen Highlands catch the rain-bearing northwesterly monsoon winds and rise high enough to condense mists. Consequently the mountain crests support ancient juniper forests. These contain many species of plants and animals that are found nowhere else (endemic species).

The mountain ranges of northeastern Turkey and Iran and the Hindu Kush in Afghanistan are generally snow-covered in winter; some peaks are under snow all the year round. Their pine and juniper forests gradually give way to alpine meadows. These habitats, too, contain many unique plants that have evolved in isolation, cut off from neighboring areas by the intervening desert. In much of Afghanistan, however, the mountains support little vegetation, or none at all.

CONSERVATION AND TRADITION

Turkey, Iran and Israel have possessed national parks, nature reserves and other forms of protected areas for quite a long time, but the practice is less established in Arabia. In general, for a region of its size and relatively low population (or perhaps because of them), the Middle East does not enjoy many protected areas of internationally recognized standards.

The International Union for Conservation of Nature and Natural Resources (IUCN) lists only 46 national parks or equivalent protected areas in the Middle East that meet its criteria of legal status and management objectives. These are distributed between Iran (24), Turkey (15), Israel (5), Oman (1) and Saudi Arabia (1). Together they total some 38,400 sq km (15,000 sq mi) – hardly more than half a percent of the entire area of the region. Even these figures are based on the perhaps optimistic assumption that Iran's protected areas, which were subjected to heavy pressure from grazing (and in some cases military use) after the Islamic revolution of 1979, continue to meet international standards.

The Islamic tradition of conservation

Throughout the Middle East, however, traditional mechanisms for protecting wildlife and natural habitats exist, derived from the Koran and Islamic law (*sharia*). The seemingly meager number of protected areas is augmented by Islam's practice of setting aside certain sites for reasons not directly concerned with nature conservation. An example of such site protection is the *hema*. Under Islamic law this is a reserve established by the ruling authority for the public good; thousands of them exist throughout the Arabian Peninsula, though the advent of modern agriculture and the declining power of the local emirates have caused their numbers to fall. Because *hemas* were set up by and for local communities, they have proved to be enduring sites, where wild animals and plants have been able to survive as the surrounding areas have gradually deteriorated.

In southwest Saudi Arabia *hemas* are still an important part of the traditional land management system. They vary in size from 10 to 1,000 ha (25 to 2,500 acres), averaging about 250 ha (620 acres), and usually occupy part or all of a wadi system or a hillside, where their boundaries can be clearly defined; otherwise stone markers or a wall may be used. *Hemas* are of several types: there are those where stock grazing and grass cutting are prohibited but the cutting of fodder is per

Components of the ecosystem
1 Sodom apple plant
2 Desert grass
3 Seeds and detritus
4 Spotted sandgrouse
5 Dorcas gazelle
6 Desert jerboa
7 Desert locust
8 Feces
9 Dung beetle
10 Egyptian vulture
11 Horned viper
12 Fennec fox
13 Scorpion

Energy flow
⇛ primary producer/primary consumer
⇛ primary/secondary consumer
⇛ dead material and detritus/consumer

A dune ecosystem has little material to decompose. Many animals adapt to these limited resources by adopting nomadic feeding patterns.

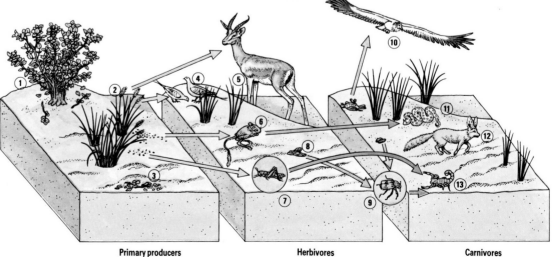

Primary producers   **Herbivores**   **Carnivores**

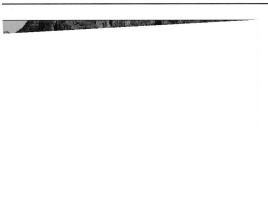

A lone acacia tree growing by the shore of the Dead Sea survives by tapping supplies of underground water. It is adapted to withstand drought, and its thorns give it protection from grazing animals.

Zigana Gorge, near Erzurum in northeast Turkey. The river greens a narrow strip on each bank, as does the dry stream bed, providing a contrast with the sparse vegetation of the surrounding countryside.

The cedars of Lebanon, mentioned in the Bible, were prized in ancient times, and the Romans created a mountain reserve for them. Today only about 400 trees survive on Mount Lebanon, but considerable forests remain farther north in the Taurus mountains.

mitted during drought years; where stock grazing and cutting are permitted on a seasonal basis; where stock grazing is restricted to certain animals only; and areas that are set aside for bee-keeping or are protected to preserve trees.

Other traditional forms of conservation in Arabia include protected zones bordering streams, ponds, lakes and other bodies of water. Here, building and farming are prohibited, and a series of injunctions concerning people's responsibilities toward the exploitation of captive and game animals is enforced. Around the sacred towns of Mecca and Medina there are large sanctuaries called Holy Harams within which the disturbance of wildlife and destruction of natural vegetation are strictly forbidden. Finally, individuals or corporations may make a charitable endowment (*waqf*) of land as a tribute to God, which is held in trust for the benefit of future generations.

Harnessing tradition for the future

It is only quite recently that Islam's traditional respect for wildlife has been employed to help to establish the protected areas the Middle East needs so badly. In Oman, for example, the local

CONSERVATION IN WAR ZONES

Few regions of the world have experienced more conflict than the Middle East has in recent decades. Inevitably the protected areas and wildlife habitats suffer. For example, the repeated bombing of Iran's Kharg Island oil terminal by Iraq during the war of 1980–88 released large amounts of oil into The Gulf, fouling the waters and reefs . Iraq's use of chemical weapons is claimed to have destroyed valuable wetlands, including about 100,000 ha (250,000 acres), or 25 percent, of Shadegan marsh in the southwest.

Afghanistan, in the late 1970s, sought to establish a network of protected areas in the country with the help of the Food and Agriculture Organization

(FAO) of the United Nations, the World Wide Fund for Nature (WWF) and American specialists. Unfortunately, after an initial declaration of five wildlife sanctuaries, plans had to be shelved in 1979 following the outbreak of civil war and the Soviet invasion.

Not all the news from the war zones is bad, however. Lebanese villagers, despairing of the inactivity of a powerless government, took the initiative in 1987 to declare the war-torn country's first protected area at Bentael, 38 km (24 mi) northeast of Beirut. The site covers some 800 ha (2,000 acres) on the slopes of Mount Lebanon and supports mixed oak and conifer forest with important bird, mammal and reptile populations.

people, traditional custodians of the endangered Arabian oryx, have been involved in a successful scheme to reintroduce it to the wild in the Jiddat al Harasis and to ensure its future protection.

In Saudi Arabia the government has found it almost impossible to establish formal protected areas by the classic Western means of legislation and protection. However, under the auspices of

Prince Saud al Faisal, the National Commission for Wildlife Conservation and Development has managed to create wildlife sanctuaries under the *hema* tradition, though on a much larger scale. The Harrat al Harrah sanctuary, declared in northern Saudi Arabia in 1986, is a super-*hema* of 40,000 ha (100,000 acres), whose boundaries are carefully respected by the local Bedouin herdsmen.

A VARIETY OF PRESSURES

After the last glaciers retreated from the northern latitudes of the Middle East about 10,000 years ago, various types of steppe and wooded habitats developed. The beginnings of agriculture some 7,500 to 4,500 years ago led to the clearing of extensive native forests: most of the comparatively treeless habitats that now predominate in the Middle East are a result of human activity over the last few thousand years, combined with an increasingly warm and dry climate. In areas where crop production was possible, the inhabitants plowed natural or seminatural pastures and grew wheat, barley, oats and rye. These cereal grasses replac-

The flowering desert The growth of annual plants in the desert is triggered by rainfall. They germinate, flower, fruit and set seed very quickly, and their seeds lie dormant for long periods of drought. Perennials have bulbs that sprout quickly when rain falls.

A scrub desert The shrubby bushes are spaced widely apart so that their extensive root systems can trap moisture from the sparse rainfall without competing with each other.

ed wild species and formed pseudosteppe habitats that were used by many steppe birds and animals. Such habitats are still found in many parts of the Middle East.

As human populations became more settled, pressures increased on both the natural and seminatural steppe and on other habitats used for raising domestic animals and growing crops. The need for increased food production to meet the dramatic increase of population in the 20th century has led to rapid agricultural mechanization and the intensive use of chemicals, both of which are hostile to wildlife. The future of most habitats and their associated wildlife throughout the Middle East, including those in many nature reserve areas, is under threat.

The destruction of the steppe

A further serious development has been the gradual degradation of steppe vegetation through overgrazing and the cutting of shrubs for fuel. This is a particular problem in Iran, where the number of sheep has increased enormously since 1979; they now amount to more than 50 million head, three times the estimated number that Iran's grazing lands can carry. As a result, shepherds take their flocks to graze in formerly well-protected nature reserves, which then become rapidly degraded.

Throughout the Middle East governments have encouraged nomadic tribes to settle. This means that the land around villages is destroyed by overgrazing: in the United Arab Emirates the original vegetation within a 2 km (1.2 mi) radius of all the settlements in the grazing lands has been destroyed. Whereas grazing range (and thus flock size) was once limited by the distance from the nearest waterhole, tankers now take water out to the herds, increasing pressure on the land still more.

Once the vegetation cover has been lost, the land quickly degrades to desert through a combination of wind erosion and loss of surface moisture. Desertifica-

Early-morning birdwatchers observe waterfowl from a reed boat on a lake in Iran. The Middle East contains some large wetlands, which attract huge numbers of migratory birds seeking rest, food and shelter on their long journeys between Europe, Asia and Africa.

tion of steppe in arid areas is probably irreversible without a favorable change in climate, and it continues to occur at an alarming scale. In the Middle East as a whole, the key to the conservation of natural and seminatural steppe habitats lies principally in finding solutions to the problem of overgrazing.

Where grazing animals have been excluded from degraded areas by fencing them off, the vegetation and wildlife can make a remarkable recovery in a matter of only a few years. This has happened, for example, at the Azraq/Shaumari Wildlife Reserve in Jordan. However, such a solution would not be practical over the very

extensive areas needed to maintain viable populations of large birds such as Arabian bustards, or mammals such as gazelles. Moreover, the fences would be greatly resented by herdsmen, who would see better ranges on the other side. Instead, long-term policies are needed to tackle ancient cultural attitudes that measure wealth in the number of sheep and goats owned, so that new, higher-yielding breeds can be introduced, and numbers limited to sustainable levels.

Spurred on by the ever-increasing loss of their wildlife and wild places, Oman and Saudi Arabia are leading the way in developing a modern network of protected areas. Oman has taken a particular interest in its terrestrial resources, Saudi Arabia in the marine ecosystems of its Red Sea and Gulf coasts. Both countries have already established one reserve each (Asir National Park in Saudi Arabia and Wadi Serin-Jabal Aswad Reserve in Oman) and both have plans to set up about 40 more during the 1990s.

The threat to the coast

As in many parts of the world, coastlines are under pressure. The Iran-Iraq war led to the despoliation of large stretches of coral reef in The Gulf, caused by oil leaking from damaged installations. Tourist developments, particularly in Turkey, also threaten to drain marshes and destroy many coastal habitats. In the late 1980s the last nesting site in Turkey of the green and loggerhead sea turtles was saved from a proposed tourist development. Similar plans were also putting Turkey's first bird reserve, Kus Cenneti National Park, at risk.

AN ISLAND UNDER THREAT

For a biologist the island of Socotra, lying about 225 km (140 mi) off the Horn of Africa, is a place of fascination because of its high numbers of endemic species. The plants of this ruggedly mountainous semiarid island of 3,100 sq km (1,200 sq mi), which belongs to Southern Yemen, are related to those of the northeast African highland and the steppe region of Somalia. Altogether, 216 species are endemic to Socotra and its offshore islets, including 85 that are considered to be under threat. Many of the native plants have economic or medicinal value. For example, juice from Perry's aloe and resin from the dragon's-blood tree are used in oriental pharmacies, and the only living wild relative of the pomegranate occurs on

the island. Socotra supports several endemic species of birds, all named after the island: the Socotra grass warbler, sunbird, mountain bunting, sparrow and chestnut-winged starling.

There is very little recent information about nature conservation on Socotra, as visits by naturalists have been sporadic and brief. The lowland vegetation appears severely overgrazed, and the plants of the coastal belt are under serious threat, but evergreen thickets do still occur at heights of between 800 and 1,000 m (2,600 and 3,300 ft) above sea level. There is an urgent need for a thorough survey of the whole island and the introduction of appropriate measures to protect the island's surviving habitats.

Asir National Park

In the southwest of Saudi Arabia lies the province of Asir, which contains the country's most impressive landscape and wildlife. Here the jagged scarp of the Asir mountains plunges steeply down 3,000 m (9,800 ft) to meet the narrow coastal plain, known as the Tihamah, that borders the Red Sea. The area contains a wide variety of habitats: burning beaches; hot and humid plains dotted with shrubs, palms and acacia trees; and cooler, gentler upper ridges clothed in gray-green juniper woodlands. Seasonal streams flow during the inter-tropical monsoon, which just touches these mountains between June and September.

These habitats, especially the upland juniper forests with their curtains of lichens, are important for water catchment and soil conservation. They also have scientific and recreational value. Their continued integrity has been ensured by including a large part of the area within the boundaries of Saudi Arabia's first national park. Established in 1978, Asir National Park covers 415,000 ha (over 1 million acres) extending from close to the provincial capital of Abha on the escarpment to the coast at Al Shuqayq.

The variety of wildlife found within the park reflects the diversity of available

A rich wildlife Against the backdrop of the Asir mountains a river winds its way across the Tihamah, the narrow coastal plain dotted with shrubs, palms and acacia trees, that divides the mountains from the Red Sea. The shrubs provide food and shelter for many small mammals – hyraxes, gerbils and jerboas – which are hunted by larger predators. Thousands of birds migrate along the corridor of the plain on their route from Europe to eastern Africa, and the park is an important sanctuary for several species of endangered Arabian birds.

Rocky summits fall sheer away to high mountain valleys studded with junipers. The Asir mountains are high enough to catch the rain-bearing northwest monsoon winds, and in the late summer the rivers, dry for most of the year, are in full spate. The variety of scenery within the park, plunging from impressive mountain peaks to the coast within a narrow space, attracts many visitors. Litter and increased motor traffic are causing problems for the wildlife.

Dwarfed by the mountains, a baboon perches high on a rock in Asir National Park. It is one of many species living in the park that have their origins in Africa, bearing witness to the land bridge that once existed between Africa and Arabia.

habitats and the closeness of the African continent, to which the land was once joined. Many of the mammals – jackals, hyraxes, baboons, gerbils, jerboas and some of the bats – are of African origin. Rarer species such as wolves, leopards, hyenas, ibex and mountain gazelles are also present. Among the plentiful bird life there is a strong population of the locally endangered African bearded vulture, while other birds of African origin found in the park include the Nile Valley sunbird, gray hornbill, Bateleur eagle and Abyssinian masked weaver.

Until fairly recently the park also supported a small number of Arabian bustards, a species of large running bird that is patchily distributed across the desert scrublands of northern Africa to the Tihamah. Unfortunately, because of persecution by hunters and the loss and degradation of the bustards' natural habitat in the Tihamah as a result of agricultural expansion, these birds are now almost extinct in Saudi Arabia.

The southwestern tip of Saudi Arabia, including Asir National Park, is situated on a major route used by birds migrating from Europe to eastern Africa. Tens of thousands of shorebirds, birds of prey, warblers and swallows pass through Asir each spring and autumn, swelling the park's inhabitants from about 80 resident species (including several confined to Arabia) to more than 300.

The breathtaking wooded scenery and pleasant climate, combined with the good populations of large mammals and the seasonal turnover of birds, attracts foreign and domestic visitors to Asir National Park. At the park's headquarters, near Abha, the provincial administration has set up a carefully designed reception center where the rapidly growing numbers of visitors can learn about the ecology and wildlife of the park and gain a deeper appreciation of the importance of conservation in arid zones. The park has proved so popular that most management effort is now expended on catering for the recreational demands of the visitors, and on dealing with problems of litter, overcrowding and car parking, rather than on maintaining the quality of the vegetation and monitoring the park's animal populations. However, once Saudi Arabia achieves its aim of extending its network of protected areas, it seems hopeful that the pressure on Asir National Park will ease.

DESERTS AND WETLANDS

THE ADVANCE OF THE DESERT · THE ADVENT OF CONSERVATION · TROUBLE AHEAD

Some 2,000 years ago northern Africa was the principal wheat-producing area of the Roman Empire. But most of this once green and fertile region is no longer so today. From earliest times humans have been instrumental in shaping the landscape, in combination with a series of climatic changes. The felling of trees and overgrazing of vegetation by the flocks belonging to the region's predominantly nomadic and seminomadic peoples have led to the spread of the desert, which millions of years ago was confined to small areas in the east on the border with Arabia. Today the barren wastes of the Sahara desert extend over most of the region. Yet this apparently inhospitable landscape supports many unique plant and animal species. It is in need of protection if it is not to become increasingly barren.

COUNTRIES IN THE REGION

Algeria, Chad, Djibouti, Egypt, Libya, Mali, Mauritania, Morocco, Niger, Somalia, Sudan, Tunisia

Major protected area	Hectares
Air and Ténéré NNR	7,736,000
Banc d'Arguin NP RS	1,173,000
Bardawil RS	60,000
Boma NP	2,280,000
Bouche du Baoule NP BR	350,000
Chréa NP	26,500
Dinder NP BR	890,000
Djebel Bou Hedma NP BR	16,488
Djurdjura NP	18,550
El Kala NP part RS	76,438
El Kouf NP	32,000
Forêt du Day NP	10,000
Gebel Elba CA	480,000
Ichkeul NP RS BR WH	12,600
Ifrane (proposed) NP	10,000
Inner Niger Delta part RS	1,700,000
Massa (proposed) NP	72,000
Merja Zerga BiR	3,500
Nefhusa PA	20,000
Ouadi Rimé/Ouadi-Achim TFR	8,000,000
Ras Mohammed NMP	17,100
Sanganeb (proposed) WH BR MNP	100
Simien NP WH	17,900
Tassili N'Ajjer NP BR WH	300,000
Taza NP	3,000
Toubkal NP	36,000
Zakouma NP	300,000

BiR=Biological Reserve; BR=Biosphere Reserve; CA=Conservation Area; MNP=Marine Nature Park; NMP=National Marine Park; NNR=National Nature Reserve; NP=National Park; PA=Protected Area; RS=Ramsar site; TFR=Total Faunal Reserve; WH=World Heritage site

THE ADVANCE OF THE DESERT

During the Miocene period (from 25 to 6 million years ago), a lush tropical environment of laurels and cedars stretched right across northern Africa, interrupted by small areas of desert in the east. Toward the end of this period the climate began to change and the eastern regions became even drier. As forests gave way to grassland, animals and plants adapted to these more arid conditions were able to move across the land bridge that then existed between Africa and Arabia. The expansion of the desert and the drifting apart of the African and Eurasian continents some 15 million years ago gradually closed this migration corridor, with the result that Africa remained isolated throughout the ice age that subsequently gripped Europe.

When, some 10,000 years ago, the first pastoralists migrated across northern Africa with their flocks of sheep and cattle, the center of the Sahara was still covered with vegetation. Rock paintings at Tassili n'Ajjer in southeast Algeria that date from this period depict herds of elephants, ostriches and gazelles roaming the vast open plains. Later paintings at the same site show the wildlife replaced by herds of cattle. This increased human and livestock activity, together with the changing climate, led to the desiccation of the Sahara. The desert started to spread between 8000 and 2000 BC, a spread that continues today.

Important wildlife refuges

Because the land bridge between Africa and Eurasia existed until relatively recent times, northern Africa is the meeting place of plant and animal species from both continents. The isolation of various mountains and marshy places by the encroaching desert has led to the evolution of many local, or endemic, species, found only in the area.

Although it is dominated by the Sahara, which covers approximately 8 million sq km (3 million sq mi), a third of the land area of Africa, the region nevertheless contains an impressive variety of unspoilt ecosystems. Moving north, the rainfall decreases and becomes increasingly seasonal, and the thorny acacia savanna of the Sahel on the southern edge of the Sahara gives way to clumped grass habitats and desert with little plant life

except in oases. The desert itself consists of gravel, pebbles and sandy soils, while the terrain features plains, high plateaus and depressions, and is rich in specialized animals and plants. Its habitats are typified by those of the Ouadi Rimé/Ouadi-Achim, Tassili n'Ajjer and the Air and Ténéré reserves, all of which harbor endangered species such as the leopard, the addax and the slender-horned gazelle.

Many of the desert mountains are important wilderness refuges: half the plant species in the mountains of Somalia in the east of the region are endemic, and

Date palms in the Sahara desert Underground water is a lifeline in the desert, and is tapped by plants that give food and shelter to people and animals. Although these palms are apparently growing out of the sand, their roots reach deep down to a reservoir where impermeable rocks have trapped the scant rainfall.

Lowgrowing cushion plants on the peaks of the High Atlas Mountains are able to resist the chilling, drying winds and are quickly covered by snow for extra protection. The diversity of species decreases with altitude: there is less vegetation, lower productivity and fewer food supplies to support the wildlife.

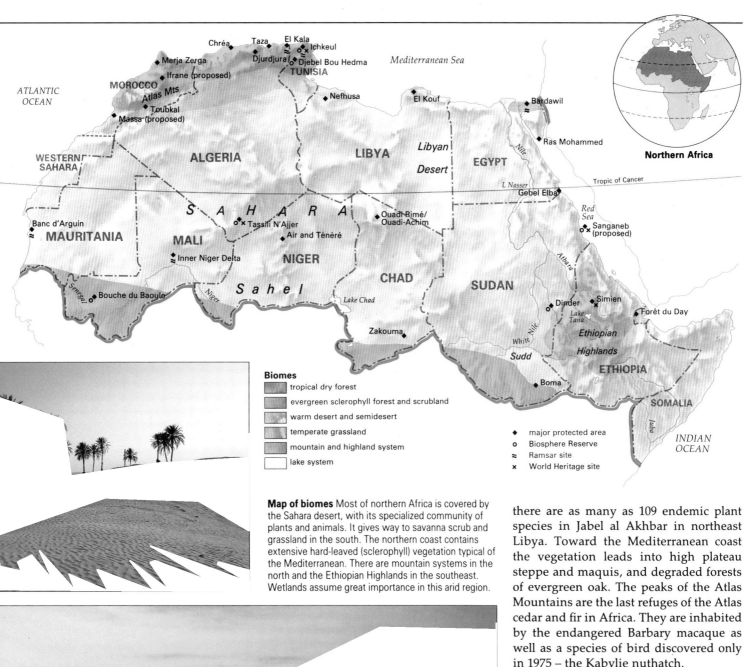

Biomes

- tropical dry forest
- evergreen sclerophyll forest and scrubland
- warm desert and semidesert
- temperate grassland
- mountain and highland system
- lake system

- ◆ major protected area
- ○ Biosphere Reserve
- ≈ Ramsar site
- × World Heritage site

Map of biomes Most of northern Africa is covered by the Sahara desert, with its specialized community of plants and animals. It gives way to savanna scrub and grassland in the south. The northern coast contains extensive hard-leaved (sclerophyll) vegetation typical of the Mediterranean. There are mountain systems in the north and the Ethiopian Highlands in the southeast. Wetlands assume great importance in this arid region.

there are as many as 109 endemic plant species in Jabel al Akhbar in northeast Libya. Toward the Mediterranean coast the vegetation leads into high plateau steppe and maquis, and degraded forests of evergreen oak. The peaks of the Atlas Mountains are the last refuges of the Atlas cedar and fir in Africa. They are inhabited by the endangered Barbary macaque as well as a species of bird discovered only in 1975 – the Kabylie nuthatch.

Water in the desert

In the few places where water is available in the desert it provides an important haven for all kinds of wildlife. Egypt has some of the greatest areas of coastal wetlands fringing the Mediterranean; more than a million birds winter in the Nile delta. At the southern extreme of the region, in Mali, the inundation zone of the river Niger is of supreme importance for birds migrating from Europe and Asia to Africa, and is visited by up to 885,000 overwintering European ducks. The extensive Sudd marshes of the Sudan are probably unparalleled on the continent as a wetland habitat.

The Red Sea coast boasts some notable marine wildernesses, supporting rich ecosystems of mangroves and coral reefs that flourish in the remarkably clear water. Dramatic examples are protected in the Egyptian Marine Conservation Park of Gebel Elba and the Sanganeb Marine National Park project in the Sudan.

THE ADVENT OF CONSERVATION

Although the vast tracts of northern Africa are among the least densely populated areas of the world, the cities, mainly on the coast, are some of the most crowded. The countries of the region were slow to realize the need to establish protected areas in this apparently limitless barren wilderness. Yet the advent of mechanized transport, changing lifestyles and, above all, the need to feed more than 200 million people, is now putting enormous and increasing pressure on land resources, and there is an urgent need to protect the environment. In a region that is widely troubled by civil war and political unrest, and where millions of people have been displaced by fighting and famine, it is perhaps not surprising that conservation legislation has been slow to arrive, and is not always adequately enforced.

The region's earliest legislation on the environment was formulated in 1884 by the forestry service of Tunisia, which passed a series of measures concerned with hunting restrictions. In 1923 the French colonial administrators of Algeria created the first national park in northern Africa at Cheilor Dar el Oued, now called the Taza National Park. This set aside land to preserve the dramatic scenery of the native forests and the coast for recreational use, and to protect the habitat of the Barbary macaque.

Many countries in the region did not possess forest or park authorities until the mid-1960s or later. It was only in 1979 that

Egypt established its Wildlife Service. Other countries are restructuring or completely replacing obsolete colonial legislation dealing with reserves, much of which lapsed years ago. Even today there is still no protection in Western Sahara and only token protection in the region's smallest country, Djibouti.

A harsh environment – yet the desert supports a surprising range of animals. Small mammals – gerbils, jerboas and several species of mice – shelter by day in rock crevices or burrows in the sand to emerge at night, when they are preyed upon by larger carnivores.

To date, approximately 285,000 sq km (110,000 sq mi), just 2 percent of the land surface of northern Africa, are protected by some form of park or reserve legislation. In many cases the protected areas have undefined boundaries, are inadequately managed, or exist only on paper.

Privately owned protected areas are almost unheard of in the region, except in Egypt, where many of the sites originated as the property of private research institutions. Most of the protected areas elsewhere are state-owned and managed.

The reserves of the future

Many existing parks and reserves support large groups of people who live there either permanently or seasonally, by hunting, grazing or cultivating the land. Their numbers are unrestricted, and so pressure on the reserves increases. The park authorities in northern Africa are

BANC D'ARGUIN – A WETLAND GEM

One of the largest and most outstanding wetlands in Africa is the Banc d'Arguin on the western Atlantic coast of Mauritania. The Banc d'Arguin National Park, created in 1976, covers just over 1 million ha (2.5 million acres) of the coastal area between Nouadhibou, on the border with Western Sahara, and the capital Nouakchott. It contains a wide variety of undamaged ecosystems, providing a unique meeting point between the Sahara desert and the Atlantic Ocean.

A number of important species and wildlife communities are found in the area, including 80,000 ha (200,000 acres) of sea grass meadows, as well as remnants of ancient mangrove vegetation

and the largest known colony of the now highly endangered Mediterranean monk seal. Nearly 3 million European birds spend the winter here. Many species of fish thrive in the sea grass beds, four species of marine turtle lay their eggs on beaches in the park, and several species of dolphins are found in the waters off the coast.

About 500 Imraguen, a local people who live in the park, fish along the coast using traditional methods unchanged since they were first recorded by European explorers in the 15th century. The park is patrolled by wardens mounted on camels – a unique feature that adds to the interest of this remarkable wilderness.

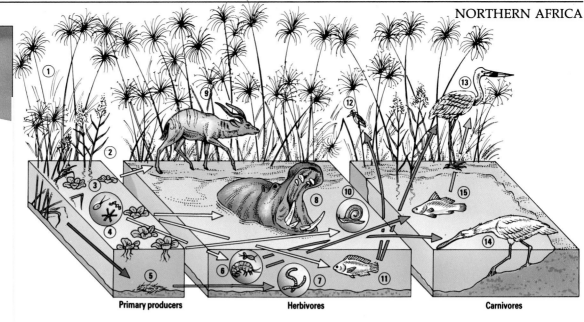

Primary producers Herbivores Carnivores

Components of the ecosystem
1 Papyrus
2 Reeds
3 Water cabbage
4 Algae
5 Detritus
6 Crustaceans browsing on algae
7 Swamp worm
8 Hippopotamus
9 Sitatunga
10 Water snail
11 Nile perch
12 Malachite kingfisher
13 Saddlebilled stork
14 African spoonbill
15 Tilapia

Energy flow
primary producer/primary consumer
primary/secondary consumer
secondary/tertiary consumer
dead material/decomposer
death

A swamp ecosystem showing the relationship between primary producers (algae and plant detritus), decomposers (invertebrates), grazers and carnivores.

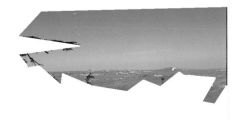

A solitary mangrove tree on the Red Sea coast of Ethiopia. The spreading roots of mangrove trees trap mud. This is colonized by other plants, and so new land is formed. The swamps, used as spawning grounds by many marine species, are rich in invertebrate life.

A forest in its final stages A few cedars are all that is left of a forest that once covered the Middle Atlas Mountains. Overgrazing by livestock has reduced the tree growth so that the remaining trees become stunted and distorted by the cold wind. Many of these conifer forests have been felled for timber.

becoming increasingly aware of the need to create multi-use protected areas that seek to establish a balance between the needs of the wildlife and of people.

One enlightened approach to this problem has been made in south Sudan with the passing of a law that permits inhabitants of the parks to capture certain species of wild animal for food, provided that only traditional forms of hunting are used, and outlaws the use of unlicensed firearms. The region contains a number of habitats that have been shaped over the centuries by human activity. These areas possess a rich variety of wildlife, and in some protected areas efforts are being made to maintain them by the controlled introduction of livestock. This should ensure that traditional land-use continues, despite the accelerating changes in lifestyle and technology.

Most such multi-use protected areas are still only at the planning stage. One example is the wetland area of the proposed Massa National Park in Morocco. As well as providing an important refuge for migrating birds, this will protect many endemic species of plants, which are closely related to those of the Canary Islands. The rare Dorcas gazelle has already been successfully reintroduced to the area. National park protection status is urgently needed to safeguard this valuable wetland site against threats from drainage schemes and tourism.

Even already established parks, such as the El Kala National Park in Algeria, have constantly to balance the needs of humans and wildlife. El Kala is one of the three most important wetland areas of the Mediterranean coast. Part of it is threatened by drainage and agriculture, while the forests, which harbor the rare caracal and Barbary stag, are being encroached on by grazing livestock. Some 100,000 people visit El Kala each summer.

The future plans of Algeria and Tunisia are typical of much of the region in that lists of new sites have been proposed, which should ensure a significant increase in the total area under protection. Whether the organizational structure to maintain these areas can be set up and sustained is another matter.

TROUBLE AHEAD

For centuries, the greatest threat to northern Africa's natural wildernesses has been the gradual degradation of the land through overgrazing and indiscriminate felling of woodland. Today, land use in the region is still based on traditional subsistence methods of agriculture and pastoralism, but these have increased in scale as human populations have swelled.

The population of most of northern Africa is increasing at a frenetic rate, doubling every 25 years on average. The need for more food has led to overstocking of the land. The result has been a series of environmental crises, particularly in the Sahel region of Ethiopia and Sudan where there are 123 million head of cattle, sheep and goats (60 percent of all the livestock of northern Africa). Attempts to restrict livestock are hindered by Islamic tradition, which measures wealth by the number of cattle, sheep and camels a person owns, and protects customary rights to grazing land.

In much of the region forest and other woods are being felled for fuel and timber. In the arid lands of northern Africa, this depletion of the vegetation cover, compounded by climatic changes, leads to soil exhaustion and erosion. Sand dunes are formed, and the desert advance continues, rendering new areas of land barren. Elsewhere the land is threatened by mineral extraction, including oil, especially in Algeria, Egypt and Libya, and by "package tour" tourism along the Mediterranean coast of Tunisia.

The stricken Sahel Foragers in Mali in search of firewood tear off branches from a stunted tree. The stony ground is already bare; without tree roots to hold it together the soil will soon erode or blow away in the wind, and the inexorable process of desertification will continue to spread across the land.

Threatened wetlands

To combat drought and famine in the region, a number of major irrigation schemes have now been undertaken with international assistance. Many wetland areas have been drained and reclaimed, and their waters used for agricultural intensification, irrigation and for schemes for hydroelectric power.

The largest of these projects was the damming of the Nile at Aswan, which created 400,000 ha (1,000,000 acres) of newly cultivable land by irrigation. The decline in fresh water supplies and silt deposition led to the salinization of the irrigated lands and marshes downstream. This had serious consequences for the Nile wildlife: water organisms gradually died out, causing a collapse of the Nile delta fish population. The loss of fertility and the resultant need for chemical fertilizers accelerated the decline in the number of birds. The loss of fresh silt has also led to erosion of the coastline, threatening the delta marshes around Lakes Burullus and Manzala, a haven for over 650,000 wintering waterfowl.

In the more agriculturally productive Mediterranean coastlands, both dryland and wetland habitats, such as that of Ichkeul (a Biosphere Reserve and Ramsar wetlands site), are threatened by water runoff containing agricultural fertilizers, pesticides and other pollutants.

The almost continuous political instability and civil war in the region jeopardizes even successful wildlife programs and threatens existing protected areas. During Algeria's independence wars in the 1950s and 1960s, for example, the wildlife of its wilderness areas was plundered to support guerrilla nationalist forces. Even today, extensive damage is evident in the Atlas Mountains, caused by the defoliant chemicals used by the French army against the guerrillas.

Positive programs

Most conservation projects are as yet still undertaken by state organizations. But realization of the need to involve the rural community is gradually growing. Many modern programs encourage public participation in an attempt to eliminate poaching and the over-exploitation of vegetation. One such project is the Gebel Elba conservation area in Egypt, a site of diverse habitats such as wooded savanna and acacia thornbush, threatened by livestock grazing. Elsewhere, schemes are under way to replenish depleted habitats. At the Djebel Bou Hedma oasis in Tunisia, for example, trees have been planted and animals, such as the endangered addax and scimitar-horned oryx, reintroduced.

Education is another way by which public awareness of conservation may be raised. Though some school programs and wildlife societies do exist, they are still relatively new concepts in the region. The Wildlife Clubs of South Sudan (WCSS), first established in 1979, have been particularly active in campaigning against the Jonglei Canal, which threatens the Sudd marshes. Most wildlife societies have tended to focus on natural history rather than conservation. An exception is the recently formed Moroccan Society for the Protection of the Environment, which campaigns on a number of environmental issues such as the threat to the ozone layer and urbanization.

Attempts are being made to develop facilities to encourage wildlife tourism. In Tunisia, for example, visitor centers and ecomuseums are being set up in national parks. A few established parks, such as that at Chréa in Algeria, already have well-developed programs to provide recreational facilities and give information on wildlife conservation, a precedent that may well be followed in other protected areas throughout northern Africa.

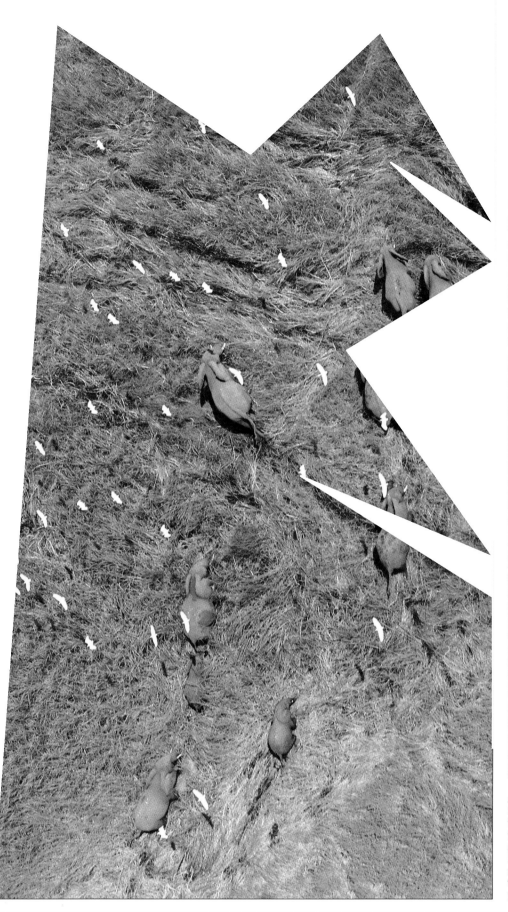

THREAT TO THE SUDD

In southern Sudan the headstreams of the White Nile flow into a great swamp, the Sudd, a vast expanse of swaying papyrus, lush riverside grasslands, and colorful spreads of water hyacinth. Covering 129,534 sq km (50,000 sq mi), the Sudd is the largest inland wetland in the world. It is famous as a refuge for migratory birds. Residents include the shoebill stork, sacred ibis and a major population of the whale-headed stork. The wetter parts of the swamp are home to hippos and crocodiles. The grasslands provide cover and food for giraffes and antelopes, including sitatunga, the world's largest population of Nile lechwe, and some 500,000 Sudanese tiang. The local people also graze their animals here.

In the early 1980s work began on a 350-km (210-mi) long canal that would link the towns of Jonglei and Malakal at either end of the great loop of the White Nile. This scheme to reclaim the swamp for agriculture would have a drastic effect on one of Africa's last great wetlands, and would devastate the pastoral economy of the local Nile people. There were even fears that changes in the ecology of the area might lead to an extension of the desert. Although the prolonged civil war in Sudan brought a halt to the project, it was not officially cancelled and the threat to the Sudd swampland remains.

A floating mat of papyrus forming an island in the middle of the Sudd, with patches of free-floating water hyacinth. It is a highly productive ecosystem.

A herd of elephants runs through the reedbeds of the Sudd marshes. Cattle egrets, which have been following them to feed on the insects they disturb, take flight. The tall reeds of one of Africa's largest wetlands provide an ideal habitat for many grazing mammals, including sitatunga and Nile lechwe, and for waterbirds such as storks and ibis. The marshes are a refuge for thousands of migratory waterfowl, which feed on its populations of fish, insects and amphibians.

Air and Ténéré

Niger, one of the poorest countries in the world, supports one of the largest conservation areas in Africa, the newly established National Nature Reserve of Air and Ténéré. The entire site of 77,360 sq km (29,860 sq mi) represents some 6 percent of Niger's total land area. The official legislation regarding the protected area has yet to be enacted, though the technical and management structures to administer it are already being installed with the assistance of the World Wide Fund for Nature (WWF).

Desert vegetation

The area encompasses a variety of arid habitats, from the waterless shifting dunes of the Ténéré desert (with recorded extremes of temperature that range from 50°C/122°F in summer to below zero in winter), to the humid mountain habitats of the Air mountains. These have been identified as one of the last strongholds for the wildlife typical of the Sahara and Sahel, and are among only a few sites where reasonable numbers of endangered large mammals can still be found. They are of importance, too, because their unique plants have affinities both with the vegetation of the Mediterranean and with that of the upland regions of east and west Africa.

The principal Saharan vegetation of the area includes spiny shrubs such as scattered acacia, characteristic of sandy and stony soils. Remnants of a plant life typical of the Mediterranean and Sudan, including wild olive, are found in wetter sheltered areas such as dry river beds, or wadis, and mountain ravines. Where the Sahel gives way to the Sahara grasses of various types predominate.

The Air and Ténéré Reserve is a refuge for the endangered addax and scimitar-horned oryx, and has significant populations of Barbary sheep, cheetahs and various gazelles. It also shelters more than 139 species of birds, a little under half of them migrants from Europe and Asia.

Nearly 3,000 people live in the region. Most of them are Tuaregs who invaded the area in successive waves during the last 2,000 years. These traditional semi-nomadic pastoralists raise camels and goats and tend irrigated gardens in the mountains. Camel caravans still travel between the Air settlements and the oases that lie beyond the Ténéré desert, but they are gradually giving way to faster motorized transport.

The Tree of the Ténéré, now destroyed, was once a famous sight. It is marked on the map as the only landmark for hundreds of kilometers, and the bones of camels lie where the exhausted animals collapsed in its shade. But a tourist drove into it, and the remains of the tree are now in a museum.

Scattered acacia trees are the only vegetation to be seen in the bare landscape of the Air and Ténéré National Nature Reserve.

Tamgak Gorge in the Air Mountains, with fresh water and palm trees. Mists that form in the higher humidity of the mountains condense and collect as water on the ground, allowing a variety of plants to thrive.

The aims of the reserve

It is hoped that the Air and Ténéré Reserve will act as a model for those trying to integrate conservation of natural resources with preserving the traditional lifestyles of the local population. Its regulations aim to be flexible enough to accommodate the needs of the Tuaregs and give protection to the reserve's valuable wildlife resources.

The people who live in the reserve traditionally respect important plant and animal species such as the acacia and the ostrich, but an influx of people from outside, who are not always bound by the same conventions, has led to the over exploitation of the vegetation and depletion of the reserve's wildlife by hunting. Management of resources will therefore need to ensure that traditional lifestyles are supported within the reserve. A series of protected zones is planned to minimize tourist pressure on the wildlife and to limit such activities as mining or inappropriate agricultural practices that would prove harmful. It is hoped that much of the decision-making will eventually be transferred to the local population.

EQUATORIAL ECOSYSTEMS

FROM JUNGLE TO SAVANNA · CONSERVATION ON THE EQUATOR · POPULATION PRESSURE

Central Africa encompasses mangrove swamps and mountain heaths, semi-arid savanna and dripping rainforests, soda lakes, seasonal rivers and isolated island communities. Within this wide range of habitats are some of the world's most famous and most spectacular landscapes and wildlife parks, such as the flat acacia grasslands of the Serengeti, with their huge herds of grazing animals, and the dank, forested slopes of the Ruwenzori mountains in Uganda, inhabited by gorillas and giant flowering plants. Yet equatorial Africa has urgent conservation problems. Its unique ecosystems are seriously threatened by the poverty and population pressures of the countries in this region. Immediate and imaginative solutions are needed if they are to be preserved for future generations.

COUNTRIES IN THE REGION

Benin, Burkina, Burundi, Cameroon, Cape Verde, Central African Republic, Congo, Equatorial Guinea, Gabon, Gambia, Ghana, Guinea, Guinea-Bissau, Ivory Coast, Kenya, Liberia, Nigeria, Rwanda, São Tomé and Principe, Senegal, Seychelles, Sierra Leone, Tanzania, Togo, Uganda, Zaire

Major protected area	Hectares
Aldabra Atoll (Seychelles) WH SNR	35,000
Amboseli NP	39,206
Benoue NP BR	180,000
Bia NP BR	7,700
Dja HR FoR BR WH	500,000
Djoudj NP BS RS WH	16,000
Fazao-Malfakassa NP	192,000
Ipassa-Makokou BR	15,000
Kahuzi-Biega NP WH	600,000
Kainji Lake NP	534,082
Korup BR NP	126,000
Manovo-Gounda-St Floris NP WH	1,740,000
Masai Mara NR	151,000
Mount Nimba INR BR WH	17,130
Ngorongoro CA BR WH	828,800
Niokolo-Koba NP BR WH	913,000
Pendjari HZ BR	880,000
Po NP	155,500
Queen Elizabeth NP BR	220,000
Serengeti NP BR WH	2,305,100
Tai NP BR WH	350,000
Tsavo NP	2,082,114
Virunga NP WH	780,000
Volcanoes NP BR	13,000

BR=Biosphere Reserve; BS=Bird Sanctuary; CA=Conservation Area; FoR=Forest Reserve; HR=Hunting Reserve; HZ=Hunting Zone; INR=Integral Nature Reserve; NP=National Park; NR=National Reserve; SNR=Strict nature reserve; WH=World Heritage site

FROM JUNGLE TO SAVANNA

From Guinea, on the Atlantic coast of west Africa, across to the mountains of the Rift Valley in the east, Africa's central zone is blanketed with evergreen lowland rainforest, which grows densely in the equable, moist climate. Although Africa's equatorial rainforest – some 19 percent of the world's total – has relatively few animal species compared with other rainforests, it nonetheless supports many different animals. Gorillas and giant forest pigs, okapis and antelopes such as bongoes and duikers forage on the forest floor; monkeys, squirrels and birds live among the upper tree branches. Nomadic pygmy peoples pursue a traditional way of life, hunting forest animals and birds and gathering plant foods.

On the high mountains of eastern Africa lowland tropical forest or savanna gives way to montane forest and then to unique mountain heath where giant forms of plants such as lobelia and St John's-wort have evolved.

The golden grasslands

At the northern and southern fringes of the rainforest prolonged dry periods are encountered, and the forest becomes less humid. Its dark green cover gradually thins to olive-green and brown woodland and to golden savanna. Here the trees are smaller, with wider canopies, and the bushes thornier. Because of the drier conditions, fires frequently break out; some savanna plants even rely on the heat of the flames to prepare their seeds for germination. While the variety of animal species in the woodland is less than in the forest, it is compensated for by their large numbers. These animals are mostly browsers such as elephants, giraffes, kudus (a large antelope) and bush pigs, and they feed off the leaves of the trees.

Farther east, where the rainfall is even lower and less predictable, pale-gold grassland scattered with trees and shrubs stretches as far as the eye can see. Fires are common here, as well, and play an important role in stimulating the fresh growth of grasses.

Huge herds of grazing animals live on the grasses – among them the buffaloes, kongoni (hartebeest), Thomson's gazelles, topis, wildebeests and zebras for which eastern Africa is famed. These grazers are hunted by predators such as cheetahs,

A typical savanna landscape The mix of grasses and acacia trees supports vast herds of grazing and browsing mammals and their predators. The trees thin out as the climate becomes more arid.

Giant groundsels grow as tall as 5 m (16 ft) on the high mountains of eastern Africa. The rosettes of cabbage-like leaves are covered with fine, silvery hairs that trap moisture, provide insulation at night and keep out excessive ultraviolet radiation.

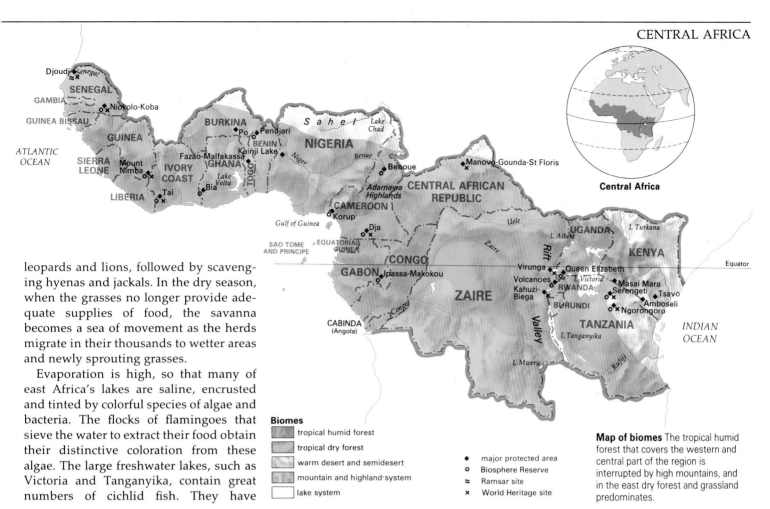

Central Africa

leopards and lions, followed by scavenging hyenas and jackals. In the dry season, when the grasses no longer provide adequate supplies of food, the savanna becomes a sea of movement as the herds migrate in their thousands to wetter areas and newly sprouting grasses.

Evaporation is high, so that many of east Africa's lakes are saline, encrusted and tinted by colorful species of algae and bacteria. The flocks of flamingoes that sieve the water to extract their food obtain their distinctive coloration from these algae. The large freshwater lakes, such as Victoria and Tanganyika, contain great numbers of cichlid fish. They have

Biomes

- tropical humid forest
- tropical dry forest
- warm desert and semidesert
- mountain and highland system
- lake system

- ◆ major protected area
- o Biosphere Reserve
- ≈ Ramsar site
- × World Heritage site

Map of biomes The tropical humid forest that covers the western and central part of the region is interrupted by high mountains, and in the east dry forest and grassland predominates.

evolved in isolation into many different species unique (endemic) to each lake.

Equatorial archipelagos

On the dry Cape Verde Islands off the Atlantic coast of western Africa, forests were once the dominant vegetation, but these have largely been felled for fuel. The coastal plains and rocky mountains are now covered mostly with scrub and grasses, some species of which are endemic to the islands.

By contrast, the 60 or so islands of the more humid Seychelles in the Indian Ocean support both moist and dry tropical forest, including six endemic species of palms. Also restricted to these islands is the magnificent coco de mer, which produces the largest seed of any plant species. However, much of this forest has been replaced with coconut and cinnamon plantations. The trees throng with bird life, the most famous example of which is the paradise flycatcher, found only in the Seychelles.

Most of the mangrove swamps that formerly fringed the islands' shores have been destroyed, and with them the Nile crocodile that inhabited them. But the warm, shallow waters support a marine ecosystem of sea grass meadows grazed by, among other animals, dugongs, turtles and at least 800 species of fish. There are many corals, including 45 that are unique to the Seychelles.

CONSERVATION ON THE EQUATOR

The first protected areas in central Africa were set up early in the century as big game reserves by the former European colonial powers. Many later became national parks: the first was Albert National Park (now the Virunga National Park of Zaire) in 1925. These reserves were mainly in savanna habitats rich in game. Because they were not heavily hunted, forests generally remained unprotected, and only recently have reserves been established in forested areas as, for example, in Cameroon.

The earliest protected areas were usually created without referring to the needs or wishes of the indigenous peoples, many of whom were nomadic herdsmen. These people suddenly found their traditional grazing lands removed and ethnic boundaries blocked by the reserves. Conservationists now try to ensure that local communities benefit from protected areas. This means integrating the areas into an overall plan with zoning for different uses. For instance, a totally protected nature reserve might be surrounded by a buffer zone for sport hunting, with adjacent areas set aside for subsistence hunting and the gathering of

firewood. Other areas may be zoned for activities that have a greater impact, such as subsistence agriculture, commercial agriculture and logging.

Examples of such "biosphere reserves", as they are called, are the Serengeti National Park in Tanzania and the Korup National Park, covering an area of Cameroon's rich rainforest. In Korup improved strains of crops are being grown to increase yields, and forestry is being developed to provide a sustainable supply of timber. By integrating the conservation and rational use of natural resources into the community's develop-

ment there is a much better chance that longterm conservation will succeed.

The need for scientific research and constant monitoring of the environment is recognized in the region. Most of the national parks and wildlife agencies have a research branch that provides a scientific basis for their management policies. There is cause for optimism in the example of Tanzania, whose Mweka College of African Wildlife Management has trained almost all the wardens in eastern Africa and continues to educate future park staff from all over Africa. It is playing a key role in changing attitudes toward

A mangrove swamp, consisting of several species of salt-tolerant trees, lines the shore of Mahé in the Seychelles. Many mangrove forests have been felled for firewood, and marine parks have been established in the Seychelles to protect them. Mud that accumulates among their dense roots provides a rich habitat for invertebrates such as fiddler crabs and for mudskippers.

Lake Naivasha, one of the lakes of the Rift Valley, is an immensely productive ecosystem. Its papyrus swamps shelter many fish, amphibians and insects. Although the native topminnow has been replaced by introduced tilapia and bass, the waters of the lake are fished by many waterfowl that migrate from Europe and Asia.

Components of the ecosytem
1 Thorn (acacia) tree
2 Zebra
3 Grass
4 Elephant
5 Blue wildebeest (brindled gnu)
6 Giraffe
7 Eland
8 Lappet-faced vulture
9 Lion
10 Spotted hyena
11 Carcass

Energy flow
⇨ primary producer/primary consumer
⇨ primary/secondary consumer
⇨ dead material/consumer

A savanna ecosystem showing the relationship between grass-eaters and predators. The grass-eaters are nomadic, depending on seasonal rainfall and local waterholes.

Primary producers/Herbivores   Herbivores   Carnivores/Scavengers

conservation, and has provided a model for other similar institutions, such as the Garoua College in Cameroon and Kenya's Naivasha Institute.

Managing the parks

Successful management brings its own problems. When, in the early 1960s, rinderpest (a disease of cattle and related species) was eradicated from the herds of cattle grazing outside the Serengeti National Park in Tanzania, the disease also ceased to attack the wildebeest and buffalo populations in the park. As a result, their numbers increased dramatically, putting great pressure on the grassland. Elephants have moved into many parks in increasing numbers because of growing pressure from poachers in unprotected areas. They strip the bark off trees and push them over, reducing wood cover, so that systematic culling is needed to protect the habitat.

Tourism has become a growing industry throughout central Africa but particularly in the east. It brings much-needed foreign currency to the region's hard-pressed economies but exerts great pressure on reserves – the very resources it seeks to exploit. Land is lost to provide hotels, service buildings and access roads, and the numerous safari vehicles break down the soil and ruin the ground cover. As the population grows, agriculture and grazing encroach on the land in and around the parks.

But the principal problem that faces reserves is lack of resources. In eastern Africa governments have long recognized the value of tourism, and wildlife and national parks departments are strongly supported. Zaire, too, has an expanding program. But elsewhere nature conservation tends to be given a low priority.

Parks departments are starved of funds and are ineffectively managed.

Insufficient funding also hampers the fight against poaching, though governments, responding to international pressure, are now starting to take a stand against ivory hunters. Right across the continent elephants have been slaughtered both inside and outside protected areas. Rhinos, eliminated from many areas, are now teetering on the brink of extinction. The fight against poachers, many of them armed with automatic weapons, has assumed an almost paramilitary scale. The park authorities try to match the poachers' tactics with aircraft fitted with special radar equipment, and even missiles, but few can afford to finance this size of operation.

The disruption caused by wars and civil strife, common in the region, spills over into the protected areas. During the 1960s the Virunga National Park in Zaire lost vast numbers of its big game to the

rebels, mercenaries, government soldiers and poachers. The Ugandan national parks suffered similarly in the political unrest of the 1970s and early 1980s.

Conserving island life

On the denuded Cape Verde Islands and in the Seychelles reforestation programs have been implemented to provide much needed ground cover and a managed source of fuel for the islanders. The hope is that this will reduce pressure on the natural vegetation.

Since 1970 the Seychelles government has established several land-based and marine national parks and has even set aside whole islands as strictly protected nature reserves. Aldabra, home of the giant tortoise, was the first coral atoll to be designated a World Heritage site. But a possible retrograde step may be the opening of the islands' waters to other countries for tuna fishing. The effects this will have on marine life remain to be seen.

COMPETING FOR LAND

Southwest Kenya contains two reserves that shelter most of Kenya's wildlife, the Maasai Mara and Amboseli. But – in a country that has one of the highest rates of population growth in Africa, nearly all concentrated in the west and southwest – there is great shortage of land. The Masai people, nomadic cattle grazers who traditionally run their herds over this land, are no longer able to live in harmony with the animals on and near the reserves. Cattle compete with wild animals for the waterholes, which are quickly drunk dry, and for grass. Disease is spread among the herds. As a result of population growth, the land cannot support the cattle that each male Masai traditionally owns.

The Masai have consequently taken up farming and have fenced off areas that were previously open rangeland.

The Kenyan government has tried to discourage the Masai from enclosing the land by giving them a share in the profits gained from tourism. However these funds are not always fairly distributed, so the incentive is not strong enough. In addition the Masai are now dividing the land into group ranches, some no larger than 2 ha (5 acres). Such restricted areas will soon become overgrazed and barren, no longer providing a basis even for settled farming. The resulting patchwork of farms and grazing land will only increase the confrontation between people and wildlife.

POPULATION PRESSURE

Throughout Africa human populations are growing rapidly: by the middle of the next century the number of people on the continent will have more than tripled. The pressure on the wild lands and protected areas will reach critical levels.

Cutting down the forests

The consequences of rapid population growth are already apparent in tropical Africa. In this century alone the area of rainforest, which once covered about 22 million sq km (8.5 million sq mi), has shrunk by at least 50 percent. Uganda, Ghana and Ivory Coast have lost much of their original forest; by the end of this century it will have disappeared completely. Despite this, the Ivory Coast and Nigeria continue to fell between 5 and 6 percent of their forests annually.

Changes in climatic patterns have been recorded in southwest Zaire, where the Mayumbe forests have been destroyed, and in Ivory Coast, where rainfall has decreased by a quarter over the last 20 years. Many experts suspect that deforestation is responsible.

Forests and woodlands protect water catchment areas and stabilize stream flow, so preventing soil erosion. Without tree cover, degradation of the land soon sets in. The loss of the forests also means

An oasis for wildlife Remote mountains are safe from the pressures of human activity. Many plants and animals have adapted to the extreme climatic conditions above the treeline on Mount Kilimanjaro, lying on the Equator, where the intense heat of cloudless days is succeeded by chilling night frosts.

that many species of plants and animals are disappearing.

Forests are essential resources. They provide the local people with meat, wild fruits and vegetables, with medicines and honey, and with wood for fuel and building. There is now a growing realization among doctors and scientists that drugs derived from forest species have a wider medical application as well.

An additional problem is the potentially high return to be made from cash crops, especially tempting in view of the high level of international debt incurred by many African countries. Natural vegetation is cleared to make way for growing single cash crops. This exhausts the soil's nutrients, and leads to soil erosion as the vegetation cover is lost.

The need for education

The most important goal of conservationists must be to try to stem the high rate of population growth. In the short term, however, much can be achieved by educating people about the risks of environmental degradation. Some public education programs have been very successful. In Rwanda the Mountain Gorilla

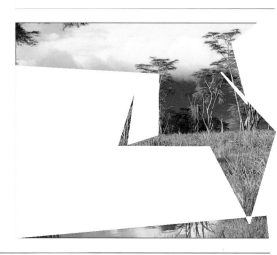

POVERTY AND ENVIRONMENTAL DEGRADATION

Uganda was once dubbed "the pearl of Africa" because the pleasant climate and fertile soil brought prosperity to its people. But the years of political turmoil between 1971 and 1985 so impoverished the country that rural communities were forced into non-sustainable use of the land, causing widespread environmental degradation.

For example, a farmer desperate for cash traps and shoots all the animals in the nearby forest to sell the meat. Had he killed only enough for the needs of his family, he could have harvested the animals in perpetuity. Instead he has sacrificed a long-term resource for only short-term gain. With no source of meat for his family he ends up poorer than before he started. So he turns to felling trees for cash. This causes the nearby stream to run empty in the dry season, and his crop yields fall. The farmer has become locked into a downward spiral of poverty as the overuse of natural resources causes further environmental degradation and yet further poverty.

Between 1966 and 1981, in most parts of Uganda, soil fertility, pasture quality, firewood supplies and the number of trees and game animals declined. The area of natural forest was reduced by about 17 percent.

A river in full spate after the rains carries a heavy load of red soil from the slopes of Mount Kenya. Tropical ecosystems are able to cope with seasonal deluge and runoff, providing that the natural checks to soil erosion are not destroyed by careless management of the land, particularly destruction of the tree cover.

Pristine rainforest, such as this in Zaire, is becoming rare outside protected areas. Rainforests are able to support only a small human population without loss to the habitat. Only with careful planning and firm policing can the surviving forests cater for the needs of growing populations without damaging the diversity of wildlife.

A vital but scarce commodity Waterholes on the savanna plains attract animals from great distances when the surrounding land has become parched. Conservation of these wetlands, and the ensured continuation of access to them, is necessary to support migratory species as well as the way of life of nomadic pastoralists, whose flocks compete with wild animals for scant water supplies.

Project tells local people about the importance of the gorillas and their forest habitats. A mobile film and lecture unit tours villages and schools, and many local people are employed as porters, guards and guides, or are involved in making and selling souvenirs and refreshments to tourists visiting the gorillas' forest haunts. Recent surveys have shown a marked swing in the attitudes of local farmers, who are no longer pressing as strongly as before for more of the forest reserve to be cleared for agriculture.

Other projects aim to reduce pressure on the forest by providing alternative sources of wood. The Kibale Forest Project in Uganda and the self-help Green Belt Movement of Kenya have set up tree nurseries managed by villagers, planted trees for fuel and taught the people how to grow crops while retaining tree cover.

Education needs to be extended to the decision-makers. Many politicians and civil servants in Africa have grown up in the cities and see wild animals and wildernesses simply as resources to be exploited ruthlessly, or as symbols of the country's backwardness, providing obstacles to progress. But even with more conservation education, finance remains a problem. Although there are international groups involved in funding, such as the World Wide Fund for Nature, and international agreements to deal with problems such as elephant poaching by banning the trade in ivory, the rich governments of the world are going to have to invest far more aid in conservation if the protected-area systems of tropical Africa are to survive.

Serengeti National Park

Serengeti, in Tanzania, is one of the world's most famous national parks. It contains the largest concentration of grazing mammals and their predators in Africa, providing the stage for some of the most impressive wildlife scenes found anywhere on this planet.

In the early years of this century Serengeti was known for its abundance of lions, but so many were shot by big game hunters that the area was made a game reserve in 1929. In 1951 it was declared a national park. Thirty years later the Serengeti's unique international importance was formally recognized when, together with the adjoining Ngorongoro Conservation Area, it became a World Heritage site.

The national park forms less than half of a much larger ecosystem, which includes the Ngorongoro Conservation Area, two game reserves, three game controlled areas and the Masai Mara National Reserve. Serengeti itself covers some 23,000 sq km (8,880 sq mi).

The Serengeti's southeastern plains are covered by grassland. The rest of the park consists of open acacia woodland. Both habitats support a community of large plant-eating mammals such as antelopes, elephants, giraffes, rhinos and zebras, which in turn support numerous large predators, including cheetahs, leopards, lions, hyenas, jackals and wild dogs. Nearly 500 species of birds, 45 species of mammals and 55 species of acacia have been recorded in the park.

Around 2 million large plant-eating mammals exploit the savanna. Competition between them is reduced because they eat different kinds of plant foods. Some are browsers and feed on the leaves and twigs of trees and shrubs; others are grazers, preferring to eat grass; and some both browse and graze. Diet also varies with body size. Among the browsers, dik-diks nibble at the trunks, duikers forage a little higher, black rhinos reach higher still and giraffes can browse up to 5.4 m (18 ft) off the ground.

The grazers also have different food preferences. At the beginning of the dry season zebras open up the dense pastures of coarse grass by eating the tips and stems of the grass and exposing the leaves lower down the plant. Wildebeest then move in and graze down the sward, exposing the small, protein-rich shoots and herbs that are eaten by the little Thomson's gazelles. Only about 70 per-

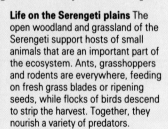

Life on the Serengeti plains The open woodland and grassland of the Serengeti support hosts of small animals that are an important part of the ecosystem. Ants, grasshoppers and rodents are everywhere, feeding on fresh grass blades or ripening seeds, while flocks of birds descend to strip the harvest. Together, they nourish a variety of predators.

Herds of wildebeest and zebras graze the Serengeti, one of the last great concentrations of big animals found anywhere on earth. The herds of grass-eaters make annual migrations around the plains as the shifting rains stimulate plant growth in different areas. The ecosystem of the plains is not static. Variations in climate and other factors such as fire are continually altering the balance between woodland and grassland.

The unending seas of grass Grant's zebras stand belly-deep in long grass. Their stomachs are adapted to cope efficiently with a diet of coarse, low-quality grass; by consuming the tougher parts of the plants, they make the rest available to antelopes, which need higher quality grazing. Wildebeest chew down the sward and are followed by gazelles that nibble the new shoots. Without the continual grazing of these herds woodland would reestablish itself over much of the plains.

cent of the plant is consumed, leaving enough for regrowth. Many of these grazers migrate over large distances during the year in search of better grazing.

Fires sweep through the park during the dry season. These can alter the balance between woodland and grassland. Controlled burning programs early in the dry season can help prevent fiercer fires later on. During the 1970s the rainfall pattern changed: more rain fell in the dry season. This led to more frequent fires, because the thick grass sward provided a greater supply of fuel.

The improved grass growth also helped wildebeest and buffalo numbers to increase, putting pressure on the grazing. At the same time the number of elephants in the park increased. All these factors caused the park's woodland areas to decline. But in the late 1970s ivory poachers began to kill large numbers of elephants, and the woodlands started to regenerate. Poaching remains a major problem, with heavily armed gangs ranging over vast areas of remote countryside.

New threats are still arising. Some of the villagers in northwest Tanzania are calling for part of the Serengeti to lose its protected status to provide extra land for cattle grazing. There are also proposals to build a railway across the Serengeti plains. This would cut across the migration routes of the wildebeest, possibly destroying one of the last great natural wonders of the world.

The Mountains of the Moon

The misty, lichen-festooned forests of the Ruwenzori mountains, the so-called Mountains of the Moon, are typical of the forests found on high tropical mountains in many parts of the world. The gnarled, twisted trees are draped in *usnea* lichens, their roots hidden by thick, soft cushions of moss. What is particular about the high-altitude vegetation of these and other eastern African mountains are the giant, flowering plants that have evolved to survive the extremes of day and night-time temperatures. Their rosettes of stiff leaves are really giant buds that close up at night around the delicate growing shoots: the temperature inside these buds may be several degrees higher than that of the surrounding air. A thick cork bark and dense sheaths of dead leaves also help to insulate the plant. Similar strategies have been adopted by some plants – the espeletias – that are found high in the Andes, providing an example of how unrelated species may evolve similar adaptations to similar environments, a process known as convergent evolution.

Giant lobelias and groundsels among the high-altitude forests of the Ruwenzori mountains.

ANCIENT WILDERNESS

COOL COASTS AND ARID DESERTS · A PATCHY NETWORK OF PROTECTION · A BRIGHTER FUTURE

Despite its rapidly increasing population, southern Africa still has large stretches of unspoiled wilderness, which have not yet been tamed. The habitats are particularly diverse, ranging from the tropical rainforest of Angola, inhabited by gorillas, chimpanzees and forest elephants, to the sand dunes and gypsum flats of the foggy Namib Desert, characterized by a rich and unique beetle fauna. Mangrove swamps and coral reefs flourish in the warm waters off the east coast, while the cooler Atlantic waters support fur seals, penguins and a host of seabirds. The plants and animals of Madagascar, which has been separated from the mainland for 30 million years, constitute a unique collection of species that have survived here since ancient times, but have long since vanished from the rest of Africa.

COUNTRIES IN THE REGION

Angola, Botswana, Comoros, Lesotho, Madagascar, Malawi, Mauritius, Mozambique, Namibia, South Africa, Swaziland, Zambia, Zimbabwe

Major protected area	Hectares
Bikuar NP	790,000
Cape of Good Hope NR	7,675
Central Kalahari (Kgalagadi) GR	5,1809,000
Chobe NP	998,000
Etosha NP	2,227,000
Gile NP	210,000
Hwange (Wankie) NP	1,465,100
Iona NP	1,515,000
Isalo NP	81,540
Kalahari Gemsbok NP	959,000
Kasungu NP	231,600
Kisama NP	996,000
Kruger NP	2,000,000
Lake Malawi NP WH	9,400
Luando INR	828,000
Macchabee/Bel Ombre (Mauritius) NR BR	3,611
Mana Pools NP WH	219,600
Marion Island NR	None given
Montagne d'Ambre NP	18,200
Moremi NP	390,000
Namib–Naukluft NP	4,976,800
Nosy Mangabe SR	520
Pongola NR	6,222
St Lucia GR RS	12,545
South Luangwa NP	905,000
Victoria Falls/Zambezi NP	58,300
Zinave NP	500,000

BR=Biosphere Reserve; GR=Game Reserve; INR=Integral Nature Reserve; NP=National Park; NR=Nature Reserve; RS=Ramsar site; SR=Special Reserve; WH=World Heritage site

COOL COASTS AND ARID DESERTS

Continental southern Africa's varied topography, the size of the region and the different ocean currents that wash its coast account for an extreme variety of climates and habitats. The east coast is bathed by the warm tropical waters of the Mozambique Current. Rainfall is high, and the lush mangrove swamps and forests of the coastal plains are alive with troops of vervet monkeys, mongooses, rodents, and a rich and colorful bird life. Extensive marshy river deltas fan out across the wide coastal plain, attracting vast herds of buffaloes, elephants and waterbuck, and thousands of wildfowl.

Moist evergreen temperate forests grow on the seafacing slopes of the mountains that fringe the narrower southern coastal plain, where rain falls throughout the year. In the extreme southwestern tip of Africa the rainfall is restricted to the winter, and here a unique plant community, the fynbos, is found. The species of this Mediterranean-type heath and shrubland are unparalleled in diversity and beauty; they include colorful proteas, heathers, gladioli and irises.

Along the west coast the cold waters of the Benguela Current well up from great depths, bringing nutrient-rich waters to the surface. This is the breeding ground of the abundant shoals of pilchards and anchovies that support several hundred thousand fur seals, jackass penguins, gannets and other marine vertebrates. Mangrove swamps are rare along this coast, and there are no offshore corals. The Benguela Current gives rise to great banks of fog, which frequently shroud the coast. They provide a vital source of water for the Namib Desert, which is one of the world's few fog deserts.

The parched hinterland

The steep escarpments that rise between the coasts and the interior plateau are covered with mountain heath and scattered shrubs, and with isolated patches of moist forest. The latter provide a habitat for many birds that are unique to the

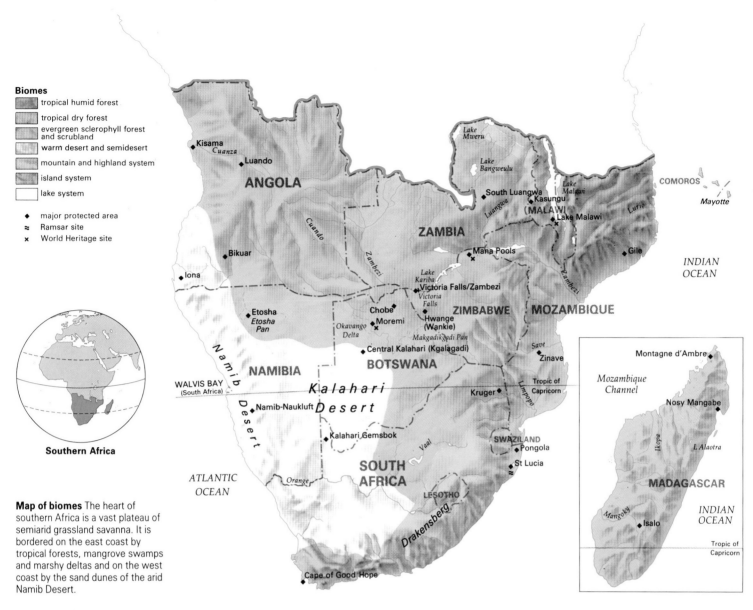

Biomes

- tropical humid forest
- tropical dry forest
- evergreen sclerophyll forest and scrubland
- warm desert and semidesert
- mountain and highland system
- island system
- lake system

- ◆ major protected area
- ≈ Ramsar site
- × World Heritage site

Southern Africa

Map of biomes The heart of southern Africa is a vast plateau of semiarid grassland savanna. It is bordered on the east coast by tropical forests, mangrove swamps and marshy deltas and on the west coast by the sand dunes of the arid Namib Desert.

locality (endemic species). The wind-swept highlands of Lesotho, rising to 3,700 m (11,000 ft), are often snowcovered in winter and bear communities of heathland and alpine plants similar to those of the east African mountains.

On the vast interior plateau the nature of the wilderness is determined by the amount of rainfall received. Grassland and thornbush predominate in the east where rainfall is seasonal and exceeds 1,000 mm (39 in) a year. The habitat supports large grazing and browsing mammals such as elephants, rhinos, giraffes, springbok and various antelopes, as well as their predators, which include lions, cheetahs and leopards.

Desert or semidesert conditions prevail in the drier west, covering much of South Africa, Botswana and Namibia. However, some oases do exist in this arid environ-

Mediterranean-type heathland Diverse species of flowering shrubs and heathland plants, called fynbos, cover the mountains and coastal plains of the western and southern Cape. The heathland plants of the fynbos flourish on the nutrient-poor soils.

ment. Major rivers drain into the Etosha Pan in Namibia, the Okavango swamps (one of the largest inland deltas in the world) and the vast Makgadikgadi Pan in Botswana. Great herds of antelope and zebra migrate into these areas during the wet season.

Endemic island tree ferns on Réunion in the Indian Ocean. Tropical forests have a complex system of layers within their dense canopies of leaves at the treetops down through luxurious undergrowth to the leaf litter of the forest floor.

Farther north, toward the Equator, lie tropical rainforests. These are particularly extensive in the western part of the region, where the rainfall is heavier.

Island life

The ecosystems of the subantarctic islands of Prince Edward and Marion, off the southeast coast of Africa, have little in common with those of the African mainland. These cold, wet and constantly windy outcrops have acquired distinctive communities of plants and animals. There are large breeding colonies of elephant seals and fur seals on the coastal plains, with even greater numbers of penguins, petrels and albatrosses.

By contrast, the tropical islands of the Comoros, Mauritius and Réunion have evergreen forests, plantations and grasslands. The high altitude of the Comoros encourages the growth of giant heathers similar to those of the high mountains on the continent. Wildlife here and in Madagascar includes tenrecs and lemurs, long since extinct on the mainland.

A PATCHY NETWORK OF PROTECTION

The history of conservation in Africa is an ancient one. For centuries, traditional hunting areas were recognized and strictly protected by African peoples. It was not until 1894, however, that the first legally protected area was officially proclaimed – the Pongola Government Game Reserve was created in what was then the South African Republic.

National parks and wildlife reserves in the region now account for more than 503,000 sq km (194,200 sq mi), nearly 8 percent of the land area. Botswana has the highest allocation of reserves; some 18 percent of the country's 575,000 sq km (222,000 sq mi) is protected. In addition hunting is controlled in many areas throughout the region, and there are also a number of privately owned reserves. As a result examples of nearly all of southern Africa's diverse ecosystems are under potential protection.

Gaps in the system

Despite the encouraging extent of the protected area network, it has many gaps and weaknesses. One of the most serious problems is the invasion of alien species that threaten indigenous lifeforms. For example, Australian acacias and hakeas have formed dense woodlands in the fynbos of the southwest Cape. They grow higher than the short native heath, choking it and depriving it of light and nutrients. Of the 7,000 plant species recorded in the fynbos, more than 150 are threatened by invasive plants.

A similar invasion threatens Marion Island, one of the world's last great remote wildernesses. In the last century sealers brought in alien weeds and mice. In 1948 cats were introduced to control the mice. By the 1970s the cat population had expanded to such a degree that the cats were preying on burrowing petrels and other ground-nesting birds; every year more than 2,000 cats were killing some 450,000 birds. Since then, feline flu and intensive hunting have reduced the cat numbers to below 300, and efforts to exterminate them continue.

Other islands in the region – the Comoro Islands, Réunion and St Helena – have yet to establish any reserves. Most of the forest on these islands has been felled, and on some of them soil erosion is severe. They are in urgent need of effective forest protection, as well as incentive programs to change traditional farming methods among a population that is growing rapidly and putting added pressure on the land.

In Angola and Mozambique, civil war has brought wildlife conservation to its knees. Initially, lack of law enforcement and the ready availability of modern firearms and vehicles as a result of the conflicts provided rural peoples with the means of poaching. Then the guerrilla armies themselves took part in organized poaching and later the international ivory cartels moved in. The elephant and rhino populations of these war-torn countries have suffered enormously as a result. The wars have disrupted the management of parks and reserves, many of which now have no staff to run them.

In the extreme southwest corner of Angola lies Iona National Park, a sanctuary of 1,515,000 ha (3,742,000 acres) that occupies a vast area of desert dunes, gravel plains and mountains.

The cold, bleak coastline, shrouded by fog, is inhabited by fur seals and jackass penguins, which provide occasional quarry for brown hyenas and blackbacked jackals. Inland, the fog-moistened dunes are populated with sand-diving lizards and beetles, burrowing geckos, side-winding adders and sand-swimming lizards, which use a variety of strategies to escape the intense heat of the midday sun (at least 50°C/122°F).

The gravel plains are scattered with stands of *Welwitschia mirabilis*, a strange plant with a giant root and two extremely long leaves that absorb the dew. These are interspersed with hardy shrubs, acacias, dwarf baobabs and grasses. Herds of gemsbok, zebra and springbok wander the plains, following the erratic rainfall. The arid savanna and succulent vegetation of the mountains in the interior support a number of small antelopes.

The Himba/Herero peoples who live in the park are pastoralists who share the resources of this harsh land with the wildlife. Recently, however, the park has suffered from the fighting on the Angola/Namibia border. It is to be hoped that once the conflict is over, the machinery to protect and manage Iona will be restored.

Management under pressure

The Kisama National Park in Angola typically lacks the resources needed to sustain management. This area of nearly 1 million ha (2.5 million acres) of unspoilt grassland, floodplain and forest was once home to massive herds of red buffalo, eland and roan antelope, as well as several hundred elephants. For many years is was also occupied by 25,000 head of cattle.

Since Angola's independence oil wells have taken the place of the cattle herds. Although these pump money into a war-ravaged economy, they pollute the once-pristine streams and mangrove swamps. In addition a dense grid of bulldozed roads has ripped into the forests and swamps. Erosion scars now cut deep into the escarpments, and layers of sediment choke the stream beds and lagoons.

One of the greatest threats to protected areas comes from the conflict between people and wildlife as human populations and their livestock increase. For example, in Botswana the European Community has funded the attempt to control foot-and-mouth disease – more than 1,200 km (745 mi) of fences were erected across Botswana to prevent infected cattle from moving into disease-free areas. Little consideration was given to the environmental impact of the scheme. As a result, the migration routes of the numerous hartebeest, eland and wildebeest were disrupted, and tens of thousands of these antelopes were isolated from grazing and water. Estimates suggest that more than a quarter of a million antelopes died along the fences between 1979 and 1984.

The Karoo National Park Karoo is the Hottentot word for the semiarid deserts of southern Africa. Succulent plants, typical of the habitat, survive the heat and aridity by conserving moisture. Other plants have life cycles that are geared to the occasional violent storms.

Where buffaloes roam The Mana Pools National Park in Zimbabwe is a classic savanna woodland with forests of acacia and mahogany. The woodland provides food for a rich variety of wildlife including elephants, rhinos and waterbuck.

A fog desert ecosystem Plants in the desert are sparse and usually have little foliage above ground, though they may have extensive root systems. Consequently productivity above ground is very low and windblown debris is the main food source for desert animals.

Components of the ecosystem

1	Welwitschia mirabilis	7	Namib clown dune cricket
2	Short grasses	8	Sidewinder snake
3	Seeds	9	Barn owl
4	Darkling beetles	10	Sand lizard
5	Ant	11	Spotted eagle owl
6	Gerbil		

Energy flow

⇒ primary producer/primary consumer
⇒ primary/secondary consumer
⇒ secondary/tertiary consumer
⇒ plants produce seeds

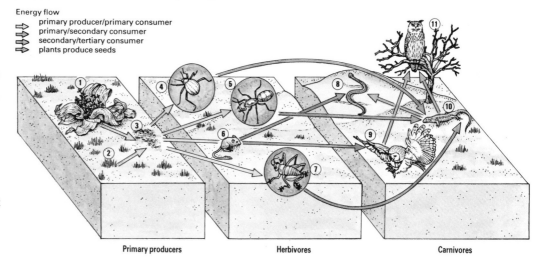

Primary producers　　　Herbivores　　　Carnivores

A BRIGHTER FUTURE

The number of national parks and reserves in southern Africa continues to rise despite the pressures posed by agriculture, industry, mineral extraction, war and a rapidly soaring rate of population growth. However, many that once had good records of conservation exist today in name only, threatened as they are by all kinds of human activities carried on within their boundaries.

But the prospects for the parks in the region are not entirely bleak. Much conservation work has been achieved in the past decade, especially in Malawi, Zimbabwe, Swaziland and South Africa. Many new national parks have been established, existing reserves extended, and the theory and practice of wildlife management advanced. Even the effects of civil war, environmentally disastrous though they are, have not been entirely without benefit to wildlife. Huge areas of rainforest, threatened by deforestation for coffee plantations, have reverted to a near pristine state, while timber extraction, mineral exploitation and other potentially damaging processes have been halted in large areas of the region.

The success stories
In the last few decades the management of African wildlife areas has become highly skilled, combining modern research techniques with traditional hunting and tracking lore.

The Kruger National Park in South Africa was one of the first parks to develop such techniques, pioneering the use of remote infrared satellite imagery to study the way vegetation patterns alter in response to grazing and browsing. It also used radio-tracking devices to follow key species and carried out a program to vaccinate rare antelope species against anthrax. Current research projects include studies of the ecological effects of fire on the environment and the water requirements and physiology of large mammals.

The park's use of technology has undoubtedly aided wildlife management. By developing immobilizing drugs that enable the capture and translocation of large mammals, it has been possible to reintroduce six species that had disappeared from the park: black and white rhinoceroses, oribi, gray rhebuck, red duiker and Lichtenstein's hartebeest. The park relies to a great extent on tourism for its income, increasing the employment opportunities for local people.

Elsewhere in the region the local people living close to protected areas have become involved in the economy of the wildlife and tourist industry. Successful models have been developed in Zimbabwe where communal areas are divided into zones for livestock grazing, cultivation, residential purposes and wildlife management. A district wildlife committee elected by villagers monitors game populations, sets hunting and culling quotas and organizes hunting safaris for tourists. Each village committee is paid for the animals that are

The roof of Africa The Maluti Mountains are just one of Lesotho's many ranges. The mountain terrain is unsuitable for most game animals, but supports a wealth of bird life including the lammergeyer or bearded vulture.

Remnant tropical rainforest amid montane grassland in the mountainous Nyika National Park in Malawi. Malawi has a good record for establishing conservation areas and parks in recent years. Game animals such as antelopes, buffaloes, elephants and lions are plentiful.

Island under threat The habitats of Madagascar range from dry desert to tropical forest, and support some unique species that have evolved in isolation. However, the island is under pressure from the increasing population, who want land for cash crops and cattle ranching.

shot in its district. The village committees also manage forest areas, grazing and water, and decide how best to use the income from hunting. More projects of this kind would help to ensure the future of southern Africa's fragile habitats.

Increasing public awareness

Public support for protected areas has increased dramatically. For example, when the South African government announced its intention to construct an emergency landing strip on Marion Island, conservationists around the globe protested and an independent environmental impact assessment team was appointed. It recommended that the project should be cancelled, and Marion should be permanently protected as a strict nature reserve.

International bodies such as the World Wide Fund for Nature (WWF), the International Union for Conservation of Nature and Natural Resources (IUCN) and the United Nations Environment Program (UNEP) are working to establish an adequate network of protected areas that will preserve viable populations of all species by the year 2000. A study carried out in South Africa indicated that breeding populations of more than 90 percent of the country's native birds and mammals and 80 percent of amphibians, reptiles and plants are to be found within the network of protected areas. It is to be hoped that such a degree of protection will one day be achieved for the whole of the southern African region.

MADAGASCAR – ON THE CRITICAL LIST

Separated from the mainland for 30 million years, Madagascar's distinctive habitats have evolved in isolation. As a result, they support an extraordinarily high number of endemic species – as many as 95 percent of the island's mammals and over 7,000 species of flowering plants. In the east, the Madagascan rainforest contains a rich plantlife, including a variety of spectacular orchids and the Madagascar periwinkle. The tree canopy is shared by red-ruffed lemurs, bats, and birds such as cuckoo rollers and helmet birds. The mountains that rise above the rainforests are covered in lowgrowing mist forest. Here, the moss on the trees is so wet that tadpoles swim along it.

Farther west deciduous woodlands extend to the coast, supporting ringtail lemurs and sifakas (an endemic species of lemur) among other animals. To the south the desert of thornbush and baobab is inhabited by lizards, radiated tortoises, sifakas and tiny lemurs.

About 80 percent of the island has been deforested, and as a result some 340,000 sq km (102,000 sq mi) have now developed into degraded coarse grassland and badlands. The island's indris (a kind of lemur) are dying out and the virtually extinct aye-aye, a primate unique to Madagascar, is protected only on Nosy Mangabe island. Although the government has been encouraging local village councils to educate the villagers in habitat management, it still needs considerable financial support from other countries if it is to ensure the survival of its unique natural heritage.

The Okavango Delta

Rising in the highlands of Angola, the Okavango river never reaches the sea. It dies in the dry expanses of Botswana's Kalahari Desert. On its journey it creates a vast inland delta, one of Africa's most diverse wilderness areas.

The delta's core covers some 160,000 ha (400,000 acres), but after good seasonal rains it increases to 200,000 ha (550,000 acres). This annual flooding of the Okavango keeps the system flushed with fresh water. The water moves slowly through the delta, arriving in its southerly floodplains in the dry season. Its sluggish speed means that on its journey 95 percent of the water is lost to evaporation. However, a few main channels carry water more swiftly through the delta to a point 165 km (102 mi) to the south, where it is trapped by a series of faultlines and redirected to the vast Makgadikgadi Pan, where it evaporates.

The juxtaposition of wetland and dryland gives the Okavango its unique character and accounts for the diversity of its wildlife. Large numbers of birds, including kingfishers, cormorants, fish eagles and the African fishing owl, fish the permanent water and dense reedbeds of the upper reaches. The large mammals are mainly well-adapted swamp specialists such as sitatunga and lechwe antelopes, crocodiles and hippos.

Farther south the islands are larger. Their dry and sandy interiors are fringed by forests, which are home to small mammals such as mongooses, genets and civets, and many birds.

In the delta's lower reaches the reedbeds give way to arid thornscrub and grassy floodplains. When grazing outside the Okavango is scarce, the plains are visited by migrating populations of buffaloes, elephants, zebras, 14 species of antelopes and their predators – lions, leopards, cheetahs, wild dogs and hyenas.

The fringes of the delta are also inhabited by cattle-herders such as the Tswana and Herero peoples. Cattle were formerly unable to live in the heart of the delta because of the presence of disease-bearing tsetse flies. However, this natural check has been virtually removed by new spraying techniques. Cattle are now spreading farther into the delta and reducing the range of antelopes such as the sitatunga, whose numbers are consequently on the decline.

Exploiting the Okavango's resources

Always important to local hunters, the Okavango is now the base of a growing tourist and hunting industry. In 1987 this industry was worth $12 million to Botswana, and accounted for nearly half the area's employment.

Aware of the fragile nature of the delta's river systems, the Botswana government has frozen development here – including a scheme to extract water from the delta to supply a diamond mine on the edge of the Makgadikgadi Pan – until completion of a land-use planning study.

The only legally protected part of the Okavango delta is the Moremi Wildlife Reserve. This was the first wildlife sanctuary in southern Africa to be created by a native people (the Tswana) on their own land, and meets the IUCN's criteria for a World Heritage site. The WWF have urged Botswana to join the Ramsar Convention, drawn up in 1971 to protect the world's wetlands. This would provide some leverage against territorial disputes over water. All decisions are in the hands of the local people, who have shown restraint in developing the Okavango and a healthy concern for its future.

The Makgadikgadi Pan stretches across the Kalahari Desert. After violent storms and flooding the salt pans are filled with water, attracting huge flocks of colorful flamingoes and waterfowl. Migratory herds of game follow the rains and, in turn, attract hyenas, jackals, cheetahs and lions.

Crossing the delta A herd of red lechwe antelope cross one of the numerous islands on the great floodplain of the Okavango delta. Lechwe are the most aquatic of the grazing antelopes and they have developed special adaptations to cope with life in the swamps and floodwaters. Their hooves, for example, splay out widely to prevent the animals from sinking into the mud.

African wilderness The Okavango delta is the greatest single wilderness in southern Africa. The rivers and wooded islands have an abundance of wildlife including huge herds of buffalo and antelope, along with crocodiles, hippos, turtles, eagles, snakes and frogs.

Desert survival

Despite their harsh environments, deserts are able to support a surprising range of species. The Namib Desert is among the driest places on Earth: yet even here plants have adapted to survive, gaining much of their moisture from the frequent fogs that roll in from the ocean. Some of the plants absorb the fog directly through their leaves; others have grooved leaves that channel the water down toward their roots.

Some Namib animals use a very similar technique. Tenebrionid beetles live in the sand; when the fog rolls in they climb to the crest of the dunes, duck down their heads and raise their abdomens in the air. Condensing fog droplets trickle down grooves in the beetles' wingcases to their mouths. Dune beetles dig narrow trenches in the sand to collect water. These scavenging beetles – which live off the dead remains of insects and plants blown in on the wind – form the basis of a food chain. Most desert predators are nocturnal, spending the day in cool burrows or under rocks and emerging to hunt in the cooler night-time temperatures.

A salt pan in the Namib Desert is one of the most inhospitable places on Earth. The dead trees have been poisoned by accumulating salt, deposited by rivers that have dried up, evaporating in the intense heat.

SANCTUARIES OLD AND NEW

THE GREAT VARIETY · A LONG HISTORY OF PROTECTION · PEOPLE AND PARKS

The great physical contrasts of the Indian subcontinent have given rise to an enormous variety of habitats that in turn support a rich diversity of plants and animals. Thousands of species of flowering plants survive the low temperatures of remote valleys and mountain slopes, which in winter are snow-covered; the succulents that thrive in hot, scrub desert areas support gazelles and the livestock of pastoral nomads; and moist tropical forests and wet grasslands provide ideal cover for the world's largest cat – the tiger. Sadly, much of this diversity has been lost with the advance of agriculture and towns in what is one of the most densely peopled regions of the world. However, some habitats have long been protected by the sacred value accorded to wildlife by the ancient Hindu scriptures.

COUNTRIES IN THE REGION

Bangladesh, Bhutan, India, Maldives, Nepal, Pakistan, Sri Lanka

Major protected area	Hectares
Bandipur NP	87,420
Chitral Gol NP	7,780
Corbett NP	52,082
Dachigam NP	14,100
Gir NP	25,871
Hazarganji-Chiltan NP	13,166
Kanha NP	94,000
Kaziranga NP	42,996
Keoladeo NP	2,873
Khunjerab NP	226,913
Kirthar NP	308,733
Lal Suhanra NP BR	31, 441
Manas WS	4,385
Margalla Hills NP	14,786
Namdapha NP	198,524
Nanda Devi NP	63,033
Nilgiri (proposed) BR	552,000
Nokrek NP	6,801
North Island S (proposed) BR	49
Royal Bardia NP	96,800
Royal Chitwan NP WH	93,200
Ruhuna NP	97,881
Sagarmatha NP WH	114,800
Sinharaja FoR BR WH	8,864
Sundarbans WP NP WH	133,010
Wagomuwa NP	37,063
Wilpattu NP	131,879

BR=Biosphere Reserve; FoR=Forest Reserve;
NP=National Park; S=Sanctuary; WH=World Heritage site;
WP=Wildlife Park; WS=Wildlife Sanctuary

THE GREAT VARIETY

The habitats of the Indian subcontinent are diverse. They include the world's highest mountain range in the north, tropical forests, wet and dry grasslands, warm desert in the northwest, mangrove forests in the east and coral reefs in the surrounding oceans, with rich marine life and palm-fringed beaches.

Forests once covered much of the subcontinent, providing food and shelter for many large mammals, including Indian elephants, deer, leopards, tigers, langurs (monkeys), and a host of birds. Much of the original forest has been felled for fuelwood, timber, and to clear the land for agriculture. In India forest covers only 11 percent of the country, but in Bhutan, where development has not proceeded apace and conservation of forests is given high priority, 53 percent of the forest cover remains.

The diversity of plant and animal species on the subcontinent is a legacy of its turbulent geological past. When the Indian subcontinent collided with the mainland of Asia some 40 million years ago, pushing up the Himalayas as it did so, plant and animal species from both land masses migrated into new habitats, diversifying as they spread.

Contrasts in the north

Despite the inhospitable climate of the northern mountains, a variety of plants and animals manage to live there. The eastern Himalayas, for example, with about 4,000 species of flowering plants, have one of the richest collections of plant species in the world. There are semi-evergreen forests of rhododendrons, laurels, maples, alders, oaks, birch and conifers below the treeline. In the western Himalayas mixed oaks and conifers predominate, with deodar (Indian cedar) on drier aspects and chir pines and sals (a native broadleaf) farther down. These forests and the alpine meadows support many species of grazing and predatory mammals, including leopards and both brown and Himalayan black bears.

Adjacent to the Himalayan foothills is a belt of marshy jungle and wet grasslands known as the terai. Here, sals, bamboos and tall grasses are grazed by mammals such as Indian rhinoceroses, wild water buffaloes, gaurs and several species of deer, and they also provide cover for tigers.

In search of grazing A Rajput shepherd with his flock among the sand dunes of the Thar Desert, part of the wide desert belt that stretches from the Sahara to the Gobi. Overgrazing and clearance of tree cover for fuel means that the Thar Desert is spreading at a rate of 8 km (5 mi) a decade.

Coniferous forests that cling to the upper slopes in Sagarmatha National Park begin to thin out near to the treeline, at about 4,000 m (13,100 ft). Mount Everest lies within the park's boundaries.

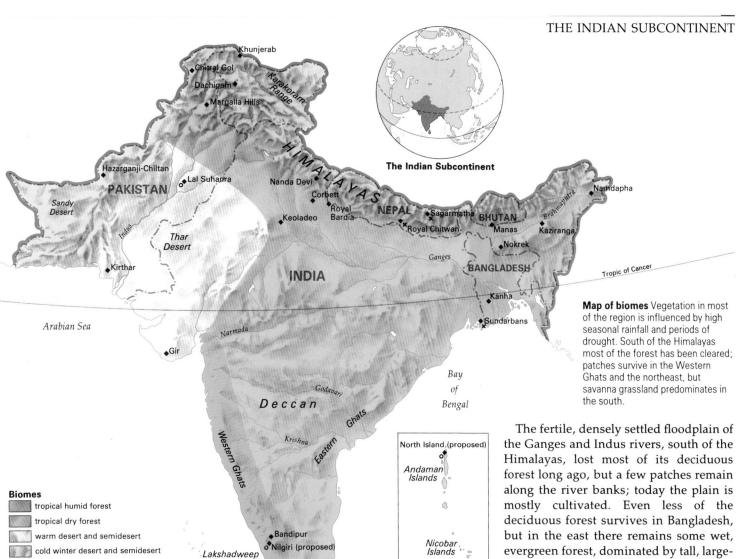

The Indian Subcontinent

Biomes

- tropical humid forest
- tropical dry forest
- warm desert and semidesert
- cold winter desert and semidesert
- mountain and highland system
- island system
- lake system

- ◆ major protected area
- ○ Biosphere Reserve
- × World Heritage site

Map of biomes Vegetation in most of the region is influenced by high seasonal rainfall and periods of drought. South of the Himalayas most of the forest has been cleared; patches survive in the Western Ghats and the northeast, but savanna grassland predominates in the south.

The fertile, densely settled floodplain of the Ganges and Indus rivers, south of the Himalayas, lost most of its deciduous forest long ago, but a few patches remain along the river banks; today the plain is mostly cultivated. Even less of the deciduous forest survives in Bangladesh, but in the east there remains some wet, evergreen forest, dominated by tall, large-buttressed dipterocarps, with a crown of branches at the top.

The rivers Ganges, Brahmaputra and Meghna drain into the Ganges delta, a vast, lowlying swampy area with extensive mangrove forests – the Sundarbans – and grassland, providing a rich habitat for mammals, birds and reptiles.

To the west lies the Thar Desert. On its rocky hills and plains grow scattered bushes and low trees of gum arabic, spiny acacia, succulents, the ubiquitous khejri (an evergreen tree found only in this area) and tamarisk. The plants support grazing mammals such as wild asses, blackbucks and chinkara. Birds that can tolerate the arid conditions include the great Indian bustard, the houbara bustard and the imperial sandgrouse.

The peninsula

Peninsular India consists of the Deccan plateau, a dry area covered with short-grass savanna and a few palm species unique to the area (endemic species). The mountains of the Western and Eastern Ghats run parallel to each coast, and are clothed in moist, evergreen forests. Lion-tailed and bonnet macaques, bats, giant squirrels, mongooses, hares, several species of deer, gaurs, tigers, wild boars and elephants dwell in the forests.

A LONG HISTORY OF PROTECTION

The protection of nature has a long history in the Indian subcontinent. Several thousand years ago sacred groves were established by hunter-gatherer communities; limited use of their resources was sanctioned only during times of calamity such as fire and drought. Sacred groves are widespread in India – more than 400 are known of in the western state of Maharashtra – and in Nepal. The Vedas, ancient Hindu scriptures, contain directives to protect the environment and all forms of life. As early as the 4th century BC, the establishment of forest reserves was being advocated in the Arthasastra, a manual of statecraft.

Much later, princes and kings preserved many areas specifically for hunting. Today, these form the basis of a number of existing national parks and sanctuaries, such as Chitral Gol in Pakistan, Keoladeo and Ranthambor in India, Royal Chitwan in Nepal and Manas in Bhutan. Many more protected areas, established over the last hundred years or so, began as forest reserves to safeguard timber and water resources. Examples include all three wildlife sanctuaries in the Sundarbans of Bangladesh, as well as Sinharaja in Sri Lanka.

Legislating for conservation

Legal provision for protected areas first appeared in Sri Lanka in 1885, following which two game reserves were established – Ruhuna in 1900 and Wilpattu in 1905. Both sites were given national park status in 1938, the same year that Corbett (then named Hailey) National Park was established in India. A dozen more reserves were created at the same time, and others soon followed. By the middle of the century Sri Lanka possessed some of the best wildlife conservation areas in the region. Today, just over a tenth of the island's land area has been granted official protection.

In both India and Pakistan, where conservation is the responsibility of the provincial governments, legislation providing for the establishment of protected areas was adopted in the 1970s. Subsequently there has been an enormous growth in protected areas. Conservation laws were passed in Bangladesh and Nepal in 1973. Bhutan as yet lacks com-

parable conservation legislation, but its protected areas, notified under the Forest Act of 1969, cover a fifth of the country.

Conservation is in its infancy in the Maldives, although a Nature Conservation Act has recently been drafted and a network of reserves proposed. The islands of the Maldives and the Chagos, which contain the largest expanse of undisturbed coral reefs in the Indian Ocean, are of international importance. For the moment, conservation on the Chagos islands is assured: the United Kingdom–United States naval base on Diego Garcia restricts access to the islands to military personnel only.

Forms of protection

Most of the protected areas in the subcontinent are national parks and sanctuaries.

Both protect wildlife, but whereas traditional uses of natural resources, such as grazing and the collection of timber, fuelwood and other forest products, are usually prohibited in national parks, they may be allowed in sanctuaries, provided they do not hinder conservation. The reverse is the case in Bangladesh and Pakistan, however, where settlement and grazing by livestock are prohibited in sanctuaries but not in national parks.

Game reserves established for hunting purposes are found in all countries in the region except Sri Lanka. Sri Lanka has several strict nature reserves – areas of outstanding ecological importance, to which access is permitted for scientific purposes only. Two of them form the cores of the Ruhuna and the Wasgomuwa National Parks. Nepal is the only other

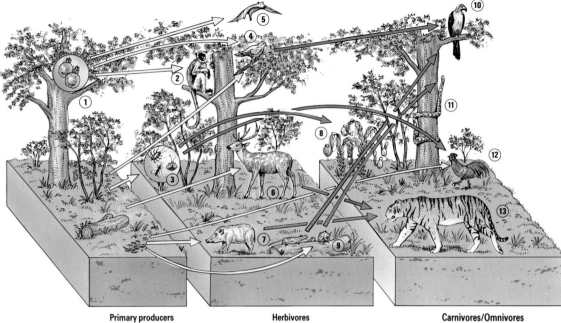

Primary producers   Herbivores   Carnivores/Omnivores

Components of the ecosystem
1 Sal trees
2 Hanuman langur
3 Insects
4 Green pigeon
5 Indian flying fox
6 Chital deer
7 Pygmy hog
8 Pitcher plant
9 Spiny mouse
10 Crested hawk eagle
11 Indian python
12 Red jungle fowl
13 Tiger

Energy flow
⇒ primary producer/primary consumer
⇒ primary/secondary consumer

A royal hunting ground
Ranthambor National Park formerly belonged to the Maharajahs of Jaipur, whose forts lie scattered through its forests. Today it is a sanctuary for the tigers that were once hunted in the royal chase.

Villagers remove excess vegetation in Keoladeo National Park. Water buffalo and cattle, which would keep it in check, may not graze in India's national parks.

The mangrove swamps of the Sundarbans supports the largest tiger population in the subcontinent. The tigers prey on the numerous spotted deer (chital) of this fertile ecosystem.

A monsoon forest ecosystem
supports a wide variety of species. Many forest soils are lacking in nutrients. Some plants, such as the pitcher plant, have adapted to the lack of nitrogen by becoming carnivorous, ingeniously trapping insects to make good the deficiency.

THE SUNDARBANS

On the world's largest delta, at the mouth of the Ganges river, lies the Sundarbans. It is one of the world's most extensive mangrove forests, covering some 1 million ha (2.5 million acres). In India 130,000 ha (321,230 acres) of this area is officially closed to all kinds of exploitation and forms the core of a much larger tiger reserve. In Bangladesh 32,386 ha (80,026 acres) are protected in three separate sanctuaries.

The vegetation in the Sundarbans contains some 334 plant species, including the elegant sundari tree from which the place takes its name. The unique mixture of species includes some related to the plants of Southeast Asia, Polynesia, Ethiopia and even the New World. The Sundarbans is the only mangrove forest in the world inhabited by tigers: it supports the largest tiger population (600–700) in the subcontinent. The man-eating Sundarbans tiger poses a very real threat to human life.

The Sundarbans is a valuable source of timber, fuelwood and honey, and its waters contain an abundance of fish and crustaceans, vital for the local economy. It is under increasing pressure from overexploitation of timber, cultivation and poaching, particularly on the Bangladesh side. Largescale irrigation projects in the Ganges floodplain have reduced the flow of fresh water, resulting in saline conditions that may have caused vegetation to die off. The longterm survival of this unique ecosystem depends on the careful management and balanced use of land and water resources farther upstream.

nation with legal provision for strict nature reserves, but none has yet been established there.

In Bangladesh most of the remaining wildlife is found in forest reserves, but the country's existing network of protected areas is not comprehensive. Low priority has been accorded to wildlife conservation owing to economic constraints but, like Nepal and Pakistan, Bangladesh is in the process of developing a national conservation strategy. In India there are plans to increase the number of protected areas to cover some 150,000 sq km (58,000 sq mi), amounting to nearly 5 percent of the country.

All the countries of the region, except Bhutan, are party to the World Heritage Convention; World Heritage sites have been declared in India, Nepal and Sri

Lanka. A number of important wetlands have been designated under the Ramsar Convention of 1971, signed by Pakistan, India and, more recently, Nepal. The Man and the Biosphere Program launched under the auspices of UNESCO is not yet far advanced in the subcontinent: Biosphere Reserves exist only in Pakistan and Sri Lanka. India is in the process of creating the Nilgiri Biosphere Reserve in the southwest of the peninsula, and has identified 12 other potential sites.

Moist evergreen forests grow in the Anamalai Sanctuary on the hills of the Western Ghats in southern India, where rainfall is high. They are rich in species; monkeys, squirrels and bats live in the tree canopy, and several species of deer, gaurs and elephants browse the lower branches and understorey.

PEOPLE AND PARKS

Extensive changes to the environment of the Indian subcontinent have had severe repercussions. Soil salinity and erosion have turned more than a third of India into a wasteland. In Pakistan, land irrigated by the Indus is becoming increasingly uncultivable because of waterlogging and excess salt. It is argued that deforestation of the Himalayan watersheds has caused rivers and dams to silt up, leading to greater flooding on the Indus and Ganges river plains. There is an urgent need for land-use planning to reconcile production requirements with habitat protection.

The cost of development

Development schemes pose a great threat even to protected habitats. Some 4,200 ha (10,400 acres) of alluvial grasslands in Corbett National Park were inundated in 1974 as part of the Ramganga River Project. The creation of a reservoir, while attracting large numbers of water birds, severely reduced the local populations of deer and cut elephants off from an important traditional migration route. The main problems, however, are fires, which in one year swept across nearly 30 percent of the park, and infestation by exotic weeds such as lantana and cannabis.

Dam projects are becoming increasingly controversial. For example, the decision to go ahead with a hydroelectric project in Silent Valley in southwestern India, site of one of the last stands of virgin tropical evergreen forest in the subcontinent, led to a fierce environmental debate. The project was finally cancelled in 1983 and the area is now a national park and part of the proposed Nilgiri Biosphere Reserve.

Encroachment on protected areas is a widespread problem. Occasionally it has even led to violent confrontation. In 1982 the exclusion of several thousand cattle and water buffalo from Keoladeo National Park, a manmade wetland of World Heritage status, kindled local resentment and ended in a forced entry into the park in which eight people were killed. In the absence of grazing livestock, water plants are allowed to grow unchecked and the wetland now provides little open water for migratory ducks.

Protected areas attract an increasing number of visitors. While tourism often contributes significantly to the national and local economy, it can also erode local culture and lead to inflation, inappropriate development of tourist facilities, overcrowding and increased dumping of garbage. In Nepal, where tourism is the major source of foreign exchange, garbage became such a problem at the base-camp on Mount Everest that in 1984 an expedition was organized to clean up the site. More than 1,800 loads of refuse were collected and removed.

Embracing the trees

One of the most important developments has been the growth of voluntary conservation organizations. Out of the growing ecological crisis in the Himalayas arose the now world-famous Chipko, or "Em-

Grasslands predominate in the dry countryside of the Manas river plain in Assam in northeast India. The habitat supports a variety of species, including two rare mammals, the hispid hare and pygmy hog. However, burning by farmers to improve the grazing for livestock is putting pressure on the surviving wildlife of the area.

Gal Oya National Park in Sri Lanka was created as a water catchment area for the large reservoir that lies at its heart. Most of the park is savanna grassland with patches of forest, and it is a sanctuary for elephants and the many birds that nest around the reservoir.

THE MAHAWELI PROJECT

A multibillion-dollar development program has been set up in Sri Lanka to harness the country's largest river – the Mahaweli – and its tributaries, to generate hydroelectricity and irrigate agricultural land. The affected area covers some 420,000 ha (1 million acres) in the interior of the island, including some of Sri Lanka's finest tropical, dry, mixed evergreen forest, as well as grasslands, swamps and riverside forests. It is rich in plants and animals, with many endemic species, and supports about a third (some 800) of the nation's elephants as well as other threatened mammals, birds and reptiles. To mitigate the project's impact on the wildlife, a system of national parks and sanctuaries has been established within the area. Almost as much land is now under protection as is developed for agriculture and settlement.

The "Accelerated Mahaweli Project" has been heavily criticized. The scheme involves the compulsory resettlement of thousands of local people. Increasing soil salinity is already a problem and, since less than 10 percent of the upper water catchment area is forested, eroded soil is likely to silt up the reservoirs more quickly than anticipated. Environmentalists have pointed to an alternative, less harmful strategy for increasing food production by revitalizing the ancient network of 15,000 or so water storage tanks. The land could be irrigated from them and freshwater fish farmed in the tanks.

brace the Trees" Movement, with its guiding Gandhian philosophy of nonviolent resistance. Its first action took place in April 1973, when villagers demonstrated against contractors felling trees in Mandal forest.

The movement spread like wildfire, and in 1981 won its demand for the protection of Himalayan forests when a 15-year ban was placed on commercial felling in the hills of the state of Uttar Pradesh. The ban was followed in 1983 by a national moratorium on the felling of trees above an altitude of 1,000 m (3,300 ft). The Chipko Movement has also stopped clearfelling in the mountains of the Western Ghats in southern India, and in the Vindhya Range in the northern Deccan. It has pressed for an environmentally sound national forest policy that is more sensitive to people's needs.

An enlightened forest policy already exists in Bhutan, where a high conservation value is placed on forest resources and they are not simply exploited as a source of revenue. All forest land above 2,700 m (8,860 ft) or on slopes steeper than 60 degrees is protected. Grazing is not allowed in these forests, and the practice of shifting cultivation is additionally prohibited on slopes of 45 degrees or more.

The growing awareness of the need to change the emphasis from "people versus parks" to "people and parks" is apparent in one of India's latest conservation initiatives, Project Snow Leopard. A number of reserves will be established specifically to protect this endangered species and its habitat, but the people living in the reserves will not be moved out – instead they will be actively involved in the project.

Royal Chitwan National Park

Extending across the floodplains of the Narayani and Rapti rivers in Nepal, Royal Chitwan National Park, which was a royal hunting reserve from 1846 to 1951, is a remnant of the patchwork of forest and grassland that once covered the terai, or lowlands. About 70 percent of the park is covered with forests of sal, the dominant tree of the terai. Vegetation in the rest of the park is a constantly changing mosaic of grassland and riverside forest, which is in various stages of development (or succession) as the result of fire, and of flooding and riverbank erosion during the heavy rains of the monsoon season.

But the most outstanding feature of Chitwan is the diversity of its animal life – it has some 35 species of large mammals and 489 bird species. In recognition of this diversity and of the minimal disturbance caused to the ecosystem by human activity, Chitwan was designated a World Heritage site in 1984.

Chitwan's wildlife includes a number of threatened species, such as tigers, sloth

A ride on an elephant is the best way to see the wildlife in Royal Chitwan National Park. Although affording an unforgettable experience for visitors and providing a major source of revenue, tourism is a mixed blessing. It upsets the delicate balance between the local population and the ecosystem.

bears, gaurs, hispid hares, Ganges river dolphins, mugger crocodiles, gavials (or gharials) and Indian pythons. Chitwan and the Kaziranga National Park in Assam in the northeast of the subcontinent are the last strongholds of the greater one-horned or Indian rhinoceros, and a number have been successfully moved from Chitwan to Nepal's Royal Bardia Wildlife Reserve. The gavial, similar to a crocodile, but with a long, slender snout, has been reintroduced into the park's rivers: these are reared in captivity from eggs taken from the wild, then released into the park's rivers. Chitwan has a fine tradition of scientific research.

Until recently, few people lived in the terai because of the high risk of malaria. Following the virtual eradication of the

disease in the 1950s, settlers began to clear the sal forests for cultivation. By 1959 some 12,000 people had moved into grasslands formerly grazed by rhinoceroses, and many more had moved in illegally. That same year the government, concerned about the threat to wildlife, established the Mahendra Deer Park, which was the basis of the national park, designated in 1973. Agricultural encroachment continued, and in 1963 22,000 people were moved out of the Rapti valley under a new land settlement scheme.

Chitwan now has a reputation for being Nepal's best-managed national park. Poaching, formerly rife, has been brought under effective control, as shown by the increase in numbers of some of Chitwan's endangered species, notably the tiger and rhino. The tiger population more than doubled from about 25 in 1974 to over 60 by 1980, more than half of which live within the park. Similarly, the rhino population increased in 20 years from 100 in 1968 to about 370 in 1988.

Jungle and mist at dawn against the snow-capped Himalayas. The marshy jungles of Chitwan, a World Heritage site since 1984, are a refuge for several threatened mammal and reptile species. Much of the park's scientific research is carried out with the Smithsonian Institution of the United States.

Deciduous sal trees withstand the seasonal heavy rains and periodic droughts of the monsoon belt. They are the characteristic tree of the terai lowlands and cover much of Royal Chitwan National Park.

The increased numbers of animals have, however, intensified problems in the areas of human settlement around the edge of the park. Damage to crops, mostly by rhinos, wild boar, spotted deer (chital) and parakeets, can amount to the total destruction of the season's crop in some villages. Moreover, tigers and, to a lesser extent, leopards prey on livestock and, occasionally, villagers.

To the east of the park, in Padampur Panchayat, the problem was solved by resettling 7,000 people from 10 villages, a move that met with local support. However, it is neither politically nor economically feasible to relocate any of the remaining 310 villages that surround the park.

Another source of conflict is tourism. Chitwan is Nepal's most popular national park, receiving thousands of visitors each year. The tourist industry has brought some jobs to villagers (and others work in the park), but most are worse off because the cost of living has been inflated.

The greatest benefit that the park bestows on the district is in conserving soil and water, but the villagers regard the grasslands as the park's most valuable asset since they build their houses from grass. The grasslands, maintained for centuries by cutting and burning, are harvested in January. Grass-cutting also benefits animals such as rhinos and deer, as the flush of growth that follows cutting and burning provides good forage. This intimate relationship between wildlife conservation and local people is a model to be encouraged in other areas where the two may conflict.

NATURE TAMED AND UNTAMED

THE EVER-CHANGING WILDERNESS · THE GREAT PANDA EMERGENCY · THE HUMAN IMPACT

China, with Taiwan, Hong Kong, and Macao, covers nearly 7 percent of the world's land area. This vast region is one of great physical beauty and great natural variety. Its habitats range from mangrove swamps and coral reefs in the tropical south to the coniferous forests of the subarctic north. Their wide variety has given rise to a great richness and diversity of animals and plants. Long human impact – China has enjoyed the longest continuous civilization of any region of the world – has had an immeasurable effect on the Chinese landscape. Today the density and numbers of its human population press heavily on its wildlife and its surviving wildernesses. Yet despite this, the region still possesses some of the most remote and unspoiled places to be found anywhere on the planet.

COUNTRIES IN THE REGION	
The People's Republic of China, Taiwan	
Major protected area	**Hectares**
Alligator Sinensis NR proposed WH	100
Arjin Shan NR	45,000
Baihe NR	20 000
Changbai Mountain NR BR	217,235
Dinghu NR BR	1,200
Dongzhai Port NR	2,500
Fanjingshan Mountain CA BR	41,533
Fujian Wuyishan NR BR	56,527
Labaihe NR	1,200
Lunan Stone Forest SZ	1,437
Mai Po Marshes NR BR	380
Mount Emei SZ	11,500
Mount Huanghai SZ	15,400
Mount Taishan WH	42,600
Poyang NR	384,100
Shennongjia NR	90,000
Wanglang NR	27,700
West Lake SZ	7,000
Wolong NR BR	200,000
Xilin Gol Natural Steppe PA BR	1,078,600
Yushan NP	105,490

BR=Biosphere Reserve; CA=Conservation Area; NP=National Park; NR=Nature Reserve; PA=Protected Area; SZ=Scenic Zone; WH=World Heritage site

THE EVER-CHANGING WILDERNESS

China's wilderness areas show enormous contrasts, from the icy peak of Mount Everest (Gomolangma) at 8,848 m (29,087 ft) to the bottom of the Turfan depression, which at 154 m (505 ft) below sea level is the world's second-lowest landmark and the country's hottest spot. The tropical evergreen and seasonally wet monsoon forests of coastal and lowland China in the warm south give way to deciduous broadleaf temperate forests at middle latitudes, and these in turn are replaced by coniferous forests growing in the frozen ground (permafrost) of the cold north. Rainfall decreases markedly toward the northeast. There are high cold deserts on the northern edge of the Plateau of Tibet and deserts of sand and fine, wind-blown loess in the Gobi and Takla Makan farther north.

Landscapes change dramatically across China. From the vast Plateau of Tibet the land descends through great hill ranges, where large rivers have carved spectacular gorges, to the flat, swampy land of the coastal plains, with lakes and reedbeds spreading out between low ranges of

Tropical rainforest at Xishuangbanna in Yunnan Province in the south of China – one of the few places in China where the habitat survives. It provides a refuge for tigers, gaur, elephants and peafowl. However deforestation and overhunting are posing serious problems for conservation, and land is being lost to cultivation and rubber plantations. Xishuangbanna has been made a priority target for the Chinese government's conservation program.

gently rolling hills. These fertile lowlands have been greatly modified by human activity and support dense populations. Here, there is little forest left and the wildlife is much depleted, but the area's mountains and deserts are still largely undeveloped and remain essentially wild.

Landscapes and wildlife
China's considerable variety of landscape and climate has resulted in a great diversity of plant and animal life. Northeastern China is dominated by the Da Hinggan Mountains, clothed in coniferous forests. These support northern Asiatic animals such as sable, moose and the last Manchurian tigers. In the more temperate regions of the lower Huang (Yellow) river valley, the mixed deciduous forests are home to pheasants, badgers, deer, foxes and other woodland animals.

The arid lands along the Mongolian border in the north and the deserts and

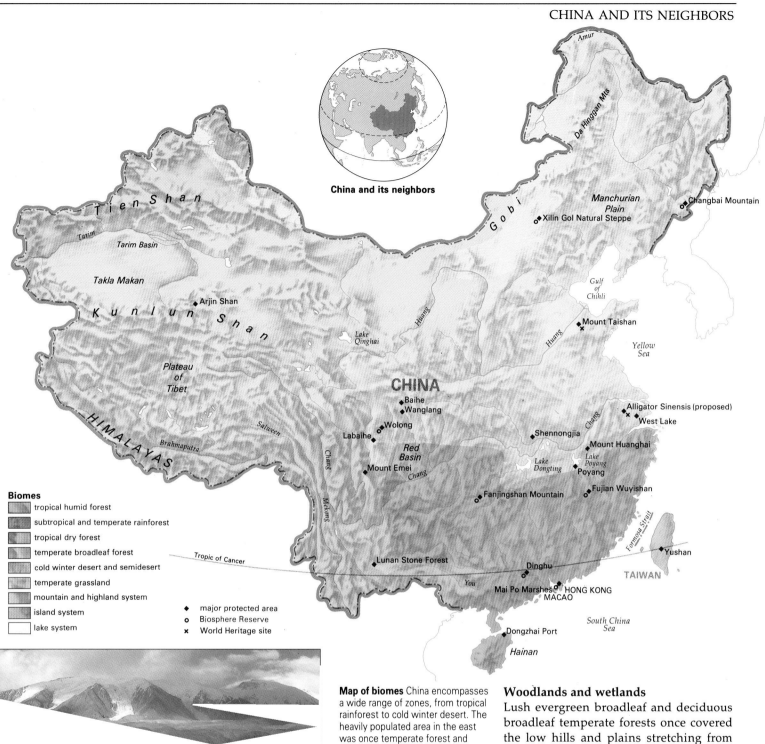

China and its neighbors

Biomes

- tropical humid forest
- subtropical and temperate rainforest
- tropical dry forest
- temperate broadleaf forest
- cold winter desert and semidesert
- temperate grassland
- mountain and highland system
- island system
- lake system

- ◆ major protected area
- ○ Biosphere Reserve
- × World Heritage site

Map of biomes China encompasses a wide range of zones, from tropical rainforest to cold winter desert. The heavily populated area in the east was once temperate forest and grassland, but little natural vegetation remains after centuries of agriculture.

High, dry grassland in the remote northwest supports large grazing mammals, now becoming restricted to domesticated yaks and goats pastured by nomads.

Woodlands and wetlands

Lush evergreen broadleaf and deciduous broadleaf temperate forests once covered the low hills and plains stretching from the Red Basin to the mouth of the Chang river. Most of the forests were destroyed long ago, but the patches that survive are inhabited by pheasants, monkeys and deer. Many important wetlands remain, where alligators and freshwater dolphins live. Lake Poyang is the wintering ground of almost 80 percent of the world's white cranes. Giant pandas live along the narrow strip that divides this area from the mountainous country to the west.

Evergreen tropical forests grow in the far south, in southern Yunnan, Hainan, southern Taiwan and along the south China coast. The lush mangrove forests that once fringed these coasts still remain in places, and offshore there are coral reefs to be found.

mountain ranges of the northwest are mostly covered in scrub by vegetation; surprisingly, this supports large mammals such as camels, wild asses, wild sheep, wild horses and wolves. Vegetation is also sparse on the cold, high Plateau of Tibet and areas to the northeast, some 3 million sq km (1.2 million sq mi) at more than 4,000 m (13,000 ft) above sea level. This is the home of wild yaks, snow leopards, Tibetan antelopes,

gazelles and wild sheep, animals typical of central Asia. The many lakes and salt marshes provide wintering grounds for rare black-necked cranes.

In southwestern China the sweep of the mountains is dissected by the steep valleys of the Salween, Mekong and Chang (Yangtze) rivers. This area is still largely forested, and shelters creatures such as the takin (a rare goat antelope), red panda and snub-nosed monkey.

THE GREAT PANDA EMERGENCY

The first nature reserves in the People's Republic of China were established in 1956 amid the surge of interest in the country's natural resources that followed the establishment of the Republic. However, the Cultural Revolution of the late 1960s and early 1970s was a black period for conservation, when research came to a complete standstill.

A single event did much to reverse the trend. In the mid-1970s the bamboo in the Min range northwest of the Red Basin flowered and then died. This is a natural phenomenon that affects most temperate bamboo species about once every 40–60 years. It posed a dangerous threat to the giant pandas living in the mountains.

Giant pandas are almost totally dependent on bamboo. Those patches that did not flower immediately did so over the next few years. Agricultural encroachment on the mountains' lower slopes had eliminated many other bamboo species with different flowering cycles, which would have provided the pandas with alternative supplies. Hundreds of giant pandas were forced to search for new foods; many died of starvation. The crisis caused great concern in China and launched a huge conservation program.

A priority was the establishment of a more extensive network of reserves to protect areas of good panda habitat. A huge survey of the panda range was undertaken. Two reserves – Baihe, which had originally been established for golden monkeys, and Labaihe, set up for takin – were redesignated panda reserves, and a further nine panda reserves were declared between 1975 and 1979, including the 200,000-ha (494,000-acre) area known as Wolong. In 1980 the World Wide Fund for Nature (WWF) was invited to join the Chinese government in its attempts to save the panda; this cooperation continues today.

The reserves multiply

Since recognition of the panda emergency in China there has been a nationwide resurgence of interest in conservation. Reserves are now being declared so fast it is difficult to keep up-to-date lists. By 1983 China had established more than 120 nature reserves; by 1988 over 400. A further 500 are planned by the year 2000.

THE BENEFITS OF UMBRELLA SPECIES

The reserves of southern China protect far more than the pandas they were set up to preserve. The giant panda is in effect an "umbrella species", whose protection also preserves many hundreds of other important species. The mountains are home to a large number of rare animals – golden monkeys, takin, several species of rare pheasants and other birds, and a wealth of rare plants. Indeed, the greatest value of the panda reserves is probably botanical.

During the climatic fluctuations of the glacial periods of the past 2 million years, much of the plant life in Europe and Asia was destroyed through drought or cold. But the mountains of southern China remained relatively warm and moist because they intercepted the moisture-laden clouds brought by the southwest monsoon, much as they do today. The land is so

steep here that plants were able to adapt to any changes in average temperature by making small migrations up or down the mountainsides.

So the plants survived, and today there is an amazing richness of species in the panda reserves. Wolong, for instance, which ranges from the permanent glaciers of Mount Siguniang at 6,250 m (20,000 ft) to subtropical evergreen forests at 1,110 m (3,650 ft), has an estimated 4,000 species of plants. They include several hundred species that have medicinal properties or are close relatives of plants that are cultivated for medicinal use. Rhododendrons, magnolias, cotoneaster, roses and buddleia abound, along with other shrubs now found in gardens around the world. The area is also rich in wild fruit trees – apples, cherries, pears, gooseberries and many varieties of currants.

The reserves include examples of China's many types of natural landscape and their ecosystems within different geographical zones, the habitats of threatened species of plants and animals, places of outstanding geological or physical features, or simply stunning scenery. They are being set up despite other pressing calls for reorganization and modernization. A range of scientific, educational and recreational interests are taken into account when drawing up the plans for reserves.

The reserves are established by several different authorities. Over 300 are in forested areas and therefore controlled by the forestry ministry; some 30 of these are national reserves administered directly from Beijing, but the remainder were established and are managed at provincial level. The Department of Environment Protection Agency has set up reserves in less forested parts of the country, among them the Arjin Shan Reserve, which is the largest in China. Local government at town and county level can also designate reserves and natural park areas. Three reserves – Changbai Mountain on the frontier with Korea, Wolong and Dinghu – have been designated as international Biosphere Reserves.

Outside the People's Republic

Hong Kong has established some important parks, despite its small size and dense human population. Almost two-fifths of the territory falls within country parks, each park having a protected wilderness zone. Forests are gradually growing once more in the parks, after being almost totally cut for firewood during the Japanese occupation in World War II (1939–1945). In addition, the energetic WWF section in Hong Kong has established a valuable mangrove and wetland reserve at the Mai Po Marshes, which is an excellent place for viewing wetland and shore birds and has a fascinating visitor center.

Conservation also has a role in Taiwan. Four national parks account for 6 percent of its surface and a number of small coastal reserves protect mangroves and other habitats. Yushan, in the mountains, is the largest and most remote of Taiwan's parks. Here the rare Taiwan serow, a goat antelope, and Taiwan black bear can be found, together with the Formosan macaque and emperor pheasant.

The fall foliage of cotinus emblazons a mountainside in Sichuan province. Southern China is home to many species of lowgrowing deciduous shrubs, many of which have been successfully transplanted as ornamental plants in gardens around the world.

Wanglang reserve is one of a number of protected areas set up to conserve the giant panda. The depletion of bamboo supplies in the mid-1970s highlighted the plight of this rare and endangered animal, and it is hoped that these reserves will support a sufficient number of bamboo species to provide the pandas with their staple food.

Components of the ecosystem
1 Bamboo
2 Giant panda
3 Herbivorous insects
4 Red panda
5 Golden pheasant
6 Musk deer
7 Red-breasted flycatcher
8 Leopard
9 Golden jackal

Energy flow
⇨ primary producer/primary consumer
⇨ primary/secondary consumer

A bamboo forest ecosystem
Bamboo flowers only very rarely, and then dies. This often causes great shortages throughout the food chain.

Primary producers   Herbivores   Carnivores

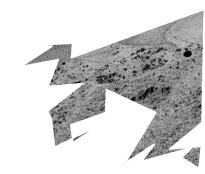

THE HUMAN IMPACT

The Chinese people have for centuries shown love and respect for natural beauty. Their poetry and paintings are full of mountains, waterfalls, lakes, flowers and wild birds and animals. Houses and gardens are filled with rock gardens, potted plants and caged birds. But there is in many Chinese – as in other people – a desire to tame nature, even to improve on it, rather than to appreciate its qualities untouched. Many Chinese would prefer to sit in a flower park or listen to a singing bird in a cage rather than walk in the wild forests and hear birds in their wild habitat. Forests are regarded as savage and primitive, and hence dirty and dangerous. Only "sacred mountains" and a few recently recognized beauty spots draw the populace out of the towns and into the countryside.

Changing attitudes

There remains a massive need to generate greater awareness of the values of natural habitats. Attitudes are changing, but rather slowly. The country's leaders have recognized the need for conservation areas, more rational exploitation of natural resources and better laws to protect rare species. But these new approaches are not fully understood by the majority of the people and conflict with traditional views and economic needs.

At one time it was every man's duty to rid the country of pests such as wolves, bears and wild cats. Now most of those

A waterfall in the Changbai Mountain Biosphere Reserve on the border with Korea. Scenic waterfalls set in rugged, rocky landscapes are commonly celebrated in Chinese art. However, the wilderness itself has not always been cherished.

species are officially listed as highly protected animals, and a heavy prison term or even death sentence awaits anyone convicted of killing them. But law enforcement remains weak and the shops are still full of the skins of rare animals.

Realizing that overpopulation is crippling the country's attempts to modernize, China's leaders have introduced some of the harshest family planning regulations in the world. Even so, the population will continue to grow for many years because a very large proportion of the population is not yet adult. Pressure for land will continue to threaten China's surviving habitats.

The new economic reforms in China have also had an adverse effect on wildlife. People have been encouraged to develop private businesses and money-making ventures in an attempt to generate greater wealth. In economic terms this is a good thing, but the exploitation of wild animals and plants has increased dramatically as a result.

The Chinese have great faith in their traditional medicines. Prescriptions are largely made up of animal and plant extracts. This, too, creates a great industry of hunting and collecting. Demand is growing while the resource base that supplies it shrinks.

China is slowly finding its way through these problems and trying to adjust to the

A valley terraced for rice in Shaanxi Province in northeast China provides the only patch of green in an otherwise barren landscape. Temperate China was once extensively forested, but the trees have disappeared to feed, house and warm a large human population, with a disastrous loss of habitats for plants and wildlife. The main concern must now be to maintain the productivity of the farmland and to prevent erosion, which will lead to further wasteland.

The Stone Forest of Kunming, in Yunnan Province, is one of China's three most visited natural sites. Its hundreds of tall limestone pinnacles were formed by the erosive effect of underground streams 270 million years ago. In many places the pillars are so dense that only mosses grow in the damp gloom, but where there is enough light and soil, lime-loving plants have taken root. Conifers crown the highest pinnacles.

THE LOSS OF CHINA'S FORESTS

Fifty years ago China's hills supported beautiful broadleaf and coniferous forests. Today huge areas bear the scars of severe erosion: hills are denuded, with little soil. There may be stunted growth where reforestation has failed.

These hills were laid waste during the brief Great Leap Forward, starting in 1958, when the population attempted to industrialize at farm level. Many villages built smelting works, quarried for iron ore and cut down trees to use as fuel in the smelting. They made some iron but failed to raise the general economic level of the country. In the process they destroyed a precious natural resource and paved the way for the political and social turmoil of the Cultural Revolution.

Fire has proved another severe blow to efforts to maintain adequate forest cover. Fires that broke out at Jeilong-jiang in May 1987, after a long, dry spell, swept out of control, destroying 600,000 ha (15,000 acres) of forest and a small town.

Demand for timber for furniture, construction, paper and fuel still places excessive strain on the remaining areas of natural forest. The rate of cutting far exceeds the rate of regrowth, though the area of replanting does now approximately equal the area officially logged. There will be a period of grave timber shortage when all the unprotected natural forests have been exploited and the new forest plantations are not yet ready for harvesting.

new possibilities offered by the recent "open door" policy with the West. International organizations such as WWF and the International Union for the Conservation of Nature and Natural Resources (IUCN) have an important role to play in conveying to the Chinese the experience of other countries in finding ways of accommodating conservation with other development objectives.

Future prospects

Already the effect of television is eroding traditional attitudes. Most nature films shown on Chinese television are imported, but a start has been made in producing domestic wildlife programs. The presence of increasing numbers of tourists from abroad may also help to develop new attitudes toward nature. Isolated for so long, young Chinese (before the events of Tiananmen Square in 1989) were keen to exchange ideas with foreign visitors.

China now has, on paper at least, a superb system of nature reserves, covering examples of most of its natural habitats and scenery. It has passed laws to protect both these areas and other natural treasures outside the reserves. However, the management skills and the determination to enforce the new laws are not so far advanced.

Money for conservation is always in short supply, and funds are not made available to back government policy statements on environmental protection. Nevertheless, compared with many other Asian countries, China has already put up large budgets for priority projects such as saving the panda, and introducing a number of reforestation schemes.

The Arjin Shan Reserve

China's largest nature reserve, the 45,000-sq km (17,400-sq mi) Arjin Shan, is one of the remotest places left on Earth. Lying on the northern edge of the high Plateau of Tibet, it covers an area larger than Switzerland, with no human settlements, although Uighur and Kazakh tribespeople sometimes graze their sheep and goats in the north. Few visitors reach this isolated and rugged reserve. Access is from the north after a four-day drive from Urumqi, capital of the autonomous region of Xinjiang. The route crosses the Tien Shan (Mountains of Heaven), then the formidable Takla Makan desert, which in Uighur means "once you get in you can never get out".

The reserve rises from some 3,100 m (10,000 ft) to 7,723 m (25,348 ft) on the peak of Wu-lu-k'o-mu-shih (Muztag). It consists of a ring of high peaks surrounding a relatively flat plain that contains several large lakes, some of them salt lakes. The spectacular mountain scenery includes weirdly pocked karst limestone landscapes, solidified molten rock formations and, in the north, hot springs. In places the prevailing winds have whipped up spectacular dunes whose outlines stand out sharply in the cold desert atmosphere. The weather is permanently cold, rising above freezing only at lower altitudes in the summer months. Precipitation is only 100–200 mm (4–8 in) a year, but enough to form shallow wetlands in flatter parts of the reserve.

The vegetation varies with altitude. Lowgrowing desert shrubs cover the valley bottoms and lower mountain slopes, with wet marshland meadows and dwarf shrub communities around the lakes, rivers and springs. Arid grassland clothes the upland plains and high, cold plateaus, and at high altitude alpine cushion plants eventually give way to bare rock and glaciers.

This superb wilderness has been called "the Serengeti of Asia": a survey of the western half of the reserve by a WWF–Chinese team revealed it to be the most important area for hoofed mammals in Asia, providing a refuge for important populations of many rare and endangered species typical of the high, cold plateaus and mountains of central Asia.

The upland plains, wild plateaus and valley basins are home to the chiru (Tibetan antelope) and kiang (wild ass). Probably no more than a few thousand kiang live in the reserve, but it is still the largest population in Asia. Both kiang and chiru tend to concentrate in large, loose herds on the open grasslands around the lakes; on the move, the herds are a magnificent sight.

The gentler mountain massifs above 4,300 m (14,000 ft) are grazed by Tibetan gazelle, wild yak – the third most abundant hoofed mammal in the reserve – and argali (wild sheep). Blue sheep (bharal) and ibex graze the higher meadows in more rocky, mountainous terrain. Wild Bactrian camels are also believed to occur in the reserve. There are six rodent species, including pikas, Himalayan marmots and steppe lemmings, which live in large colonies. A rich wildlife attracts predators. Wolves are common and brown bears, foxes, lynx and snow leopards also live here. The birds include scavenging lammergeiers (bearded vultures). The shorelines of the large lakes provide valuable feeding areas for migratory waders and nesting waterfowl.

There are only two manned posts to guard the reserve, but the area needs little protection. There are no roads, no food and limited fresh water. Transport in more remote parts is by camel or horse. The area's remoteness deters visitors and tourists, and the best prescription for management is probably simply to leave it alone. The Chinese authorities hope to make the area a Biosphere Reserve as soon as possible.

Salt mudflats – a rich habitat for migratory waders and nesting ducks – skirt the shore of Aqqikkol lake in the Arjin Shan Nature Reserve. The inaccessibility of the reserve makes it one of the most remote wildernesses left on Earth.

A large herd of Tibetan antelopes grazes on the high plain of the Arjin Shan. This desolate area supports impressive herds of many large mammals. They need great spaces in which to wander in search of food, and these are still available in this mountain fastness.

RICHES OF TROPICAL LIFE

A SPECTACULAR WILDERNESS · STRATEGIES FOR CONSERVATION · VANISHING HABITATS

Southeast Asia extends from the Tropic of Cancer to south of the Equator, and from the Himalayan foothills of Burma to the tropical islands of the Philippines and Indonesia, a distance of more than 6,000 km (3,730 mi) west to east. It encompasses a wide variety of habitats, and is the meeting place for species from two continents. Burma, Thailand, Indochina and most of Malaysia mark the southernmost limit for Asian species, and the islands of eastern Indonesia, including western New Guinea, lying on the Australian continental shelf, contain many Oceanian species. This warm, fertile region contains some of the wildest places still remaining anywhere in the world, both on land and in the seas, though they are increasingly threatened by the pressures of ever-increasing numbers of local inhabitants and tourists.

COUNTRIES IN THE REGION

Brunei, Burma, Cambodia, Indonesia, Laos, Malaysia, Philippines, Singapore, Thailand, Vietnam

Major protected area	Hectares
Alaungdaw Kathapa NP	160,667
Angkor Wat NP	10,717
Cat Ba NP	27,700
Cibodas NP BR	15,000
Cuc Phuong NP	25,000
Dumoga-Bone NR	300,000
Gunung Leuser NP BR	792,675
Gunung Lorentz NR	1,560,250
Gunung Mulu NP	52,865
Gunung Niut NR	110,000
Huai Kha Khaeng WS	257,464
Khao Yai NP	216,863
Komodo NP BR	75,000
Lore Lindu NP BR	231,000
Mae Sa-Kog Ma R BR	14,200
Mount Apo NP	72,814
Mount Kinabalu NP	75,370
Puerto Galera BR	23,545
Sakaerat RA BR	8,104
Siberut NP BR	56,500
Taman Negara NP	434,351
Tanjung Puting NP BR	355,000
Ujung Kulon NP	78,619

BR=Biosphere Reserve; NP=National Park; NR=Nature Reserve; R=Reserve; RA=Research station; WS=Wildlife Sanctuary

A SPECTACULAR WILDERNESS

Southeast Asia was created between 15 and 3 million years ago when outlying fragments of ancient drifting supercontinents collided in the vicinity of the island of Sulawesi. Molded by its geological past, at a crossroads for animal and plant migrations, the region has some of the most spectacular and diverse tropical habitats in the world. These range from muddy coastal mangroves and peat swamp forests to moss-draped cloud forests and shrubby alpine plant communities; from tall, lowland dipterocarp (two-winged fruited) forests to the palm-thick jungles of the eastern islands; from tidal wetlands to the multicolored crater lakes of still-active volcanoes; from craggy limestone hills with spearlike pinnacles and vast underground cave systems to harsh, nutrient-poor heathlands.

The region has had a long history of human settlement, and people have left their mark on the landscape. The open rolling grasslands of Thailand, east Java and the Lesser Sunda Islands, which extend from Bali to Timor, are great swathes of once forested land that was cleared for agriculture a long time ago and subsequently abandoned.

A species-rich ecosystem Tropical rainforest covers much of the region. It supports many rare and unusual plants and animals, which are specially adapted to living in a particular layer of the tree canopy, in the understorey or on the forest floor.

Mist clings to a forested mountainside Highland cloud forests contain fewer species than lowland rainforest. Their inaccessibility means that they are less vulnerable to human disturbance and encroachment.

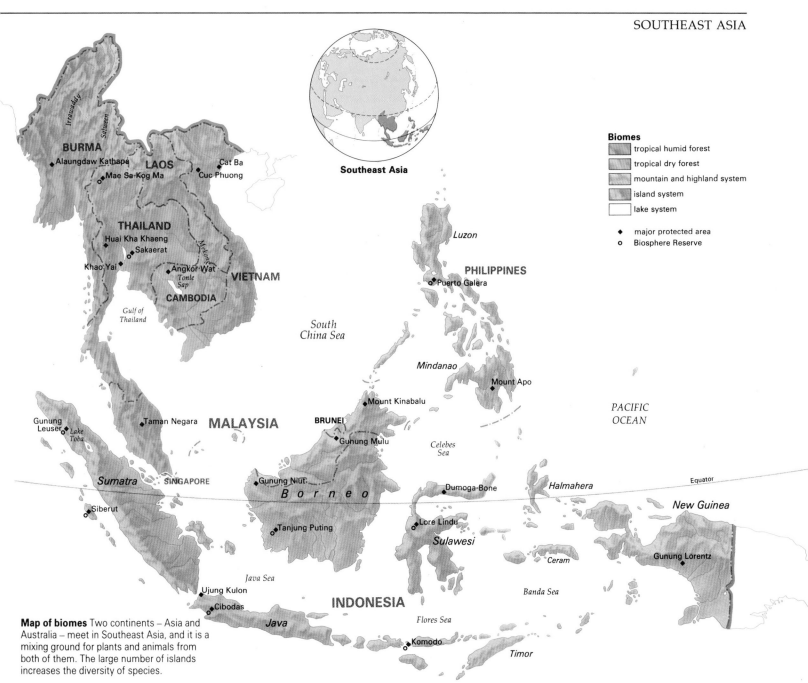

Southeast Asia

Biomes

- tropical humid forest
- tropical dry forest
- mountain and highland system
- island system
- lake system

◆ major protected area
○ Biosphere Reserve

Map of biomes Two continents – Asia and Australia – meet in Southeast Asia, and it is a mixing ground for plants and animals from both of them. The large number of islands increases the diversity of species.

The variety of terrestrial habitats is matched offshore by a marine world of superb diversity. Meadows of sea grass make lush feeding grounds for sea cows (dugongs) and turtles, while vast numbers of colorful fish and other sea creatures live and breed in the complex world of the coral reefs that fringe the islands, among the richest ecosystems on Earth. The southern seas are rich grounds for marine mammals: schools of sperm whales and dolphins, following their traditional migration routes into the Indian Ocean, sport in the straits between the Lesser Sunda Islands.

The luxuriant forests

With an equable climate, year-round sunshine and regular rainfall, Southeast Asia has ideal growing conditions for plant life. Temperate oak, teak and conifer forests in Burma give way farther south to dry deciduous forests and then to a broad band of tropical rainforest.

Seasonal monsoon rains mean that these rainforests are among the most luxuriant and species-rich habitats on Earth. Giant trees stretch upward, sometimes 60 m (200 ft) or more, and each hectare (2.5 acres) of forest may contain 100 different species of trees. Clinging to the taller trees, other plants strain to reach the life-giving light: orchids and bird's-nest ferns festoon the highest boughs; spiny rattans (climbing palms) thrust their tendrils upward; and gnarled vines and berry-laden climbers wind round the trunks. Strangling figs, which begin life high in the canopy, develop arched roots that eventually strangle the unfortunate host tree.

Distinct communities of plants and animals occupy layers of the tropical forest between the ground and the topmost tree canopy. The whole habitat is thus so complex that as yet little is

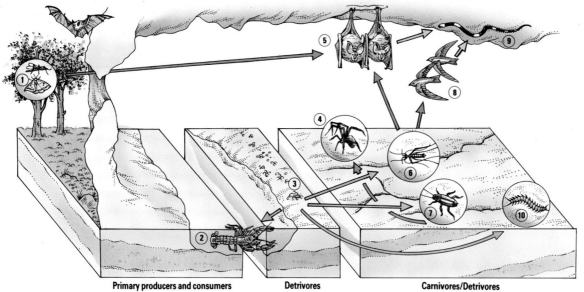

Components of the ecosystem
1 Forest insects
2 Cave crayfish
3 Bat guano
4 Cave spider
5 Roosting bats
6 Cave cricket
7 Cave beetle
8 Cave swiftlets
9 Cave snake
10 Cave millipede

Energy flow
→ primary/secondary consumer
→ secondary/tertiary consumer
→ dead material/consumer

A cave ecosystem is based on nutrients brought into the cave by bats, birds and underground streams.

Primary producers and consumers   Detrivores   Carnivores/Detrivores

understood of its tangled interrelationships or even of all the species it contains. There are many strange and unique plants and animals: the giant parasitic rafflesia, the world's largest flower, the delicate maiden's-veil fungus, and the insect-eating pitcher plants and ant plants that flourish on the nutrient-poor soils of Borneo. The forest is the home of the solitary orangutan, territorial gibbons, big-eyed tarsiers, flying lemurs and a host of other "flying" mammals, lizards and frogs that have adopted ways of gliding through the trees. Indonesia alone boasts more than 500 mammal species, including such rarities as Sumatran and Javan rhinos, tree kangaroos, babirusa hogs and pot-bellied proboscis monkeys, as well as a wealth of colorful birds, ranging from tiny iridescent sunbirds to birds of paradise, helmeted cassowaries and the magnificent hornbills that are so important in Bornean folklore.

STRATEGIES FOR CONSERVATION

The idea of conservation is not new to Southeast Asia. The first nature reserve in the Indonesian archipelago was established in southern Sumatra as early as the 7th century by decree of one of the first kings of Srivijaya; it reflected the early Hindus' appreciation of all animal life. Traditionally, communities harvested natural resources wisely and sustainably, taking only what they needed for themselves and husbanding the rest for future use. Today, however, much greater demands are being made on wilderness areas, and conservation has consequently become a matter for governments rather than individual villages.

During the period of European colonization a network of game and forest reserves was established throughout Southeast Asia. These set out to prevent overexploitation of certain species rather than give total protection. This tradition of conservation continued after independence, and emphasis has been given to improving the protected area systems of the region, and expanding them to cover the whole range of natural habitats, from coral reefs to rainforests.

Poor as it is, Burma has already established a national strategy for developing national parks. Malaysia and Thailand have some well-run reserves, while Indonesia has plans to declare approximately 10 percent of its land area as protected, which will give it one of the finest reserve systems in the region. In Vietnam, the first national park, Cuc Phuong, was declared in 1962; in Cambodia the ancient temple complex of Angkor Wat has been designated a national park. Cambodia, Laos and Vietnam are cooperating to establish international parks to protect the kouprey, a wild forest ox.

Governments have come to appreciate the need to preserve natural habitats. Many parks and reserves protect important watersheds that control the flow of water to the surrounding country. The Dumoga–Bone National Park in Sulawesi was established to protect the watershed around a large irrigation scheme. To establish the reserve cost only a fraction of the World Bank loan that subsidized the development, yet the reserve protects the forest and extends the life of the irrigation canals, so that both the people of the island and Sulawesi's unique wildlife benefit from the scheme.

Lowland swamps in New Guinea. These unspoilt wetlands are inhabited by many species of animals, and preservation of the habitat is essential to safeguard their diversity. But the surrounding land is being encroached on by coconut plantations.

THE SUMMIT OF BORNEO

Towering dramatically above the coastal plains of northern Borneo, Kinabalu at 4,094 m (13,432 ft) is the island's highest peak. Kinabalu National Park was established in 1964, and is one of the best-managed parks in the region. The mountain has one of the richest communities of plants in the world, with representatives from more than half the families of flowering plants. There are a thousand or more species of orchids, 24 species of rhododendrons, 5 of them unique to Kinabalu, and 10 species of insect-eating pitcher plants.

As the land rises, lowland rainforest gives way to montane chestnut and oak forest, succeeded by cloud forest and rhododendron thickets. On the wind-swept pavement of the granite summit a few hardy alpines struggle to survive. Tens of millions of years ago the spine of Borneo was much higher than it is today, and supported a rich variety of alpine plants; Kinabalu is a refuge for these plants of cooler climes. The mountain plants include many living fossils, such as the celery-top pine and the tree oak, the missing link between the oaks and the beeches. At higher altitudes there are several plants more typical of temperate latitudes, isolated here when the climate warmed at the end of the most recent ice age some 10,000 years ago.

The animals of the park are as varied as the plants, though less conspicuous. More than 100 species of mammals and 300 bird species have been recorded, including the orangutan, gibbon, mouse deer, clouded leopard and ferret badger. Even the elusive Sumatran rhino is rumored to have been sighted.

Sheltered crevices on the bare summit of Mount Kinabalu provide a refuge for alpine plants isolated since the last ice age some 10,000 years ago. The harsh climate allowed them to survive here.

The long-term security and survival of protected areas in densely populated Southeast Asia depends upon the support of local communities. Increasingly, conservation programs will need to establish buffer zones outside the most strictly protected areas to provide sources of income, firewood and forest products for the local people. In the protected areas at Khao Yai in Thailand and Gunung Mulu in Sarawak there has been a determined effort to employ local guides and guards, so that tourism benefits both local communities and the national economy.

Reserves under pressure

Even officially protected areas are not yet secure. Forest clearance and hunting continues unchecked in many parts of the region. Reserves are threatened by all kinds of development, from landfilling of estuaries to resettlement schemes or even the creation of prestigious golf courses. Both Taman Negara in Malaysia and Huai Kha Khaeng in Thailand will be severely damaged if plans to build dams for hydroelectric projects go ahead. Mount Apo park in the Philippines, the last stronghold of the monkey-eating eagle and other rare wildlife, has been partly reduced in status in order to provide agricultural land for settlers. Gunung Niut in Borneo is threatened by logging and by goldmining. The area around Tonle Sap in Cambodia continues to be damaged by fighting between the warring factions in that country's civil war. The list of Asian parks under threat is depressingly long. Many feature on the list of the most endangered protected areas drawn up by the International Union for the Conservation of Nature and Natural Resources (IUCN).

Many of these protected areas are of global as well as national significance, and the world community is responding to the challenge. For many years international organizations such as the World Wide Fund for Nature (WWF) and IUCN have helped to identify conservation needs and improve management in reserves in Southeast Asia. There is an encouraging trend as financial institutions such as the World Bank and – through their aid agencies – Western governments take on a more active role in sponsoring conservation activities; in this way they often protect their investments. The beneficiaries will be the forests and their wildlife, and the local and world-wide human communities.

VANISHING HABITATS

The demands of agriculture place tremendous pressures on wilderness areas in the region. Forests are being felled by farmers who cultivate the land and then move on, for plantations, and to satisfy the world's seemingly insatiable appetite for hardwoods. Tropical rainforests are disappearing at an alarming rate. Malaysia, where an estimated 230,000 ha (570,000 acres) are cut down every year, will lose all its remaining forest during the 1990s if this rate continues.

Forests have also been lost as a consequence of war, as in Vietnam where 2 million ha (5 million acres) of forest and mangroves were sprayed with herbicides such as Agent Orange during the conflict with the United States between 1964 and 1975. In Borneo large areas have been lost to forest fires, which consumed 3.6 million ha (8.9 million acres) in Kalimantan, in the center and south of the island, and another million in Sabah in the north during the drought year of 1983. Fires started by farmers practicing shifting cultivation raged out of control and were spread by the underlying coal and peat seams; they swept through forests that had already been logged, and damaged the edges of primary forest. In 1987 east Kalimantan was a sea of fire again until the flames were quenched by the late monsoon rains.

Most countries in Southeast Asia have already lost at least half their forest cover. In densely crowded Java only 9 percent of the island remains forested. Even on Borneo, renowned for its vast tracts of tall dipterocarp and swamp forests, the forest boundaries are being pushed back farther and farther inland and every major river is congested with floating logs. Until people throughout the world become aware of the threat and there is a dramatic reduction in the use of tropical hardwoods, the destructive deforestation of Southeast Asia will continue.

Where the loggers go the farmers follow. Traditional peoples living at low densities were once able to practice shifting cultivation in ecological balance with their environment. The Land Dayaks of Borneo, the Muong people of Laos and the hill tribes of Burma all cleared land, grew their crops and then let the land lie fallow for several years before returning to cultivate it again. As human popula-

tions have grown and remote areas have been opened up for new settlers, more forest areas have been cleared, often on vulnerable lands with nutrient-poor soils. These fields may provide crops for a year or two, but with regular burning they cannot return to secondary forest.

Eventually these abandoned fields become a sea of alang-alang (cogon) grasslands. These are becoming increasingly common throughout tropical Asia. The tough, tall grass is difficult to supplant, has little value except for new grazing and thrives on burning. Fires sweep through the grasslands, destroying

Cloudy moss forest at 3,500 m (10,500 ft). Northern Borneo is one of the wettest places on Earth, and in the highlands it is also very cold. The trees are stunted, being no more than 10 to 15 m (30 to 45 ft) high; as there is only a single canopy layer enough light penetrates to promote the growth of hanging lichens, mosses and other epiphytes, as well as ground plants.

Life from the ashes As lava and ash on a volcanic peak weather to a fertile soil, the radiating ridges and channels formed in the lava flows are colonized by wind-blown seeds, and vegetation begins to cover the mountainside.

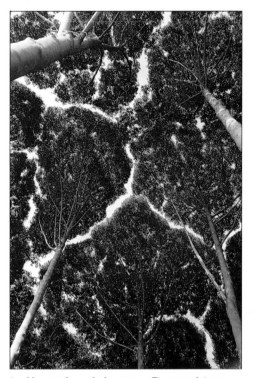

Looking up through the canopy The tops of these giant dipterocarp trees do not overlap. One explanation for this "crown shyness" is that it inhibits the spread of leaf-eating caterpillars. It also allows light to penetrate through the trees, and has led to the evolution of jumping and gliding mammals and reptiles.

WALLACE'S LINE

During the last ice age the countries of mainland tropical Asia and the Greater Sunda Islands – Sumatra, Borneo and Java – were connected by dry land; they were separated as melting ice caused sea levels to rise again. Before this wildlife was able to move along the land bridges, and the islands still have plant and animal species in common with those of the mainland, characterized by tall dipterocarp trees and by monkeys, native deer and hornbills. These habitats, together with their plants and animals, fall within the Indomalayan biogeographical region.

By contrast the islands of eastern Indonesia – Irian Jaya (western New Guinea), Kai and Aru – lie on the Australian continental shelf and belong within the Oceanian biogeographical region. Here there are mound-building birds, bowerbirds, parrots and birds of paradise, while wallabies take over the ecological niche of deer.

On the edges of these two biogeographical regions the islands of Sulawesi, the Philippines, the Moluccas and the Lesser Sundas form a mixing ground for plants and animals from both east and west. They include many species that are found only here.

It was the famous 19th-century British naturalist, Alfred Russel Wallace (1823–1913), co-publisher with Charles Darwin (1809–82) of the theory of natural selection, who first recognized the pattern of species distribution in the area. Wallace's Line runs through the islands between Bali and Lombok, Borneo and Sulawesi, and Palawan and the rest of the Philippines. Still recognized as the biogeographical boundary for many families of plants, and of birds, mammals, insects and other animals, it is a reminder of Southeast Asia's turbulent geological past.

adjacent plantations of newly planted trees and eating into the natural forest.

Sometimes the need for conservation is understood only too well by the local people but ignored by governments and business interests, more concerned with short-term profit than the long-term cost of environmental folly. The Penan of Sarawak, hunter-gatherers who harvest wild meat and a few minor forest products for their own use, are now building blockades to halt the timber trucks of the companies that are destroying their traditional lands. Local people, struggling for a livelihood from shifting cultivation, are often blamed for starting the fires that lead to forest losses, but the areas that burn most intensely are those that have already been logged.

Paying the price of "progress"

Many countries in the region are suffering the environmental consequences that follow the loss of natural landscapes – droughts where there was once rain, floods sweeping down deforested valleys, and erosion of coasts and hillsides. Lowland habitats, particularly on the fertile alluvial lands along river valleys, are the first to disappear. Swamps, mangroves and wetlands are drained for development and agriculture, forests are cleared, limestone hills are quarried for cement, coral reefs are burned for lime and damaged by blasting to kill the fish.

As these habitats are lost, so too are the benefits they provide: natural products, fish nurseries, coastline and watershed protection. Their value is often appreciated only after they are gone. Very few wildlands, apart from inaccessible mountain areas, will remain inviolate outside protected areas. The next decade will be a time for decision, a last chance to save many natural areas.

Ujung Kulon National Park

Situated on Java's westernmost point, the Ujung Kulon National Park is Indonesia's premier reserve and one of the few remaining wilderness areas on this densely populated island. The park includes the Ujung Kulon peninsula, connected by a narrow isthmus to the Gunung Honje massif, and the offshore islands of Panaitan and Peucang, with their spectacular coral reefs. Its boundaries extend seaward to embrace the island remnants of the famous volcano Krakatau in the Sunda Strait between Java and Sumatra. Protected on three sides by the sea and to the east by Gunung Honje, Ujung Kulon provides a last refuge for some of the unique wildlife on an island where 100 million land-hungry people have cleared most of the original vegetation.

The reserve was created to protect one of the last surviving populations of one of the world's rarest animals, the one-horned Javan rhinoceros, which is found in the lowlying swamps of Ujung Kulon. Other rare animals living in the park include the leopard, native Javan gibbon and troops of leaf monkeys (surili). The Javan tiger, which survived here until recently, is now extinct.

The landscape was shaped by the eruption of 1883, when the island of Krakatau was rent by a massive explosion. After the explosion came devastating sea waves, which destroyed the coastal forests of Ujung Kulon. Today these forests are characterized by dense thickets of rattan and salak palms, gingers and bamboos, so tangled that only the thick-skinned rhinos can pass with ease.

Several open grazing grounds are maintained within the park and here the Javan wild cattle (banteng) come to graze in the early morning and late afternoon. Here, too, sambar deer come to feed, as well as peacocks and the green junglefowl that are found only on Java. On the golden beaches on the south coast of the park green turtles nest, though even in this sanctuary their eggs are not secure from natural predators such as pigs and monitor lizards.

Ujung Kulon is a paradise for birds, from the forest-ranging hornbills to tiny sunbirds and the waterbirds and storks that enjoy the wetlands of the Nyiur swamps. The park's marine and river life includes the archerfish, which fells insects by shooting jets of water at them, and the beautiful and colorful animals of the coral reefs. Scenically the park is outstanding, with white coral beaches, lofty rainforest, strangling figs with cathedral-like roots and the rich plant life of the forest floor – exotic fungi, intricate palm fronds and twisting lianas that wind their way up to the canopy.

Ujung Kulon's most dramatic spectacle is the crater of Anak Krakatau (child of Krakatau). The smoking cone bears witness to the fact that the volcano is only sleeping; during its restless phases the sea and night sky are lit dramatically by showers of sparks from its glowing core.

People have probably lived on Ujung Kulon since early times. At the time of the 1883 eruption there were several coastal settlements practicing shifting cultivation

A forest stream carries away the excess water deposited by the area's heavy rainfall. By acting as a sponge, the forests form a natural reservoir, regulating the supply of water to the surrounding country. Without the forest, water flow becomes erratic.

A last refuge for wildlife Ujung Kulon's forests shelter some of Java's most endangered animal species, including the one-horned Javan rhinoceros. The park is a reminder of how the densely populated island must have been before the original vegetation was cleared.

A mangrove swamp in Ujung Kulon. These salt-tolerant trees form a very rich ecosystem in many estuaries and coastal waters of the tropics. Their leaves and twigs fall into the water and decompose to make the first link in the food chain.

and growing rice. These were swept away by the tsunami that followed the eruption. The ash from the volcano smothered the peninsula and its vegetation, making the land much less productive, so the human population of Ujung Kulon declined. The region's wildlife benefited further when the area was evacuated by government decree because of the danger from disease and from man-eating tigers.

Protection of Ujung Kulon began in 1921; it was granted national park status in 1980. The WWF has had an important role in conservation since 1965, working closely with the Indonesian Conservation Department to improve the protection and management of the reserve. Today it is probably one of the better protected Indonesian reserves, though its eastern borders have been encroached on by village farmers. Vigilance is needed against poachers (the rhino is particularly vulnerable) and collectors of turtle eggs; local fishermen not only overfish the reef areas but can cause serious damage from untended campfires.

Living in the dark

In parts of Southeast Asia huge cave systems have been formed by underground streams, which have eroded the limestone rocks. The most famous are the Niah Caves of Borneo and the Batu Caves of Malaysia.

These caves form islands that contrast with the sea of forest around them. They are an ecosystem with special characteristics, the main feature of which is the complete absence of light.

Away from the entrance no green plants can survive, so the animal community within the cave needs another source of energy and nutrients. These are provided by bats and swiftlets, which feed outside the caves and return to breed and roost. They form the basis of a food chain, being preyed on by snakes, lizards, rats and civets, while their corpses and droppings fall to the cave floor and are scavenged by a variety of animals. Cockroaches, springtails, flies and beetles swarm over the cave floor in vast numbers and are eaten by spiders, scorpions, whip-scorpions and centipedes. Many of these invertebrates are colorless and blind, color and sight being of little use in the dark. Touch and smell are the secret to successful predation and scavenging.

The vast entrance to a cave in Mulu National Park, Sarawak, Malaysia provides the link between the outside world and the specialized ecosystem within.

ON THE PACIFIC'S NORTHERN RIM

A LAND OF CONTRASTS · A LOVE OF LANDSCAPE · HOME OF THE GODS

The long, narrow islands of Japan stretching from the subarctic almost to the Tropic of Cancer, have high volcanic mountain backbones and long coastlines. The mountains in the north support coniferous forests, home to many mammals and rare birds; this blends into deciduous forest farther south and, on the southernmost islands, fringed with coral reefs and seagrass beds, there is hot, humid rainforest. The mountains in the north and east of Korea are densely forested; lowland areas to the west and south are heavily cultivated. Where the larger rivers reach the coast, extensive wetlands are important wintering grounds for birds. Both Korea and Japan have large, growing populations and the ever-increasing demand for land is pressing heavily on the remaining wilderness areas of the region.

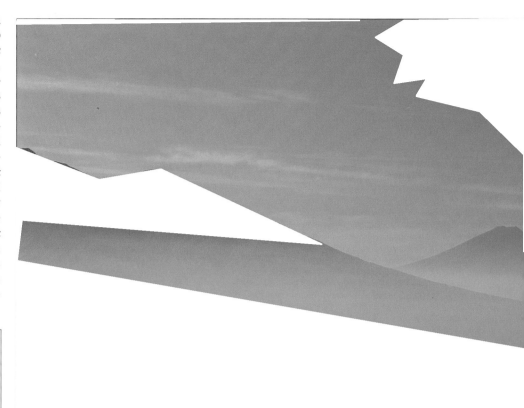

COUNTRIES IN THE REGION

Japan, North Korea, South Korea

Major protected area	Hectares
Akan NP	90,481
Aso NP	72,492
Bandai-Asahi NP	189,582
Chiri Mountain NP	44,045
Daisenoki NP	31,927
Daisetsuzan NP	230,894
Fuji-Hakone-Izu NP	122,686
Hallyo MNP	47,862
Iriomote NP	12,506
Ise-Shima NP	55,549
Kwan-Po NM	1,064
Mount Hakusan NP BR	47,700
Mount Paekedu NPA	14,000
Mount Sorak BR	37,430
Samil-Po NM	676
Seto-Naikai NP	62,957
Shiga Highland BR	13,000
Shiretoko NP	38,633

BR=Biosphere Reserve; MNP=Marine Nature Park; NM=National Monument; NP=National Park; NPA=Nature Protection Area

A LAND OF CONTRASTS

Japan may be small, but its mountains are certainly not: steep, forested slopes cut by deep gorges and waterfalls rise to an alpine zone in the northern islands, often with active volcanoes or spectacular crater lakes. The high mountain ranges form a climatic divide, which has a powerful influence on the country's habitats. Winds blowing directly from Scheria bring much snow and rain to western coasts, while the Pacific side of the country has dry winters.

The Korean peninsula is similarly dominated by great volcanic mountains, especially in the north and east, where the slopes plunge directly into the sea. The land is tilted towards the west and south; rolling hills run down to a twisting and turning coastline. Most of Korea's important wetlands are in this lowland area; off the south and southwestern shores lie numerous small, coral-fringed islands.

The sacred mountain, Mount Fuji, symbolizes the Japanese people's veneration of nature. Japan has a surprisingly rich wildlife. Its diverse habitats contain 130 species of mammals, 500 bird, 50 amphibian, 76 reptile and over 6,000 plant species.

A rich wildlife

Japan's marked contrasts in climate, altitude and landscape have produced a great diversity of plant communities: more than 6,000 species have been recorded. Forests cover 67 percent of Japan's land area and an even larger proportion of the Korean peninsula, although South Korea has suffered extensive deforestation. In Japan much of the native forest has been replaced by conifers, mostly larch; today, 40 percent of the forests are commercial plantations.

The remaining natural forest is of three main types: evergreen coniferous forest

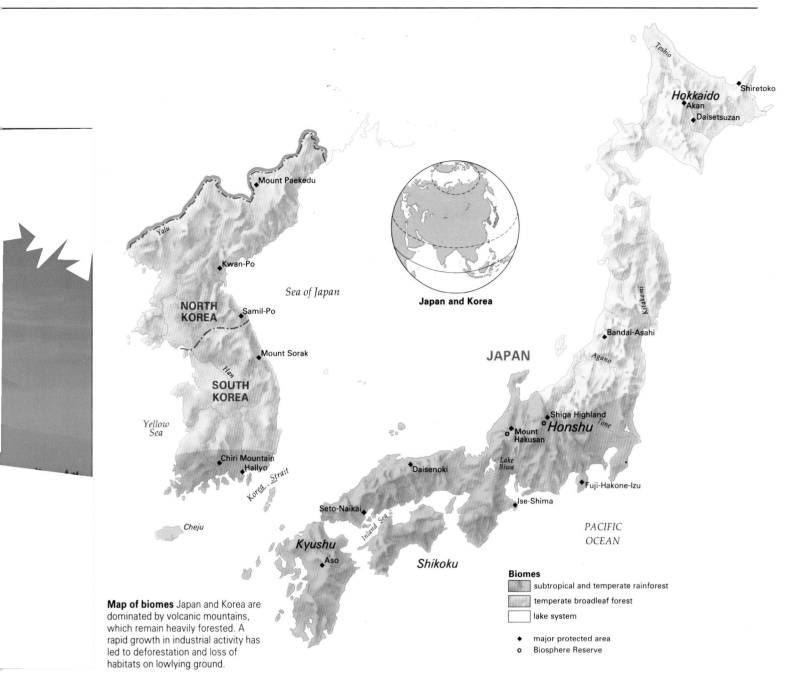

Map of biomes Japan and Korea are dominated by volcanic mountains, which remain heavily forested. A rapid growth in industrial activity has led to deforestation and loss of habitats on lowlying ground.

Biomes

- subtropical and temperate rainforest
- temperate broadleaf forest
- lake system

◆ major protected area
○ Biosphere Reserve

(which gives way above the treeline to alpine scrub, grassland and rocky desert) on the mountains of the northernmost island, Hokkaido; deciduous broadleaf forest in the southwest; remnants of evergreen broadleaf forest, with tree ferns, palms and mangroves, in the subtropical Ryukyu Islands, about 500 km (310 mi) to 1,200 km (745 mi) southwest of the mainland islands. Bamboo thickets are common in the forest understorey in many parts of the islands of Honshu, Kyushu and Shikoku, and provide good habitat for some wild mammals. A similar pattern is found in Korea, with forests of larch, Siberian fir, spruce, pine and cedar mixing farther south with deciduous species such as oak, alder and chestnut. Patches of subtropical broadleaf forest are found along the south coast.

The region is also rich in animal species. These have their origins mostly in northern Asia, Siberia, Manchuria and China (the Palearctic region) with some mixing from the Indian subcontinent and Southeast Asia (the Oriental region). Many, evolving in isolation, are found only on particular islands (endemic species): the endangered Iriomote wild cat is confined to one island only in the southern Ryukyu Islands; the Amami rabbit, the sole survivor of a family that flourished 20 million years ago, is found on two islands north of Okinawa. North Korea's inaccessible mountain forests still harbor small numbers of tigers, leopards, snow leopards, Korean pandas and the Amur goral, a species of goat antelope.

Korea's extensive coastal wetlands and the rice paddies that flank them are important overwintering and migratory stopover sites for waterfowl and waders, as well as nurseries for over 800 species of fish. Rare species breeding here include the endangered Chinese egret and the black-faced spoonbill. Many red-crowned

cranes overwinter here, and hooded and white-naped cranes and whooper swans pass through in large numbers on their way to winter on Japanese lakes. There is a well-protected colony of red-crowned cranes, once thought to be extinct, on wetlands near Kushiro on Hokkaido, which probably accounts for about one-third of the total world population.

Threatened marine life

The seas surrounding Japan and Korea have a rich marine life, ranging from the temperate species of the north to those of the coral reefs far to the south where the Ryukyu Islands are washed by the warm northward-flowing Kuroshio Current. Fringing reefs are found around most of the islands south of Kyushu. However, many are now severely threatened as silt from soil erosion caused by agricultural intensification and coastal development washes into them and chokes them.

229

A LOVE OF LANDSCAPE

There is very little wilderness to be found near Japan's centers of population, but wild places remain on remote islands. Japan's first conservation legislation was the Imperial Game Law, passed in 1892, but it was not until 1931 that a national park law, designed to protect the natural environment, came into being. This was reinforced by new legislation in 1957 that provided for a new category of reserves, natural parks.

There are now many national and natural parks. The 27 national parks account for 20,200 sq km (7,800 sq mi) or just over 5.4 percent of the land area, and with the natural parks, the total rises to just over 14 percent. There are also 23 marine parks. However, the stated aim of natural parks is "to conserve scenic areas, to promote their utilization, and to contribute to the health, recreation and culture of the people". Conservation of plants and animals does not feature at all as a consideration.

National parks in Japan are basically scenic parks – few if any qualify as national parks according to international criteria. In 1982 some 828 million people visited Japan's national and natural parks. Catering for tourists takes precedence over protection of endangered habitats, plants or animals. To meet their demands, many buildings and other facilities have been built. As a result, many parts of the natural parks have become highly commercialized leisure centers whose activities are damaging to the environment.

Legal protection

The survival of Japan's forests owes much to a policy of buying timber from abroad, notably from Indonesia and Malaysia. Although the total forest area is said not be to decreasing, the quality of the forest habitats is undoubtedly declining, and the Forest Agency has instituted conservation measures for watershed management and to ensure the protection of forest resources.

The legislation specifically restricts the grazing of livestock, lighting of fires, wood gathering, capture of wildlife, tree planting and building within national and natural parks. If any of these restricted activities are being practiced at the time of designation they have to be registered, and are permitted to continue

Autumn colors reflect the variety of tree species that make up the native woodland of the Japanese Alps on Honshu Island. Over the last 40 years much of this forest has been cleared for industrial development or replanted with conifers. Consequently many species have become rare or extinct.

A landscape of lake and forest in the northern island of Hokkaido is more similar to neighboring Siberia than to the rest of Japan. The coniferous forest of spruce and fir, mixed with some deciduous birch, oak and maple, supports populations of brown bears, Siberian chipmunks, Asiatic pika and Hokkaido squirrels.

classified as Wilderness Areas and Nature Conservation Areas.

Serious about conservation

Korea's tradition of conservation dates back to the 6th century AD. Many habitats have been under the protection of monasteries. A fine example is Mount Myohyung in the forests of North Korea, the site of many Buddhist temples, themselves important tourist attractions.

North Korea has many types of reserves, including some specifically designed to protect seabirds, plants and marine resources such as scallops, abalones, oysters and char. Many rare plants and animals, along with their breeding or wintering grounds, are given special protection by being designated as natural monuments, a practice also adopted in South Korea. The Academy of Sciences, which enjoys a status equivalent to a ministry, allocates the areas to be given protection, and the Union for Nature Conservation, an umbrella organization incorporating many small societies and groups, involves the people in conservation efforts.

South Korea's first park law was passed in 1967, and the various conservation laws and decrees were rationalized in a new national park law in 1981. Legislation of 1972 prohibits all hunting of wild animals except in the islands of Cheju off the south coast. Responsibility for conservation is shared between no fewer than five ministries, but the country has succeeded in establishing many national parks on both land and sea. Well over 2.5 percent of South Korea's land area is now under protection. With increasing income and leisure time coupled with expanding cities and worsening pollution, the popularity of South Korea's parks is increasing rapidly – well over 12 million people visit them every year.

subject to certain controls. The result is that many kinds of land use, including farms and settlements, exist within the boundaries of the natural parks.

A zoning system attempts to accommodate the needs of tourists and local residents and still protect habitats. Some 12 percent of the national park area is classified as "Special Protection Areas" where no development is permitted. The rest is open not only to human residence but also to agricultural and forestry activities, so long as they do not affect "scenic beauty". Even here, the legislation makes no specific reference to the protection of habitats or wildlife.

There is some hope for the future: in recent years public interest in nature has grown and a nature conservation law was passed by the Diet in 1972. This new law aims to extend protection to ecosystems, but so far only small areas have been

AN ISLAND OF RARITIES

The island of Iriomote, in the tropical Ryukyu Islands, is the central feature of Japan's southernmost national park. It is fringed with rich coral reefs, and the largest reef in Japan extends for 15 km (9 mi) to neighboring Ishigaki. On land there are dense subtropical forests of evergreen broadleaf trees, while mangrove forests grow along the shores. Cycads and tree ferns are also locally abundant. The land and marine animals are quite different from those of mainland Japan.

In addition to the unique Iriomote wild cat, thought by some to be a living fossil survivor of the ancestral species of the cat, the endemic Ryukyu dwarf boar lives in Iriomote's forests, and the diverse birds include a rare crested serpent eagle. Dugongs graze on sea grasses offshore, sharing the shallow feeding grounds with marine turtles. Some 250 species of corals and at least 100 species of fish live on the reefs.

The discovery of the Iriomote wild cat in 1965 aroused scientific interest in the island worldwide and stimulated concern for its lowland evergreen forests, which had already been substantially felled to make way for agriculture. Since the Iriomote National Park was created in May 1972 the park has twice been reduced in size, mainly in response to opposition from the local inhabitants.

Park policy prohibits tree felling, but registered local inhabitants have special rights that cannot be easily removed, and other people fell illegally. Public awareness of Iriomote is growing, and there is now considerable pressure to enforce the park protection laws to save the remaining reefs and forests.

A coral reef ecosystem is based on planktonic plants and animals, which support the intricate and colorful corals, known as polyps, and other invertebrates. The crevices of the reef itself shelter many of the fish that graze around the corals, as well as their larger predators such as barracudas.

Components of the ecosystem

1 Phytoplankton	13 Sea urchin
2 Algae	14 Sea slug
3 Detritus	15 Starfish
4 Zooplankton	16 Butterfly fish
5 Fanworm	17 Octopus
6 Coral shrimp	
7 Damselfish	
8 Clam	
9 Brittlestar	
10 Reef crab	
11 Corals	
12 Barracuda	

Energy flow

primary producer/primary consumer
primary/secondary consumer
secondary/tertiary consumer
dead material/consumer
death

Primary producers   Herbivores/Omnivores   Carnivores

HOME OF THE GODS

In Japan gods dwell in water, rocks and trees. This ancient belief, combined with Buddhism, introduced to Japan in the 6th century, has protected the forests for centuries. Buddhism has had an equally powerful influence on Korean attitudes to nature. Many mountains were held to be sacred; the few people who visited them went not to conquer the summit but to be purified by experiencing the benevolent presence of the gods.

Nature for the Japanese people was not something they could control: it was much bigger and more powerful than they were. They lived under the constant threat of typhoons and earthquakes, which could destroy their houses and fields within seconds. Each year Japan experiences about a thousand earthquakes that are strong enough to be felt.

In the past crops were frequently destroyed by extreme weather conditions, both of cold and drought.

As a result, the Japanese developed a love, respect and fear of nature, but gave no thought to ruling it, except in their garden and flower arrangements, where they tamed nature and lent it artificiality.

Godless destruction

This attitude had a positive effect on conservation for a long time. Almost all the areas of surviving lowland forest are attached to religious establishments such as temples or shrines, and there has been little or no change in them for many centuries. However, industrialization, along with the rise of science and technology, has changed the scene completely. As the gods of forests, mountains, rocks and water ceased to have power in people's minds, there was nothing to prevent them from exploiting nature for

Mount Seorak National Park is one of South Korea's growing number of protected areas. The pressure of population, urbanization and pollution on the landscape is matched by the public's increased desire to visit parks and wild areas.

Origami strips mark a shrine, the dwelling place of a *kami* or spirit. The Japanese people, by tradition, have a strong affinity with nature based on the ancient beliefs of the two religions of Japan, Shinto and Buddhism.

NATIONAL PARK IN DANGER

Fuji–Hakone–Izu National Park is the nearest national park to Japan's capital, Tokyo: it encompasses the dramatic scenery that surrounds Fuji – at 3,776 m (12,348 ft) Japan's highest mountain – the lakes and many hot springs of the Hakone area, and the Izu group of volcanic islands lying just off the coast. All this helps to make Fuji–Hakone–Izu one of the world's most visited parks, with consequent detrimental effect to wildlife. Trees have been felled to make room for tourist facilities, and there is serious pollution.

In 1982 proposals were made to create a bird sanctuary and marine park on Miyake Island, in the Izu Islands. However, three years later permission was granted to construct a military airport at the center of the proposed sanctuary. With its associated human activities this will undoubtedly disturb the rare and endemic birds that breed on the island. The proposed airport also poses a significant threat to the coral reefs and their inhabitants.

This incident has a wider implication than just the threatened destruction of the natural habitat on Miyake Island. If the protection afforded by the Natural Park Law can be waived so easily, then no protected area in Japan can be considered safe.

their own ends. The destruction of the forests advanced with tremendous speed. Those who still retain the old ways of thinking are pushed aside by bureaucrats in the name of progress.

The Japanese people in the past had a particularly strong affinity with beech trees, which predominated in the original forest cover, and with their associated plant and animal communities, harvesting what they had to offer through the seasons. Since beech is not regarded as a useful timber for building, most of the primary beech forests have now been felled and the cleared areas replanted with Japanese cedar, cypress and other more profitable timber. The few remaining beech forests are under great pressure. Shirakami mountain on the northern tip of Honshu has the largest surviving tract of beech forest in Japan. Part of it is included in both a national and a natural park but most of it, because it lacks spectacular scenery, is not protected under the Natural Park Law, which makes it highly likely that plans to log it and

build a road through it will be carried out without opposition.

Despite the devastation of the Korean war (1950–53), some 75 percent of the peninsula remains forested. Destruction is greatest in the densely settled lowlands of South Korea, where land is needed for agriculture, industry and houses.

Vanishing wetlands

In both Japan and Korea, wetlands have been reclaimed for urbanization and agriculture, and contaminated by pesticides, sewage and other pollutants. Japan's bird populations – especially its famous cranes – have declined drastically as a result. Many of Korea's large wetlands are in serious danger, especially in South Korea, where there are no protected wetland areas. A government program of reclamation of estuaries and shallow bays on the west and south coast, involving the construction of dikes, dams, barrages and channel diversion, threatens to destroy some 66 percent of the country's coastal wetlands. This will decimate the over-

Flowing water in the Joshin-Etsu Kogen National Park on Honshu Island. This is the third largest of Japan's 27 national parks. Its proximity to Tokyo puts its wild areas under intense pressure from tourists.

wintering and feeding grounds of countless waterfowl and shorebirds.

While Korea is expanding its protected areas and pressing for their international recognition, the situation in Japan is very different. There is now increasing concern for the natural environment, following some well-publicized cases of forest destruction. The uncontrolled disposal of chemical waste has caused widespread coastal pollution. Many small, locally based citizens' movements have organized themselves to protect particular areas, and they have been achieving some positive results. However, Japan's continued economic development, its shrinkage of open space, and its overcrowded cities and transport systems means that official planning policy still gives a low priority to conservation – which may have disastrous environmental consequences for future generations.

Shiretoko National Park

Shiretoko National Park lies on a narrow peninsula at the northeastern end of Hokkaido. Snowcovered for over half the year, this mountainous and inaccessible island, the home of the aboriginal Ainu people, was not settled by the Japanese until the end of the 19th century. Even to the Ainu, whose way of life was adapted to the harsh natural environment, the narrow Shiretoko Peninsula appeared remote and wild: its name is derived from the Ainu, meaning "land's end".

The range of volcanic mountains (only one is active) that runs the length of the peninsula rises impressively from the sea, forming steep, rugged cliffs up to 200 m (650 ft) high. Visitors come in boats to observe the numerous birds that nest along these inhospitable cliffs, particularly black-tailed gulls, pelagic shags and Japanese house martins. White-tailed sea eagles breed locally, and in the winter months magnificent Steller's sea eagles fly south from their breeding grounds in the Arctic, and can be seen perching on the ice flows. The coastal waters sometimes freeze over completely off the peninsula, forcing seabirds and mammals such as large Steller's sea lions to migrate farther south or to fish far out to sea.

The vegetation is quite different from that of the rest of Japan: creeping pine, which does not grow below 2,500 m (8,000 ft) in the central part of Japan, is found on slopes less than 1,000 m (3,250 ft) high on Shiretoko, and because of the severity of the climate many alpine plants flourish at these altitudes too. The lower parts of the mountains are covered with undisturbed forests of subarctic conifers such as pine and spruce as well as decidous maples and oaks. The lakes and swamps hidden deep in the forests support deer, foxes and brown bears.

Land's end is the Ainu people's name for Shiretoko. This remote national park, designated in 1964, is the wildest park in Japan. It has a variety of habitats including coastal cliffs, alpine slopes, subarctic forests, lakes and swamps.

Water cascades over the wooded cliffs that buttress the Shiretoko peninsula. The peninsula attracts many visitors in summer but in winter the rivers freeze solid and the land is surrounded by ice flows. Many seabirds winter inshore and the park is famous for its winter population of sea eagles.

The Shiretoko Five Lakes lie in natural hollows in the lava bed of the peninsula's volcanic plateau. The fringes of the lakes support typical marsh plants including spatterdock, buckbean and skunk cabbage. The lakes are surrounded by thick forests of birch, spruce, oak and silver fir.

At 38,630 ha (95,340 acres), Shiretoko National Park is not particularly extensive, but more than half of it is classified as a special protected area, where no development is permitted. Yet the park has been threatened by a plan to fell more than 10,000 trees by the mid-1990s.

The Forestry Bureau responsible for the proposal argued that a certain amount of logging was necessary to "make the old forest more active". Opponents said that it would lead to the destruction of one of the last real wildernesses left in Japan. It would disturb the park's many rare animals, and threaten the survival of the country's last viable population of Blakiston's fish owl. These birds played a large role in Ainu folklore. They were known as "god-birds" and regarded as the protectors of the villages.

Fighting the planners

Since the area in which logging was proposed did not fall within the special protected area, the plan did not contravene existing conservation legislation. However, so strong was the campaign mounted against it that the Forestry Bureau was forced to postpone logging indefinitely within the park. At the core of the campaign was the "100 square meter" movement, which invited members of the public to contribute money to buy land at the edge of the park. A surprising number of people responded to the appeal, many of them from large cities like Tokyo and Osaka – perhaps an indication that many Japanese people are developing an acute sense of how great the loss of nature has been, and wish to protect what is left before it is too late.

Shaping the mountains

The backbone of volcanic mountains that stretches almost the length of Japan covers nearly 80 percent of the country. The peaks rise to 3,776 m (12,460 ft), forming a barrier between the cold, rain-bearing westerly winds and Japan's east coast. In the mountains, snow-capped peaks surround spectacular crater lakes, and the steep forested slopes are cut by deep gorges and waterfalls, a source of inspiration over the centuries for Japanese artists.

Only Japan's highest peaks have been scoured by ice: unlike the European Alps, whose valleys were carved by glaciers during the ice age, these mountains and valleys were mainly shaped by rain and snow meltwater. The same water feeds the lush forests that clothe their slopes. These forests contain more species of plants, especially broadleaf deciduous trees, than the montane woodlands of Europe, many of whose plants were wiped out by the severe weather of the ice age.

The Japanese Alps, the mountain chain that covers much of the island of Honshu. Dense forests grow on the steep slopes.

FRAGMENTS OF GONDWANALAND

AN ANCIENT TAPESTRY · PROGRESS IN CONSERVATION · NATURE INVADED

Traces of the ancient, lush forests that once covered Gondwanaland, the supercontinent that millions of years ago extended across almost half the globe, still survive in parts of Australia and New Zealand. In Australia they are inhabited by marsupial (pouched) mammals of great diversity. Australasia broke away from Gondwanaland and drifted northward millions of years ago. Antarctica detached itself from the ancient supercontinent rather later. It drifted southward and became colder, losing almost all its wildlife. By contrast, many of the coral atolls and volcanic islands of the Pacific are new wildernesses, having emerged from the sea in relatively recent times. The region thus contains an astonishing variety of habitats and wildlife that have evolved in isolation from the rest of the world.

COUNTRIES IN THE REGION

Australia, Fiji, Kiribati, Nauru, New Zealand, Papua New Guinea, Solomon Islands, Tonga, Tuvalu, Vanuatu, Western Samoa

Major protected area	Hectares
Arthur's Pass NP	94,420
Cape Range NP MP	50,581
Coorong NP GR	38,987
Egmont NP	33,540
Farewell Spit RS	11,388
Fiordland NP	1,252,400
Firth of Thames RS	7,800
Flinders Ranges NP	80,265
Great Barrier Reef NP MP BR WH	20,000,000
Great Sandy NP	52,400
Kakadu NP WH RS	667,000
Kooragang NR RS	2,206
Kopuatai Peat Dome RS	9,665
Macquarie Marshes NR RS	18,200
Mount Aspiring NP	285,580
Myall Lakes NP	31,005
Nelson Lakes NP	96,110
Royal NP	15,014
Simpson Desert NP	692,680
Tongariro NP	76,500
Uluru (Ayers Rock) NP	132,550
Urewera NP	207,460
Varirata NP	1,063
Waituna Wetland SR RS	3,557
Western Tasmania Wilderness NP WH	76,355
Westland/Mount Cook NP	187,500
Wet Tropics of Queensland WH NP R	920,000
Whangamarino Wetland RS	5,690
Willandra WH	600,000

BR=Biosphere Reserve; GR=Game Reserve; MP=Marine Park; NP=National Park; NR= Nature Reserve; R=Reserve; RS=Ramsar site; SR=Scientific Reserve; WH=World Heritage site

AN ANCIENT TAPESTRY

Nearly three-quarters of Australia – its "red center" – is a vast arid or semiarid plateau. Its deserts range from the desolate and stony "gibber" deserts of Queensland and South Australia to the vast sand ridges of the central Simpson Desert. In places sparse desert grassland supports tussocky, stiff-leaved spinifex and patches of saltbush.

Fringing the deserts are habitats known as mallee – scrubland and forest of low growing eucalyptus – and mulga, which is dominated by various species of drought-tolerant acacia trees (wattles). In contrast, the Mediterranean climate of the southwestern tip of Australia supports species-rich heathlands dominated by eucalypts and banksias.

To the north of the arid heartland the climate becomes more tropical and monsoonal. The semidesert areas merge into savanna grassland grazed by kangaroos and wallabies, with wet sclerophyll forests, marshes and lakes occuring farther north. Patches of tropical rainforest survive in the northeast, especially in Queensland.

To the south and southeast the soil is extremely fertile, and the natural habitats have been greatly modified by agriculture, pastoralism and urbanization. A quarter of the island of Tasmania, however, is still wilderness. Its temperate rainforests, survivors from the ancient continent of Gondwanaland, are among the finest in the world.

Australia's rich coastal habitats range from mangrove and tidal forests in the north to lagoons, reefs and sea grass beds on the northeast coast and complex dune systems, lakes and sheltered inlets in the south and west. Coral reefs fringe many parts of the coast, the finest being the Great Barrier Reef that flanks the northeast coast. Over 1,600 km (995 mi) long, it supports a wealth of marine life.

Unique wildlife

New Zealand, which parted from Gondwanaland even earlier than Australia, still retains ancient temperate rainforests of southern beech. Its unique wildlife includes no mammals, no snakes and very few flowering plants. There is little grazing, except by introduced mammals, and the birds fill almost every ecological niche. Its habitats range from snow-

The Queensland rainforest in Australia has been nominated a World Heritage site because of its diversity of species and Aboriginal rainforest culture. Protection is backed by the federal government but opposed by the state government which favors logging and development programs.

capped mountain peaks, glaciers and tussock grassland in the south to volcanic landscapes, mud pools and geysers, and evergreen tropical forests of kauri, podocarps and tree ferns in the north.

New Guinea has perhaps an even greater range of habitats: volcanic peaks, glaciers, tundra and alpine grasslands give way at lower altitudes to tropical ecosystems – dry monsoon savanna, lush rainforests and coastal swamps and coral reefs. Matching this great diversity is a wealth of wildlife, with plant and animal species derived from Southeast Asia, Indonesia, the Philippines and Australia.

Island of diversity

Each of the Pacific islands has its own history and its own unique, or endemic, species of plants and animals, which have evolved in isolation to fill a variety of island habitats. Some of the islands like New Guinea and New Zealand were once part of the supercontinent, and the animals and plants they contained when they broke away later diversified into new species and varieties. Coral atolls and volcanic oceanic islands were created much more recently. Their stark features have been colonized by plants, carried as seeds by the wind or by birds, or washed up by storms and currents; some seeds were brought by the few animals that were somehow able to cross the ocean. Oceanic islands are important stopover and breeding sites for seabirds, and provide nesting beaches for marine turtles.

The arid heartlands of Australia are not a desert wasteland as commonly portrayed. They have thriving plant and animal communities that have specialized in living with little moisture and intense heat. Vegetation includes mulga (acacia) scrub and mallee (dwarf eucalpyt) scrub. Animal species are varied and include snakes, lizards, scorpions, centipedes, marsupial moles and even colonies of feral cats.

Biomes

- tropical humid forest
- subtropical and temperate rainforest
- tropical dry forest
- evergreen sclerophyll forest and scrubland
- warm desert and semidesert
- tropical grassland and savanna
- temperate grassland
- island system

- ◆ major protected area
- o Biosphere Reserve
- ≈ Ramsar site
- × World Heritage site

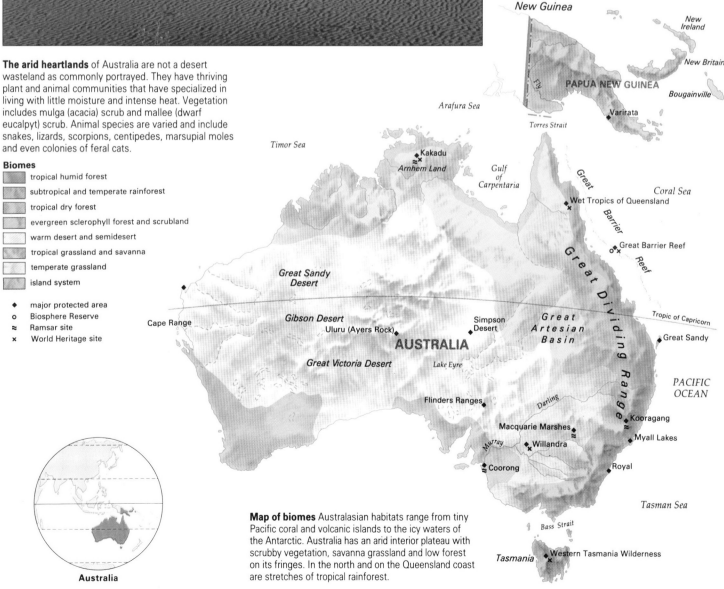

Map of biomes Australasian habitats range from tiny Pacific coral and volcanic islands to the icy waters of the Antarctic. Australia has an arid interior plateau with scrubby vegetation, savanna grassland and low forest on its fringes. In the north and on the Queensland coast are stretches of tropical rainforest.

Australia

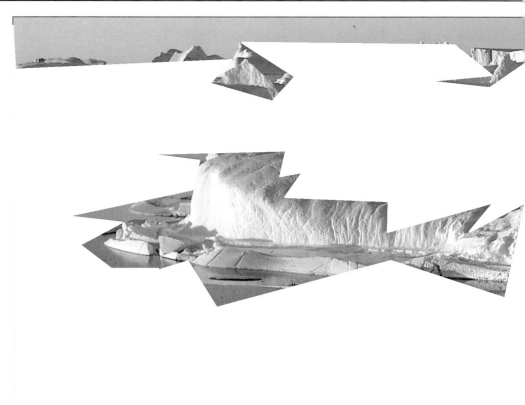

PROGRESS IN CONSERVATION

For at least the last 40,000 years the Aboriginal people of Australia have adapted to the often harsh environment of the country and maintained a deep reverence for the land. Although they imposed some changes on natural habitats, it was nothing compared to the far-reaching disturbance caused by European settlement. The first colonists felled the forests and cleared the land for sheep farming with reckless excess, and as early as the 1820s the Philosophical Society of Australia voiced its concern for the environment. Australia's first protected area was set aside in 1866, and its first national park, the Royal National Park (south of Sydney on the east coast), was created in 1879, and was only the second national park in the world.

Slow but steady progress
Since then, conservation legislation has been strengthened and the number of protected areas has grown slowly but steadily through the efforts of enlightened individuals or community groups and societies. Dust clouds over Melbourne in 1982, caused by drought and soil erosion, focused public attention on environmental issues, and this brought a change in attitude toward conservation.

Conservation is primarily the responsibility of the seven state governments through their National Park and Wildlife Service (NPWS). In addition, a federal agency, the Australian National Parks and Wildlife Service (ANPSW), administers the two national parks owned or part owned by indigenous peoples (Kakadu and Uluru) and has general responsibility for marine conservation. The Australian Heritage Commission (AHC) also plays a part in protecting Australia's natural and cultural heritage. It is estimated that under 5 percent of Australia's total land area is protected, whether for its scenic, scientific, recreational or nature conservation value: not all Australian ecosystems are yet represented in the reserves system. Nevertheless, the scientific importance of environmental protection has increased, and by the end of the 1980s Australia had some 8 areas included in the World Heritage listing. No fewer than 28 of its wetlands had been designated as sites of international importance under the Ramsar Convention on wetlands.

By contrast, in New Zealand – where preservation started even earlier, in 1840 – 17 percent of public land is within the protected reserve system, and there have been moves to extend protection to rich habitats on private land outside the parks. However, funds are lacking to protect examples of all New Zealand ecosystems, even where potential sites have been identified, and the country's unique marine and offshore island habitats are inadequately covered. The New Zealand government has its own Department of Conservation, but the National Parks and Reserves Authority oversees policy for protected areas.

Papua New Guinea's Forest division was established early in the century, but its Wildlife Division did not become active until the late 1950s, and the National Parks Board was set up only in 1969: its first park, Varirata National Park, opened in 1972. However deforestation is taking place at a frightening rate in order to meet the needs of a rapidly increasing population. Added to this, multinational corporations are exploiting the timber resources and clearing jungle for large cash crop plantations.

Protecting island ecosystems
Oceania has the world's highest proportion of threatened species per unit area or

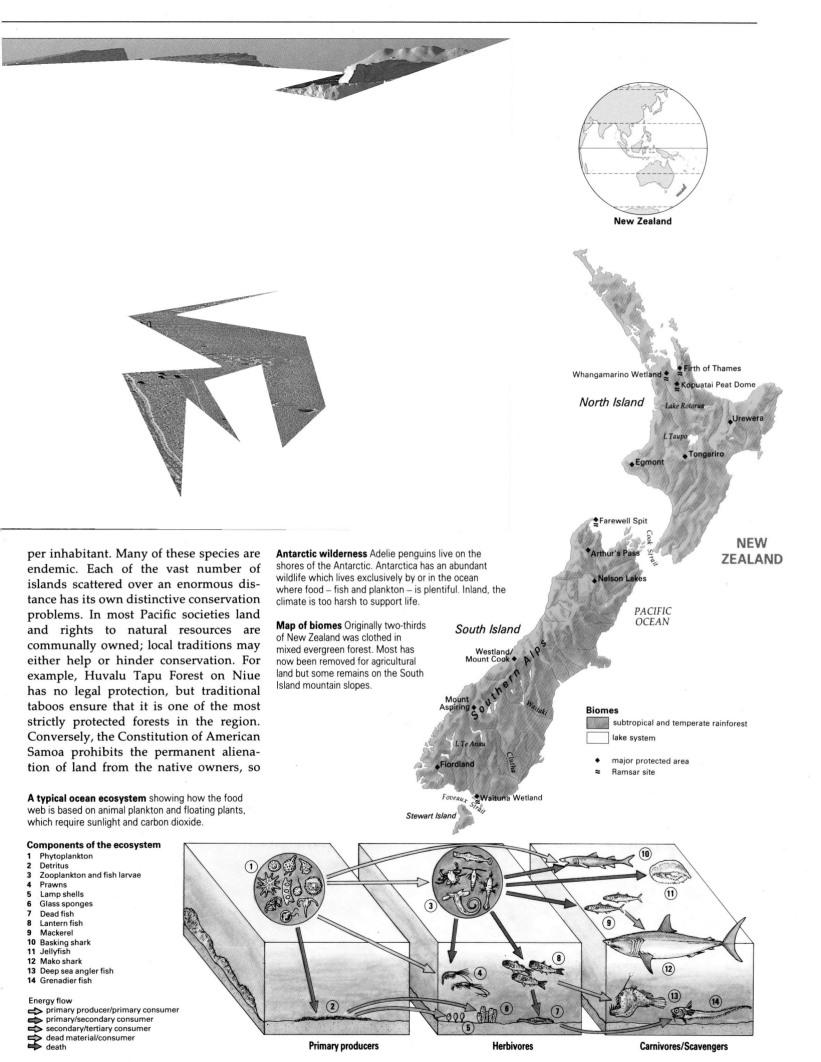

per inhabitant. Many of these species are endemic. Each of the vast number of islands scattered over an enormous distance has its own distinctive conservation problems. In most Pacific societies land and rights to natural resources are communally owned; local traditions may either help or hinder conservation. For example, Huvalu Tapu Forest on Niue has no legal protection, but traditional taboos ensure that it is one of the most strictly protected forests in the region. Conversely, the Constitution of American Samoa prohibits the permanent alienation of land from the native owners, so

Antarctic wilderness Adelie penguins live on the shores of the Antarctic. Antarctica has an abundant wildlife which lives exclusively by or in the ocean where food – fish and plankton – is plentiful. Inland, the climate is too harsh to support life.

Map of biomes Originally two-thirds of New Zealand was clothed in mixed evergreen forest. Most has now been removed for agricultural land but some remains on the South Island mountain slopes.

New Zealand

Firth of Thames
Whangamarino Wetland
Kopuatai Peat Dome
North Island
Lake Rotorua
Urewera
L. Taupo
Tongariro
Egmont

Farewell Spit
Arthur's Pass
Nelson Lakes

Cook Strait

NEW ZEALAND

PACIFIC OCEAN

South Island
Westland/ Mount Cook
Southern Alps
Mount Aspiring
Waitaki
L Te Anau
Clutha
Fiordland
Foveaux Strait
Waituna Wetland
Stewart Island

Biomes

subtropical and temperate rainforest

lake system

◆ major protected area

≈ Ramsar site

A typical ocean ecosystem showing how the food web is based on animal plankton and floating plants, which require sunlight and carbon dioxide.

Components of the ecosystem

1 Phytoplankton
2 Detritus
3 Zooplankton and fish larvae
4 Prawns
5 Lamp shells
6 Glass sponges
7 Dead fish
8 Lantern fish
9 Mackerel
10 Basking shark
11 Jellyfish
12 Mako shark
13 Deep sea angler fish
14 Grenadier fish

Energy flow

primary producer/primary consumer
primary/secondary consumer
secondary/tertiary consumer
dead material/consumer
death

Primary producers   **Herbivores**   **Carnivores/Scavengers**

protected areas can be secured only if they have been granted a lease.

In 1985 there were some 95 protected areas in the South Pacific Regional Environment Program, an international program designed to conserve the oceanic environment. But fewer than 20 percent of the major ecosystems were represented. The 1985 South Pacific Nature Conservation and Protected Areas Conference approved guidelines and standard goals for development in general, and for protected areas in particular. It was hoped that each country in the region would soon have at least one protected area. By the end of the 1980s very little progress had been made.

Success on Samoa

There are, however, some notable success stories. O Le Pupu Pu'e National Park on Western Samoa extends from the central mountains to the coast, and includes most of the ecosystems of the island of Upolu. Kiribati Wildlife Sanctuary covers almost the entire island (formerly Christmas Island), and so does the Baker Island National Wildlife Refuge. Some islands, such as Lord Howe Island and Henderson Islands, have been designated World Heritage sites.

But over the Pacific in general expansion of the protected areas is hampered by the lack of an adequate listing of ecosystems and habitats, and a shortage of suitably trained staff. Conservation has historically been given low priority in relation to development, and there are relatively few non-governmental organizations to lobby for more action.

Saltwater paperbark trees are found in the salty lakes and estuaries of west and southwestern Australia. There are over 600 species of eucalypt in Australia, found in various habitats including deserts, rainforest edges, swamps and fertile valleys.

NATURE INVADED

Because Australia, New Zealand and some of the islands of Oceania broke away from Gondwanaland early in the evolutionary history of the world's plants and animals, its species and ecosystems have evolved separately from those of the other continents. The introduction of alien species can therefore have a devastating effect on the environment, resulting in the extinction of native species and loss of habitats.

In some parts of Australia, an invasive species of mimosa is taking over riverside vegetation. In New Zealand the threat is even greater – some 70 percent of New Zealand's plant species are introduced. European pasture grasses and grazing deer have invaded the islands' tussock grasslands. The changed grassland is unsuitable for many of the original inhabitants, such as the endangered flightless rail, the takahe, a bird unique to New Zealand, which – since it lacked natural predators – lost its ability to fly. Another alien browser, an Australian possum brought in as a source of fur, has altered the species composition of the forests, threatening local populations of seed- and fruit-eating birds.

Many Australian national parks have been invaded by introduced mammals. Goats are grazing the land bare in parts of the Flinders Ranges National Park, pigs make quagmires of valuable wetlands like the Macquarie Marshes, rabbits eat crops, pasture and regenerating native vegetation, and hard-hoofed horses and donkeys destroy the soil surface of arid areas (kangaroos have soft pads).

The loss of the trees

Australasia and Oceania have supported human populations since prehistory; their traditional systems of land ownership, combined with an encyclopedic knowledge of the native plants and animals, successfully conserved rather than depleted natural resources. The opening up of the territories to European and American colonists in the 19th century disrupted this balance, and the land was exploited to produce food and also surplus crops for export.

This process has continued: natural forest cover is being destroyed to make room for plantations of coconuts and other cash crops, and to accommodate rapidly increasing populations. Soil erosion results, and causes waterways, coral reefs and lagoons to silt up. As natural habitats become islands in an

Porcupine grass or spinifex is widespread on the arid savannas of inland Australia. This tussocky grass is hardy and thrives on prolonged drought and dry salty soils. Aboriginal tribesmen burn the leaves to extract a resin, which they use in spear-making.

ever-growing sea of cultivation, more and more species are lost. Most of New Zealand's forests have been destroyed. Even today, conflicts between conservationists and timber extractors over those that survive are not uncommon, and the tropical forests of Irian Jaya and Papua New Guinea are also at extreme risk.

Australians are learning to abandon the old pioneering view of their country as a hostile land that has to be tamed, but they have yet to acquire the Aboriginal people's older wisdom in viewing the earth as their mother. Urbanization and tourism press heavily on natural environments; the collection of fallen timber for heaters and stoves, for example, depletes the habitats fringing the cities.

The ocean environment is also being threatened on a vast scale by drift net fishing – hundreds of kilometers of fine non-degradable net are strung out across the oceans behind fishing boats, creating an indiscriminate trap for marine life, whether fish, mammals or birds. Often nets are abandoned, and may drift around the ocean for decades.

THE COORONG

On the south coast of Australia, south of Adelaide, lie some 9,000 ha (22,200 acres) of coastal sand dunes that date back to the recent ice ages. Between these dunes is the Coorong – a saltwater lagoon 100 km (62 mi) long. Its complex network of waterways forms an extensive mosaic of habitats, of both salt and freshwater, that supports large numbers of birds. It is a rare wetland in an otherwise very arid continent, and Coorong National Park – a Ramsar wetlands site – is a haven for both resident and migratory species of waders, and a refuge for other waterfowl in times of drought. It also supports a large seabird breeding colony.

In the swampy mudflats and seasonal lakes of the southern part of the lagoon, active dolomite (magnesium-rich limestone) formation is taking place, a process often associated with oil-bearing rocks. The Coorong thus provides a rich field of study and research for geologists. It is also an important archaeological site, preserving the remains of an extensive Aboriginal settlement.

The Coorong is a popular recreational area for people living in Adelaide. Fishing and four-wheel driving on the beaches, which disturbs the beach-breeding birds, has brought locals into conflict with the conservationists, and many emotive public meetings have been held. Despite plans to strengthen management of the Coorong, there are fears that underfunding will hamper conservation programs.

An aerial view of the Murray River mouth and Lake Alexandrina in the Coorong National Park. This area of saltwater lagoon and swampy mudflats is a wetland of international importance protected under the Ramsar Convention.

In recent years there has been a significant increase in the populations of the fur seals that breed on the coasts of the subantarctic islands, perhaps because the overhunting of whales has reduced competition for food. Seals are now present in such numbers that they are damaging the moss cushions and eroding potential penguin nest sites. Increasing human interference threatens even Antarctica. Large populations of seabirds and marine mammals are at risk from pollution of coastal waters, and an airstrip, used to offload supplies to scientists working there, has been built in the middle of an area used by colonies of penguins for nesting.

Many small Pacific islands are themselves at risk from the oceans. Predicted increases in sea level due to global warming, already detectable in places, threaten to inundate low-lying atolls and coastal areas. The Tokelau Islands, the Marshall Islands, Tuvalu and Kiribati could all be completely destroyed.

Australia's Great Barrier Reef is threatened by plans for oil exploration, limestone mining, offshore sewage discharge, overfishing and tourism. The removal of large quantities of reef limestone on many coral-based Pacific islands has seriously impaired their sea defenses, leaving them vulnerable to storms.

Kakadu National Park

Covering 667,000 ha (1,647,490 acres) of Australia's lush, green, tropical north, Kakadu National Park contains a remarkable range of tropical habitats. Mangrove swamps and saltmarshes fringe the coast, while inland lie the floodplains of the South and East Alligator rivers, drier lowland plains with gentle hills covered in eucalypt forest, and a spectacular 500-km (310-mi) long sandstone escarpment rising to 300 m (823 ft) that forms the western edge of the Arnhem Land plateau. Joints in the sandstone have been carved into deep valleys with many permanent pools. After the monsoon rains, cascades of water plunge over the edge of the escarpment, eventually submerging the vast floodplains.

On the escarpment, low spinifex shrubland and heath-type plants merge into woodland, with tall, open forest in the ravines. In some sheltered places there are pockets of rainforest. More than half the park is covered in lowland eucalypt

The Yellow Waters Lagoon in the Kakadu National Park is fringed with paperbark trees. These wetlands are famous for their populations of water birds (there are 30 species of migratory waders) and crocodiles.

Clinging to the rocks Tenacious bushes have found a hold in the sandstone joints of the Arnhem Land escarpment. Much of this national park is Aboriginal land and the cliffs and caves possess many sacred sites and wall paintings.

From exploitation to protection Settlers were originally attracted to the Kakadu area for the rich hunting and fishing and for the mineral deposits, but it is now a national park, a Ramsar site and a World Heritage site.

woodlands which are extremely rich in species, many of them endemic. The wetlands and floodplains are a vital habitat for waterfowl, crocodiles and a great variety of fish.

Unlike many of the world's large parks, Kakadu is fortunate in enclosing several complete ecosystems, and an entire drainage basin from source to coast. It includes many Aboriginal sacred places; Aboriginal wall paintings are found on the cliffs and in the caves. European settlers began to raise livestock here at the end of the 19th century; they hunted crocodiles and the introduced Asian water buffaloes. However, Kakadu's remote location saved it from many in-

vading species, such as rabbits and foxes, which accompanied the settlers elsewhere in Australia. Consequently, a large number of its native plants and animals have survived. A plan to eradicate invasive alien mimosa species from the river basins has had considerable success, and the buffalo population has been substantially reduced.

Park and people

About half the area of Kakadu National Park is owned by Aboriginal organizations, who lease it to the Director of National Parks and Wildlife. The first part of the park was designated in 1979. In 1985 a second area was added, and in 1987 a further part was incorporated. Stages I and II have since become World Heritage sites. Stage I is also a wetland of international importance (designated by the Ramsar Convention), and Stage II has been proposed for this status.

Over 200,000 people visit the park each year, and there are plans to develop tourism and recreational facilities further to benefit local residents and increase employment in the area. Under the terms of the lease, local Aboriginals have a say in park management decisions, and the authorities are training them in the skills necessary for park management.

Fire is a problem in the park: natural fires are common in the dry season, but unseasonal burning by the settlers disrupted existing habitats. Park management aims to restore the Aboriginal burning patterns that contributed to the region's diverse habitats.

Kakadu's ancient rocks contain rich seams of valuable mineral deposits, including gold, platinum, iron, cobalt and even diamonds. Mining has been proposed in the upper catchment areas of three rivers in a part of the park not yet fully protected. It is strongly opposed by conservationists, who hope to make the whole park a strict conservation zone; further research is being carried out by the Resource Assessment Commission set up by the federal government.

The oldest forests on Earth

At the time, more than 100 million years ago, that Gondwanaland began to fragment, the world's climate was warmer and vast forests, already more than 30 million years old, extended across much of the globe. New Zealand was one of the first pieces to detach itself from the supercontinent and to start drifting northward. Seed-bearing plants (angiosperms) had not yet reached the dominant position they have in the world's vegetation today. Consequently New Zealand's temperate rainforests contain very few wildflowers.

Yet these forests are as lush as any tropical rainforest. They are dominated by the southern beech (*Nothofagus*), podocarps, evergreen broadleaf trees, and primitive tree ferns. Mosses and ferns grow luxuriantly in the heavy rainfall, blanketing branches and roots, as well as trailing lichens, lianas, epiphytic plants and climbers. Similar forests are found in Tasmania and on the west coast of South America in south-central Chile – all remnants of the forests that once covered Gondwanaland.

Luxuriant temperate rainforest on New Zealand's North Island.

GLOSSARY

Acid rain
Popular term used to describe PRECIPITATION that has become acid through reactions between moisture and chemicals in the atmosphere. It normally refers to the excessive acidity of atmospheric moisture caused by polluting emissions of sulfur dioxide and nitrogen oxides from burning fossil fuels.

Adaptation
The process by which, under the influence of NATURAL SELECTION, an ORGANISM gradually changes genetically so that it becomes better fitted to cope with its environment.

Agroforestry
Cultivation systems that integrate forestry and farming within a coherent framework. It allows a diversity of products to be produced without exhausting any particular component of the ENVIRONMENT.

Algae
Small plants that live in water or damp ground, and that do not have true stems, leaves or roots.

Alpine
The TUNDRA-like environment found in mountain areas between the TREELINE and the permanent snow line.

Amphibian
An animal that lives on land but whose life cycle requires some time to be spent in water, e.g. the frog.

Annual
A plant that grows from seed, flowers, sets seed and dies within one year.

Aquatic
Plants and animals that grow and live in water.

Arctic
The northern POLAR region. In biological terms it also refers to the northern region of the globe where the mean temperature of the warmest month does not exceed 10°C (50°F). Its boundary roughly follows the northern TREELINE.

Arid
An area that is hot and dry. Arid areas are rarely rainless, but rainfall is intermittent and quickly evaporates or sinks into the ground. Little moisture remains in the soil, so vegetation is sparse.

Biogeography
The study of the distribution of plants and animals over the surface of the land.

Biological control
The introduction by humans of one SPECIES (usually a predator or parasite) to control the POPULATION of another species. It is used in agriculture to control weeds and pests.

Biomass
The total mass of all the living ORGANISMS in a defined area or ecological system (see ECOLOGY).

Biome
The major global ecological units (see ECOLOGY) of plants and animals throughout the world, often described by their vegetation type, e.g. SAVANNA grassland.

Biosphere
The thin layer of the Earth that contains all living ORGANISMS and the ENVIRONMENT that supports them.

Biosphere Reserve
Part of a BIOME protected under the Man and Biosphere program. Each is large enough to allow its unique biological characteristics to be self-sustaining. They also provide valuable areas for research.

Boreal
The northern cold winter climates lying between the ARCTIC and latitude 50°N. The region is dominated by CONIFEROUS FOREST, also known as TAIGA.

Brackish water
Water with a detectable quantity of salt, but less than that of sea water. It is most often found in estuaries where fresh water and the sea mix.

Browser
A HERBIVORE that eats the leaves and shoots of shrubs and trees.

Buffer zone
An area adjacent to a protected area where the use of the land is managed to prevent damage to the protected area while allowing local people to continue making a living.

Canopy
The uppermost layer of a forest, where the branches spread out and the foliage restricts the amount of light reaching the forest floor.

Captive breeding
Breeding using animals brought in from the wild in wildlife parks, zoos and other places in order to increase and so conserve the POPULATIONS of ENDANGERED SPECIES.

Carnivore
An animal that eats the flesh of another animal.

Chaparral
The scrub vegetation growing in the areas of California, in the United States, where there is a MEDITERRANEAN climate.

Climax
The final stage of natural SUCCESSION in an ECOSYSTEM where there is no further net growth in BIOMASS and the ORGANISMS are in equilibrium with their ENVIRONMENT.

Cloud forest
Forest growing in moist areas at high altitudes. The trees have an abundance of EPIPHYTES growing on them.

Colonization
The establishment of plants or animals in a new environment.

Community
A collection of SPECIES occupying a common ENVIRONMENT and interacting with each other.

Competition
The struggle between individuals of the same or other SPECIES for food, space, light and NUTRIENTS, where these are inadequate to support all of them.

Coniferous forest
A forest comprising mainly coniferous trees, i.e. those that bear cones and have needle-like leaves. Most coniferous trees are EVERGREEN and are found principally in the TEMPERATE zones and BOREAL regions.

Conservation
In relation to living things, the planning for and wise use of living resources in order to maintain their sustainable use and ecological DIVERSITY.

Continental climate
The type of climate associated with the interior of continents. It is characterized by wide daily and seasonal ranges of temperature, especially outside the TROPICS, and low rainfall.

Continental drift
The complex process by which the continents move their positions relative to each other on the plates of the Earth's crust. Also known as plate tectonics.

Coral
A group of animals related to sea anemones and living in warm seas. Individuals, called polyps, combine to form a colony.

Coral reef
A barrier of CORAL formed by the accumulation of the skeletons of millions of polyps. Living polyps attach themselves to the skeletons of past generations. Reefs may be an extension of the shore (fringing) or offshore (barrier).

Crustaceans
Hard-bodied, mainly AQUATIC animals, usually with five pairs of legs, two pairs of antennae and head and thorax joined, e.g. crab, crayfish, shrimps and woodlice.

Deciduous forest
A forest comprising mainly deciduous trees, i.e. those that shed their leaves each year during the winter or the dry season.

Decomposer
ORGANISMS such as fungi and bacteria, which feed on the dead tissue of other ORGANISMS and release the nutrients held in the tissue into the ENVIRONMENT.

Desertification
The creation of desert-like conditions through the inappropriate use of the land by humans, or as a result of changes in climate.

Detritus
Debris from decayed plant or animal matter.

Dipterocarp
A family of 400 SPECIES of trees, often very tall, living in the RAINFORESTS of Asia and Africa. Many are valued for their timber and resin.

Diversity
The variety of SPECIES living in a given area.

Dormancy
A period during which the metabolic activity of a plant or animal is reduced to such an extent that it can withstand difficult environmental conditions such as cold or drought. In animals it is also known as hibernation.

Ecology
The study of ORGANISMS in relation to their physical and living ENVIRONMENT.

Ecosystem
A COMMUNITY of plants and animals and the ENVIRONMENT in which they live and react with each other.

Endangered species
A SPECIES whose POPULATION has dropped to such low levels that its continued survival is insecure.

Endemic
A SPECIES that is NATIVE to one specific area.

Environment
The living and nonliving surroundings of an ORGANISM.

Epiphyte
A plant that grows on another plant, but does not get NUTRIENTS or water from it.

Equatorial
In the region of the Equator.

Eutrophication
An excessive growth of plants in water resulting from its enrichment with minerals and nutrients. The oxygen content of the water declines, and this causes fish to die.

Evergreen
A plant that bears leaves throughout the year.

Evolution
The process by which SPECIES have developed to their current appearance and behavior through the process of NATURAL SELECTION.

Exotic
A living animal or plant that is not NATIVE to an area but has become established after being introduced from elsewhere, often for commercial or decorative purposes.

Extinction
The loss of a local POPULATION of a particular SPECIES or even the entire species. It may be natural or be caused by human activity.

Fauna
The general term for the animals that live in a particular region.

Filter feeder
An animal that feeds by filtering small items of food out of the water.

Flora
The general term for the plant life of a particular region.

Food chain
A succession of living things, each of which is the food for the next one up in line.

Food web
The complex feeding interactions between SPECIES in a COMMUNITY.

Game reserve
An area originally set aside for the management and protection of game animals for hunting. They are now usually areas where all wildlife is protected.

Garigue
A form of MAQUIS vegetation that grows on poor, very rocky soil. Named after an area north of Nimes in southern France, it consists of a sparse covering of grass, aromatic shrubs and scattered trees.

Genes
Sets of "coded" instructions that determine the inherited characteristics of an ORGANISM.

Genetic variation
The variation in form, physiology or behavior between individuals of a SPECIES that results from the immense possibilities of reassortment of the genes each one receives from its parents, and from sudden and unpredictable changes due to mutations.

Genus
(plural genera) A level of biological classification of ORGANISMS in which closely related SPECIES are grouped. For example, dogs, wolves, jackals and coyotes are all grouped together in the genus *Canis*.

Germination
The process by which a seed sprouts and is transformed into a plant. It is affected by environmental conditions such as heat and exposure to water and light. Seeds can lie dormant for several years (see DORMANCY).

Gills
The respiratory organs in many AQUATIC animals that allow oxygen from the water to be taken into the bloodstream and carbon dioxide to be released into the water.

Grazer
A HERBIVORE that feeds on grass and other herbaceous vegetation growing near the ground.

Greenhouse effect
The warming of the Earth as a result of the accumulation of gases such as carbon dioxide, which absorb the heat radiated from the surface and slow down its escape into space. The burning of fossil fuels is responsible for about 50 percent of the extra warming.

Groundwater
The water held in spaces in rocks beneath the surface.

Habitat
The locality within which a particular ORGANISM is found, usually including some description of its character, such as grassland habitat.

Halophyte
A plant that is able to grow in a very salty ENVIRONMENT.

Hardwoods
The general term for broadleaf trees; their wood is usually harder than that of CONIFEROUS trees.

Heath or heathland
An area of open land covered with lowgrowing shrubs such as heather and ling. They are common on peaty upland SOILS where they were created by the clearance of forest. Fire and grazing prevent trees from becoming established again. (See also PEAT.)

Herbivore
An animal that feeds exclusively on living plants. (See also BROWSER, GRAZER.)

Hibernation
The state of reduced metabolic activity or DORMANCY in animals during winter.

Hunting reserve
An area of land set aside for hunting, usually for royalty or other privileged persons. The area is managed to favor the quarry SPECIES.

Hydrology
The study of water and the paths that link surface water, GROUNDWATER, oceans and atmospheric water.

Ice age
Long periods of the Earth's history during which ice covered a greater proportion of the surface than it does today. Also called a glacial.

Intertidal
The zone between high and low tides.

IUCN
The International Union for the Conservation of Nature and Natural Resources, to be renamed the World Conservation Union, is a membership organization bringing together governments, nongovernment organizations and individuals to promote CONSERVATION and the sustainable use of living resources.

Lagoon
A shallow stretch of water lying behind a barrier such as a CORAL REEF, sandbar or spit.

Leach
The process by which water moves soluble minerals downward from one layer of soil into another.

Mallee
The scrubland vegetation with grass and small trees found over the SEMIARID coastal regions of southern Australia. It also describes the multistemmed habit of dwarf eucalyptus trees found in the area.

Mangrove
A dense forest of shrubs and trees growing on tidal coastal mudflats and estuaries throughout the TROPICS. Many plants have aerial roots.

Maquis
The name given to the vegetation found along the Mediterranean coast of France, consisting of aromatic shrubs, laurel, myrtle, rock rose, broom and small trees such as olive, fig and hermes oak. It corresponds to macchia in Italy and CHAPARRAL in California.

Maritime climate
An area where the (generally moist) climate is determined mainly by its proximity to the sea. The sea heats up and cools down more slowly than the land, reducing variations in temperature so the climate is equable.

Mediterranean climate
An area that has a climate similar to that of the Mediterranean region, namely warm, wet winters and hot, dry summers. The natural

vegetation is of a type that can withstand drought.

Migration
The periodic movement, normally seasonal, of animals from one region to another to feed or to breed.

Mire
An alternative name for a bog or swampy PEATLAND.

Monsoon
The wind systems in the TROPICS that reverse their direction according to the seasons. When they blow onshore they bring heavy rainfall, also known as the monsoon.

Monsoon forest
TROPICAL FOREST in MONSOON areas in southeast Asia and Africa, where the monsoon climate produces wet and dry seasons. The trees are DECIDUOUS, losing their leaves at the start of the dry season.

Montane
The zone at middle altitudes on the slopes of mountains, below the ALPINE zone.

Moorland
An ill-defined term that covers PEATLAND and HEATH and is often used to describe open, treeless hill or mountain country. Vegetational categories of grass, heather and sedge moor are sometimes used.

Mulga
A HABITAT of western and central Australia, consisting largely of low acacia trees and forming a dry woodland with patches of shrubs and a ground cover of grasses. The desert appearance of the dry season changes to a carpet of flowers after the rains.

National Monument
An area in the United States preserved by the federal government for the protection of a site or object of prehistoric, historic or scientific interest.

National park
Large areas of natural or seminatural land that are maintained in that condition for people to enjoy and for conservation purposes.

Native
An ORGANISM that lives naturally in a particular locality.

Natural resources
Materials such as minerals, oil, timber, water and other plant or animal resources that have a value to humans.

Natural selection
The process by which ORGANISMS not well fitted to their ENVIRONMENT are eliminated by predation, parasitism, COMPETITION, etc., and those that are well fitted survive to breed and pass on their GENES to the next generation.

Nature reserve
An area of land set aside where nature is managed in such a way as to protect its special features.

Nutrient
A substance or ingredient derived from the ENVIRONMENT that provides nourishment for plants and animals.

Nutrient cycle
The constant movement of NUTRIENTS between the living and nonliving components of an ECOSYSTEM.

Organism
Any living thing – animal, plant or microbe.

Omnivore
An animal that feeds on both plant and animal matter.

Ozone
A poisonous, unstable gas that is a form of oxygen with three atoms to each molecule instead of two. It is present in small amounts in the atmosphere, and is responsible for absorbing most of the Sun's harmful ultraviolet radiation.

Palearctic
A biogeographic region encompassing Europe and Asia north of the Himalayas, and Africa north of the Sahara.

Peat
An accumulation, over 40 cm (16 in) deep, of partly decomposed plant remains in wetlands. Combinations of high acidity and low temperature, NUTRIENT and oxygen levels prevent total decomposition.

Peatland
A WETLAND HABITAT, such as fen and bog, in which PEAT has accumulated. Also called mire.

Permafrost
Ground that remains permanently frozen, typically in the POLAR regions. A layer of SOIL at the surface may melt in summer, but the water that is released is unable to drain away through the frozen subsoil.

pH
A measurement on the scale 0–14 of the acidity or alkalinity of a liquid: pH7 is neutral, less than pH7 is acidic and more than pH7 alkaline. The scale is logarithmic, hence pH5 is 10 times more acid than pH6.

Photosynthesis
The making of organic compounds, primarily sugars, from carbon dioxide and water, using sunlight as the source of energy and chlorophyll, or other related pigment, for trapping the Sun's energy.

Phytoplankton
Microscopic plants in PLANKTON.

Plankton
The COMMUNITY of microscopic plants and animals that float at or near the surface of the sea or a lake.

Polar
The regions that lie within the lines of latitude known as the Arctic and Antarctic circles, 66° 32' north and south of the Equator respectively. They mark the lowest latitude at which the Sun does not set in midsummer or rise in midwinter.

Polder
An area of lowlying ground that has been reclaimed from the sea or other body of water.

Pollination
The fertilization of plants by the transfer of pollen grains from the male anthers to the female pistil. Pollen is usually transferred by wind or insects but occasionally by animals such as hummingbirds.

Pollution
The disruption of a natural ECOSYSTEM as a result of contamination from human activities.

Population
In biological terms, a more or less separate breeding group of animals. The term is also used for the total number of individuals counted within a given area.

Prairie
North American STEPPE grassland between 30°N and 55°N.

Precipitation
Moisture that reaches the Earth from the atmosphere, including rain, snow, sleet, mist and hail.

Predator
An animal that feeds on another animal (the prey).

Primeval forest
Forest that has never been significantly altered by human activity.

Productivity
The amount of weight (or energy) gained by an individual, a SPECIES or an ECOSYSTEM per unit of area per unit of time.

Rainforest
Forest where there is abundant rainfall all year round. The term is normally associated with TROPICAL rainforests, which are rich in plant and animal SPECIES and where growth is lush and very rapid.

Ramsar site
A protected area designated under the Convention on Wetlands of International Importance, convened at Ramsar, Iran in 1971.

Refuge
A place where a SPECIES of plant or animal has survived after formerly occupying a much larger area. For example, mountain tops are refuges for ARCTIC species left behind as the glaciers retreated at the end of the last ICE AGE.

Riverine forest
A thin belt of forest along a river bank in otherwise more open country. Also called gallery forest.

Salinization
The accumulation of soluble salts near or at the surface of SOIL. This can occur when water used for irrigation evaporates; eventually the land becomes worthless for cultivation.

Saltmarsh
A coastal marsh where silt has accumulated and been stabilized by the growth of salt-tolerant plants. The TEMPERATE equivalent of MANGROVE swamp.

Savanna
A HABITAT of open grassland with scattered trees in TROPICAL areas. Also known as tropical grassland, it covers areas between tropical RAINFOREST and hot deserts. There is a marked dry season each year and too little rain to support large areas of forest.

Scavenger
An animal that feeds on the remains of food killed or collected by other animals.

Sclerophyll
Hard-leaved scrub vegetation. The low trees and shrubs have tough EVERGREEN leaves. It is found where there is drought in summer.

Secondary forest
An area of forest that has regenerated after being felled. It usually has fewer SPECIES than the PRIMEVAL FOREST it replaces.

Semiarid
Areas between ARID deserts and better-watered areas, where there is sufficient moisture to support a little more vegetation than can survive in the desert.

Slash-and-burn farming
Growing crops on land cleared by felling trees and burning off the vegetation cover. Also called swidden agriculture.

Soil
The unconsolidated, weathered upper layer of rock in which plants grow. It consists of fine particles of mineral matter as well as living and dead organic material.

Soil erosion
The removal of the topsoil mainly by the action of wind and/or rain. The term sheet erosion is used when a fine layer of soil is removed from the land; in gully erosion the rain carves deep channels.

Species
ORGANISMS that resemble one another and can breed among themselves to produce similar offspring that can themselves breed with other individuals of their species.

Steppe
Open grassy plains with few trees or shrubs. Steppe is characterized by low and sporadic rainfall and experiences wide variations in temperature during the year.

Subspecies
A recognizable subpopulation of a SPECIES, typically with a distinct distribution. Sometimes called a race.

Subtropical
The climatic zone between the TROPICS and TEMPERATE zones. There are marked seasonal changes of temperature but it is never very cold.

Succession
The development and maturation of an ECOSYSTEM, through changes in the types and abundance of SPECIES. When is reaches maturity it stabilizes in a CLIMAX.

Succulent
A plant that stores water in soft tissues of the leaves or stem to withstand drought, e.g. the cactus.

Symbiosis
Two living things of different SPECIES that derive mutual benefit from their close association with each other.

Taiga
Russian name given to the CONIFEROUS FOREST and PEATLAND belt that stretches around the world in the northern hemisphere, south of the TUNDRA and north of the DECIDUOUS FORESTS and grasslands.

Temperate
The climatic zones in mid latitudes, with a mild climate. They cover areas between the warm TROPICAL and cold POLAR regions.

Territory
An area of land occupied by a single animal or group of animals and actively defended.

Terrestrial
ORGANISMS whose entire life cycle is spent on the land.

Transpiration
The loss of water from leaves that draws a column of water up the stem from the roots, where it is absorbed from the soil. It is controlled by the opening and closing of small pores called stomata.

Treeline
The limit of tree growth beyond which the growing season each year is not long enough for trees to grow.

Tropics
The area lying between the Tropic of Cancer and the Tropic of Capricorn. They mark the LATITUDE farthest from the Equator where the sun is directly overhead at midday in midsummer.

Trophic level
The level at which living things are positioned in a FOOD CHAIN.

Tundra
The level, treeless land lying in the very cold northern regions of Europe, Asia and North America, where winters are long and cold and the ground beneath the surface is permanently frozen. (See also PERMAFROST.)

Udvardy classification
A system of biogeographic units proposed by M.D.F. Udvardy in 1975 as a basis for gauging the adequacy of protected area coverage on a global scale. The Earth's natural ECOSYSTEMS are grouped into 14 BIOMES that are subdivided into 230 provinces.

Understorey
The layer of shrubs and small trees under the forest CANOPY.

Water catchment
The area of land drained by a river and its tributaries. (See also WATERSHED.)

Watershed
The boundary line dividing two WATER CATCHMENT areas. In the United States the term is also used to mean the area itself.

Water table
The upper level of ground saturated with water. If the water table is at the surface, the ground is waterlogged.

Wetland
A HABITAT that is waterlogged all or enough of the time to support vegetation adapted to those conditions.

World Heritage site
A protected area that may have natural or cultural interest, or both, and designated and listed under the World Heritage Convention of the United Nations Educational, Scientific and Cultural Organization (UNESCO).

Xerophyte
A plant that is able to withstand long periods of drought.

Zoogeographic region
A large area of the Earth's surface that has a characteristic FAUNA. The Earth is divided into six regions. The botanical equivalent are known as floristic kingdoms.

Zoogeographic realm
The ZOOGEOGRAPHIC REGIONS are combined to make three realms.

Zooplankton
The microscopic animals living in PLANKTON.

Further reading

Attenborough, David et al. *The Atlas of the Living World* (Weidenfeld & Nicolson, London 1989)
Attenborough, David *Life on Earth* (Collins, London 1979)
Attenborough, David *The Living Planet* (Collins, London 1984)
Ayensu, E.S., Heywood, V.H., Lucas, G.L. and Defilipps, R.A. *Our Green and Living World – the wisdom to save it* (Cambridge University Press, Cambridge 1984)
Brown, B. and Morgan, L. *The Miracle Planet* (Merehurst Press, London 1990)
Durrell, Lee *State of the Ark* (The Bodley Head, London 1986)
Ehrlich, A.H. and P.R. *Earth* (Franklin Watts, New York 1987)
Goldsmith, E. and Hildyard, N. (Eds) *The Earth Report – monitoring the battle for our environment* (Mitchell Beazley, London 1988)
Goudie, A.S. *The Human Impact on the Natural Environment*, 2nd edn, (Basil Blackwell, Oxford 1986)
Gribbon, J. and Kelly, M. *Winds of Change – living with the greenhouse effect* (Headway/Hodder and Stoughton, London 1989)
Lovelock, James *The Ages of Gaia, a biography of our living earth* (Oxford University Press, Oxford 1988)
Moore, D.M. (Ed.) *Green Planet – the Story of Plant Life on Earth* (Cambridge University Press, Cambridge 1982)
Myers, Norman (Ed.) *The Gaia Atlas of Planet Management* (Pan Books, London 1987)
The New Encyclopedia Britannica, 15th edn (Encyclopedia Britannica, Chicago 1989)
Jones, G., Robertson, A., Forbes, J. and Hollier, G. *Environmental Science* (Collins, London 1990)
The Times Atlas of the World (Times Books, London 1988)
The Times Atlas of the Oceans (Times Books, London 1983)
Tudge, C. (Ed.) *The Encyclopedia of the Environment* (Christopher Helm, London 1988)
Udvardy, M..D.F. *A Classification of the Biogeographical Provinces of the World* (IUCN Occasional Paper No. 18, International Union for Conservation of Nature and Natural Resources, Geneva 1975)
Wilson, E.O. *Biodiversity* (National Academy Press, New York 1988)
World Conservation Strategy (International Union for Conservation of Nature and Natural Resources, Geneva 1980)

Acknowledgments

Picture credits
Key to abbreviations: A Ardea, London; **ANT** Australasian Nature Transparencies, Victoria, Australia; **BCL** Bruce Coleman Ltd, Uxbridge, Middlesex; **BF** Biofotos, Farnham, Surrey; **E** Explorer, Paris; **HQ** Hoa-Qui, Paris; **ICCE** International Center for Conservation Education, Guiting Power; **LO** Landscape Only, London; **NFG** Naturfotografernas Bildbyra, Osterbybruk, Sweden; **NHPA** Natural History Photographic Agency, Ardingly, Sussex; **OSF** Oxford Scientific Films, Long Hanborough, Oxon; **PP** Panda Photos, Rome; **RHPL** Robert Harding Picture Library, London; **SAL** Survival Anglia Picture Library, London; **SGA** Susan Griggs Agency, London; **SWPL** Swift Picture Library, Ringwood, Hants; **VE** Victor Englebert, Cali, Colombia.

b bottom, c center, t top, tl top left, tr top right.

6–7 E/F. Boizot 8–9 Magnum/F. Scianna 10 ANT/Tony Howard 10–11t PP/A. Petretti 10–11b NHPA/John Shaw 12 Brian & Cherry Alexander 12–13 Peter D. Canty 14–15t Gerald Cubitt 14–15b E/J.P. Ferrero 15 BF/Heather Angel 16–17t OSF/Harry Taylor 16–17b Rob Cousins 17 HQ/Denis Huot 18 Office of Public Works, Dublin 18–19t VE 18–19b Fred Bruemmer 20–21t Anthony Bannister 20–21b WWF Photo Library/Ron Petocz 21 Tim Fitzharris 22 A/J.P. Ferrero 22–23 HQ/G. Boutin 23 NHPA/Stephen J. Krasemann 24l A/J.P. Ferrero 24r NHPA/View A/S 24–25 E/M. Moisnard 25 A/J.P. Ferrero 26–27t BCL/Frans Lanting 26–27b E/Jean-Paul Nacivet 27 ICCE/Philip Steele 28 ANT/Jan Taylor 28–29 LO 29 ANT/Otto Rogge 30t LO 30c SAL/Dieter and Mary Plage 30b WWF Photo Library/Nancy Sefton 31 Bob Gibbons 32 NHPA/Dave Currey 32–33 OSF/Martyn Chillmaid 33 OSF/J.A.L. Cooke 34–35t SAL/J.B. Davidson 34–35b VE 35 Dr. Alan Beaumont 36 Derek G. Widdicombe/Countrywide Photographic Library 36–37 Andres Hurtado Garcia 37 Peter D. Canty 38l SGA/Adam Woolfitt 38r BCL/Erwin and Peggy Bauer 39 OSF/Edward Parker 40t BCL/Alain Compost 40b NHPA/Anthony Bannister 41 SAL/Andy Buxton 44 Tim Fitzharris 45 BF/Heather Angel 46–47 Fred Bruemmer 47 OSF/Brian Milne 48 BCL/Jeff Foott 49l BCL 49r BF/Heather Angel 50 NHPA/Stephen J. Krasemann 50–51t, 50–51b Tim Fitzharris 52–53 Fred Bruemmer 54 OSF/Stan Osolinski 55 PP/S. Dimitrijevic 56–57 NHPA/John Shaw 57 A/Francois Gohier 58–59 Jerome Wyckoff 59l OSF/Zig Leszczynski 59r Robert Perron 60 OSF/Michael Fogden 60–61 William Ervin/Natural Imagery 61 OSF/Stan Osolinsky 62 PP/S. Pirovano 62–63 A/Wardene Weisser 63 BCL/Nicholas DeVore 64t INCAFO/A. Larramendi 64b William Ervin/Natural Imagery 65 OSF/Michael Fogden 66 PP/S. Dimitrijevic 66–67 SAL/John and Irene Palmer 67 A/Wardene Weisser 68 BCL/Jeff Simon 68–69 OSF/Stan Osolinski 69 Natural Image/Mike Lane 70–71 Brian and Cherry Alexander 72–73 OSF/Kjell B. Sandved 73 BF/Brian Rogers 74t Gina Green 74b Michael Fogden 75 BF/Soames Summerhays 76 OSF/Michael Fogden 76–77 Colin Hughes 78–79t, 78–79b, 79 SGA/Adam Woolfitt 80–81 South American Pictures/Tony Morrison 81 BCL/L.C. Marigo 82 SAL/Annie Price 83 BCL/Udo Hirsch 84–85 VE 85t Julio Etchart 85b BCL/L.C. Marigo 86 Andres Hurtado Garcia 86–87 South American Pictures/Marion Morrison 87 Hutchison Library/Choco 88–89 BCL/Erwin and Peggy Bauer 90t NFG/Tore Hagman 90b, 92, 93 NFG/Claes Grundsten 94–95 NFG/Klas Rune 95t BF/Lars Serritsler 95b NHPA/Jan-Peter Lahall/View A/S 96 NFG/Edvin Nilsson 96–97 NFG/Axel Ljungquist 97 Tiofoto/Hans Andersson 98–99 NHPA/E. Murtomäki 100 SWPL/Mike Read 101 SWPL/Martin King 102 SWPL/Mike Read 103l Jennie Woodcock 103r Bob Gibbons 104–105 NHPA/Laurie Campbell 105t Tim Woodcock 105b SAL/Tony Bomford 106 Bob Gibbons 106–107 Andrew Lawson 107 BF/Heather Angel 108–109 E/Andre Bugaud 109 E/C. D'Hotel 110t LO/Cornish 110b, 112–113 E/Pierre Tetrel 113 Bob Gibbons 114 E/Dupont 114–115t PP/R. Ricci Curbastro 114–115b OSF/G.I. Bernard 116–117, 117, 118 NHPA/Picture Box/J.H. Hopman 119 NHPA/Picture Box/Lee 120, 120–121 NHPA/Picture Box/JWE 121 NHPA/Picture Box/VIT 122–123,123l, 123r NHPA/Picture Box/Lee 124 Bob Gibbons 125 Chris Mattison 126 Bob Gibbons 127 Desert Photo/J.M. Miralles 128–129 INCAFO/A. Camoyan 129l Wildlife Matters 129r Bob Gibbons 130 INCAFO/Jose L. Gonzales Grande 130–131 INCAFO/Juan A. Fernandez 131 INCAFO/Candy Lopesino and Juan Hidalgo 132–133 Teresa Farino 134 PP/G. Cappelli 135 PP/R. Mattio 136 PP/M. De Medici 137t PP/G. Marcoaldi 137b PP/F. Pinchera 138t Sonia Halliday Photos 138b PP/A. Boano 139 RHPL 140l PP/R. Mattio 140r PP/G. Cappelli 141 PP/R. Mattio 142–143 NHPA/Silvestris 143t Zefa/Thonig 143b NHPA/Silvestris/Rauch 144–145 NHPA/Silvestris/Rosing 145 NHPA/Silvestris 146, 146–147 Dr. Hans Bibelriether 147 NHPA/Silvestris/Karl-Heinz Jorgens 148l Dr. Hans Bibelriether 148r NHPA/Silvestris/Hansgeorg Arndt 148–149 Zefa/Thonig 150–151t PP/W. Lapinski 150–151b Marek Peitrzak 152t Wlodzimierz Lapinski 152c Marek Pietrzak 153 Tim Sharman 154t SAL/John Harris 154b PP/W. Lapinski 155 Wlodzimierz Lapinski 156 A/J.P. Ferrero 156–157t INCAFO/Juan A. Fernandez 156–157b A/Liz and Tony Bomford 158 A/Masahiro Iijima 158–159 Vadim Gippenreiter 160 A/Masahiro Iijima 160–161 Zefa 162 Vadim Krohin 162–163 Vadim Gippenreiter 163 John Hartley/JWPT 164–165 OSF/Tony Allen 165t A/Masahiro Iijima 165b Doug Allan 166 ICCE/Marie Matthews 166–167 Planet Earth Pictures/Hans Christian Heap 168–169 J. Allan Cash 169t Sonia Halliday Photos 169b E/Francis Jalain 170t, 170b Werner Braun 171 A/C. Weaver 172t A.R. Pittaway 172b Peter Saunders 172–173 A.R. Pittaway 174–175t Zefa 174–175b VE 176–177 Hutchison Library/Robert Francis 177t VE 177b Klaus Paysan 178 Magnum/Steve McCurry 179l, 179r BF/Murray Watson 180t VE 180b, 181 RHPL/Jon Gardey 182 OSF/Richard Packwood 182–183 OSF/Walna Cheng 184t Natural Image/Julie Meech 184b Gerald Cubitt 186 OSF/Richard Packwood 186–187t Jacana/Denis Huot 186–187b, 187 HQ/Denis Huot 188 BCL/M.P. Kahl 188–189t HQ/Denis Huot 188–189b BCL/Michael Freeman 190–191 OSF/Richard Packwood 192 Gerald Cubitt 193 BF/Heather Angel 194–195t Anthony Bannister 194–195b Gerald Cubitt 196 NHPA/David Woodfall 196–197t Anthony Bannister 196–197b Gerald Cubitt 198–199 NHPA/Anthony Bannister 199t NHPA/Peter Johnson 199b SAL/G.D. Plage 200–201 OSF/Arthur Gloor 202 Gerald Cubitt 202–203, 204 Gertrud and Helmut Denzau 204–205 Bob Gibbons 205, 206 Gerald Cubitt 207t SAL/Joanna van Gruisen 207b SAL/Dieter and Mary Plage 208 Gerald Cubitt 208–209 A 209 A/McDougal 210 J. MacKinnon 211 Chris Fairclough Colour Library 212–213 BCL/J. MacKinnon 213 BF/Heather Angel 214, 214–215t RHPL 214–215b Chris Fairclough Colour Library 216, 216–217 WWF/Ron Petocz 218 BCL/Gerald Cubitt 218–219 Robert and Linda Mitchell 220 BCL/Brian Coates 221 Robert and Linda Mitchell 222–223t BCL/Gerald Cubitt 222–223b RHPL/Sassoon 223 Robert and Linda Mitchell 224 J. MacKinnon 224–225 BCL/Dieter and Mary Plage 225 BCL/Alain Compost 226–227 BCL/Gerald Cubitt 228–229 RHPL/Carol Jopp 230 BCL/Orion Press 230–231 RHPL 232 RHPL/Carol Jopp 232t WWF/Michèle Depraz 232b RHPL/S. McBride 234, 234–235t, 234–235b NHPA/Orion Press 236–237 RHPL/Carol Jopp 238 NHPA/ANT/ Mary and Robyn Wilson 239 Jake Gillen 240–241 NHPA/ANT/Tony Howard 242 ANT/Fredy Mercay 242–243 ANT/Otto Rogge 243 ANT/Natural Images 244t ANT/Fredy Mercay 244b, 244–245 ANT/Grant Dixon 246–247 A/J.P. Ferrero

Editorial, research and administrative assistance
John Baines, Graham Drucker, Victoria Egan, Barbara James, Shirley Jamieson, Dr Tim King, Amanda Kirkby, Sarah Rhodes, Martin Walters

Artists
David Ashby, Graham Brown, Lynn Chadwick, Mike Long, The Maltings Partnership, Oxford Illustrators, John Woodcock

Cartography
Maps draughted by Euromap, Pangbourne
Thanks to Lynn Neal, Alison Dickinson

Index
Barbara James

Production
Clive Sparling

Typesetting
Brian Blackmore, Catherine Boyd, Peter MacDonald Associates

Color origination
Scantrans pte Limited, Singapore

INDEX

Page numbers in *italics* refer to captions, maps or tables

Abbreviations: BR = Biosphere Reserve; NP = National Park; NR = Nature Reserve; RP = Regional Park; RS = Ramsar Site; SR = State Reserve; WH = World Heritage site; WR = Wildlife Refuge; WS = Wildlife Sanctuary